COMPUTATIONAL ANALYSIS OF
POLYMER PROCESSING

COMPUTATIONAL ANALYSIS OF POLYMER PROCESSING

Edited by

J. R. A. PEARSON

and

S. M. RICHARDSON

*Department of Chemical Engineering and Chemical Technology,
Imperial College of Science and Technology, London, UK*

APPLIED SCIENCE PUBLISHERS
LONDON and NEW YORK

APPLIED SCIENCE PUBLISHERS LTD
Ripple Road, Barking, Essex, England

Sole Distributor in the USA and Canada
ELSEVIER SCIENCE PUBLISHING CO., INC.
52 Vanderbilt Avenue, New York, NY 10017, USA

British Library Cataloguing in Publication Data

Pearson, J. R. A.
Computational analysis of polymer processing.
1. Polymers and polymerization—Data processing
I. Title   II. Richardson, S. M.
668.4′028′54   TP1120

ISBN-13: 978-94-009-6636-9   e-ISBN-13: 978-94-009-6634-5
DOI: 10.1007/978-94-009-6634-5

WITH 9 TABLES AND 97 ILLUSTRATIONS

© APPLIED SCIENCE PUBLISHERS LTD 1983

Softcover reprint of the hardcover 1st edition 1983

Preface

Large, fast, digital computers have been widely used in engineering practice and their use has had a large impact in many fields. Polymer processing is no exception, and there is already a substantial amount of literature describing ways in which processes can be analysed, designed or controlled using the potentialities of modern computers. The emphasis given varies with the application, and most authors tend to quote the results of their calculations rather than describing in any detail the way the calculations were undertaken or the difficulties experienced in carrying them out. We aim to give here as useful and connected an account as we can of a wide class of applications, for the benefit of scientists and engineers who find themselves working on polymer processing problems and feel the need to undertake such calculations.

The major application we have in mind is the simulation of the dynamics of the various physical phenomena which arise in a polymer process treated as a complex engineering system. This requires that the system be reasonably well represented by a limited number of relatively simple subprocesses whose connections can be clearly identified, that the dominant physical effects relevant to each subprocess can be well defined in a suitable mathematical form and that the sets of equations and boundary conditions developed to describe the whole system can be successfully discretised and solved numerically. Problems arise at each of these operations, and the objective here will be to suggest general methods for overcoming these difficulties. In particular we shall concentrate on those aspects that are peculiar to polymer processing, and on those techniques that seem to hold most promise for the future, when progressively greater accuracy will be expected of computer simulations.

It must be emphasised that the mathematical model chosen for any process should reflect the needs of the worker concerned and be selected to provide the information required in its most convenient form and at acceptable cost computationally. This implies that there is no universally suitable model for any polymer process, nor any universally suitable method of mathematical treatment. In some cases very crude methods will suffice, while in others even the most elaborate may prove unsuccessful. However, most direct engineering applications will seek to predict physical dimensions of items of equipment or of the product of the process, and so usually the aim is to achieve the greatest accuracy possible. To this extent, it may be assumed that few of the models described in this text are unnecessarily elaborate.

In writing such a text, it is always tempting to use a very didactic approach and to present material in a very careful sequential fashion, eminently suitable for teaching engineering graduates under class conditions. We have opted instead for an approach that borrows much from the case study method, with each chapter covering a particular process in as self-contained a fashion as possible. This has the advantage that an occasional reader can start at one of the later chapters, only going back to one of the earlier chapters to refresh his or her mind about certain fundamental issues. This leads to a limited amount of repetition, but far less than might be expected; indeed such as there is may well prove a help to those readers trying to master the whole contents of the text. The reason for this is that each process presents its own peculiar characteristics— geometrical, kinematical or dynamical—with attention directed at different engineering aspects, so that quite different techniques of solution may be employed.

Another advantage of having essentially independent chapters is that we have been able to get an acknowledged expert to write each chapter. This has ensured that most approaches are well represented in an authoritative fashion and that there is a welcome element of variety in the presentation. Most of the authors concerned could not have found time to write a complete text; the present arrangement has led to a more up-to-date account than any one of them could have produced on his own. By choosing a group of authors who knew one another's work and had a common background in the subject, a strong underlying element of homogeneity is to be found in the whole work, while editorial additions

have introduced valuable forward and backward references between chapters.
The structure of the book is explained in the introductory chapter. As outlined earlier, it is aimed at those engineers and scientists who have a background in mechanics, particularly fluid mechanics and heat transfer, and some experience of computing, wishing to tackle problems in polymer processing. We hope that they will find it helpful.

J. R. A. PEARSON
S. M. RICHARDSON

Contents

ix

List of Contributors

M. J. CROCHET

Unité de Mécanique Appliquée, Université Catholique de Louvain, Bâtiment Simon Stévin, Place du Levant, 2, B-1348 Louvain-la-Neuve, Belgium.

M. M. DENN

Department of Chemical Engineering, University of California, Berkeley, California 94720, USA.

R. T. FENNER

Department of Mechanical Engineering, Imperial College of Science and Technology, Exhibition Road, London SW7 2BX, UK.

S. F. KISTLER

Department of Chemical Engineering and Materials Science, University of Minnesota, 151 Amundson Hall, 421 Washington Avenue S.E., Minneapolis, Minnesota 55455, USA.

J. R. A. PEARSON

Department of Chemical Engineering and Chemical Technology, Imperial College of Science and Technology, Prince Consort Road, London SW7 2BY, UK.

C. J. S. Petrie

Department of Engineering Mathematics, University of Newcastle upon Tyne, Stephenson Building, Claremont Road, Newcastle upon Tyne, NE1 7RU, UK.

S. M. Richardson

Department of Chemical Engineering and Chemical Technology, Imperial College of Science and Technology, Prince Consort Road, London SW7 2BY, UK.

L. E. Scriven

Department of Chemical Engineering and Materials Science, University of Minnesota, 151 Amundson Hall, 421 Washington Avenue S.E., Minneapolis, Minnesota 55455, USA.

R. I. Tanner

Department of Mechanical Engineering, University of Sydney, Sydney, New South Wales 2006, Australia.

K. Walters

Department of Applied Mathematics, University College of Wales, Penglais, Aberystwyth, Wales, UK.

J. Wortberg

Institut für Kunststoffverarbeitung, Pontstrasse 49, D-5100 Aachen, West Germany.

Notation

A list of the more commonly used symbols is given below.

Flow variables

$\mathbf{D}$, d_{ij}, e_{ij}	Rate-of-strain ($\equiv \frac{1}{2}(\nabla \mathbf{v} + (\nabla \mathbf{v})^{\mathrm{T}})$)
$\mathbf{I}$, δ_{ij}	Unit tensor
I_2	Second invariant of $\mathbf{D}(\equiv 2\mathrm{tr}(\mathbf{D}^2))$
p	Pressure (positive in compression)
T	Temperature
$\mathbf{T}$, t_{ij}, P_{ij}	Total stress (positive in tension)
$\mathbf{T}^{\mathrm{E}}$, t_{ij}^{E}, P_{ij}^{E}	Extra stress (positive in tension)
$\mathbf{v}$, v_i	Velocity
$\dot{\gamma}$	Shear-rate ($\equiv \sqrt{I_2}$ for a shear flow)

Physical properties

G	Modulus
K	Consistency†
α	Thermal conductivity
γ	Specific heat
ζ	Temperature exponent of viscosity†
η	Viscosity†
λ	Relaxation time
ν	Power-law index†
ρ	Density
σ	Surface tension

† For a fluid whose viscosity η has a power-law dependence on shear-rate $\dot{\gamma}$ and an exponential dependence on temperature T: $\eta = K\dot{\gamma}^{\nu-1}\exp(-\zeta T)$.

Dimensionless groups

Br	Brinkman number
Ca	Capillary number
De	Deborah number
Gz	Graetz number
Na	Nahme (or Griffith) number
Nu	Nusselt number
Pe	Péclet number
Re	Reynolds number
St	Stokes number
Ws	Weissenberg number

CHAPTER 1

Introduction: Polymer Melt Mechanics

J. R. A. PEARSON

*Department of Chemical Engineering and Chemical Technology,
Imperial College of Science and Technology, London, UK*

Computers have been extensively used in the analysis, design and control of polymer processing operations. In this text, most attention will be paid to the flow processes themselves and not to the manufacture or operation of the processing equipment. In this context, it is the mechanics of the flow of polymer melts that forms the starting point for computational analysis and control strategies.

In all cases, a mathematical model is sought for the process in question, using velocity, stress, density and temperature as the primitive variables in a three-dimensional time-dependent flow field. A full knowledge of the flow field is regarded as a necessary and sufficient description of any melt flow process, though design criteria are usually defined in terms of certain average (or integrated) values of the primitive variables, like flow rate, pressure drop and mean temperature.

In the analytic phase of any investigation, the geometry and operating conditions (speed, temperature or pressure) of the processing equipment are considered as known, while the consequential melt flow fields are derived by computation. If, however, the computational techniques are to be used as aids to design then the procedure has to be inverted, to the extent that equipment geometry and operating conditions are the unknowns and various characteristics of the flow field are given.

The mathematical models used in later chapters will all be based on the basic conservation laws of physics, namely those of mass, momentum and energy, and the constitutive relations for the polymeric fluid in question, namely those relating density to stress and temperature; stress to density,

1

temperature and deformation or deformation rate; and heat flux to temperature, density and temperature gradient. Where relevant, phase changes (effectively involving a solid/liquid or liquid/solid transition) and chemical reaction will be separately described. Certain boundary conditions on velocity, stress, temperature and/or heat flux at solid or gaseous interfaces will be prescribed.

In most cases, dimensionless formulations will be used, leading to dimensionless groups common to many processes; these prove useful in defining regimes of operation and in selecting and justifying approximation methods.

As far as possible a uniform notation is used throughout, but this has not been insisted upon, because subsequent chapters are intended to stand very largely on their own. For convenience, Cartesian tensor forms are given in brackets below the tensor forms.

1.1 CONSERVATION EQUATIONS

For any fluid flow, it can readily be shown (see any standard text on fluid or continuum mechanics) that the mass conservation equation may be written as

$$\frac{D\rho}{Dt} + \rho \nabla . \mathbf{v} = 0 \qquad (1.1)$$

$$\left[\frac{D\rho}{Dt} + \rho \frac{\partial v_i}{\partial x_i} = 0 \right]$$

where ρ is density, $\mathbf{v}$ is velocity, t is time and

$$\frac{D}{Dt} \equiv \frac{\partial}{\partial t} + \mathbf{v} . \nabla \qquad (1.2)$$

$$\left[\frac{D}{Dt} = \frac{\partial}{\partial t} + v_k \frac{\partial}{\partial x_k} \right]$$

is the material time derivative; ∇ is the gradient operator in a coordinate system $\mathbf{x}$. For almost all of the applications in this book, ρ is taken to be constant and so the incompressibility condition, or continuity equation,

$$\nabla . \mathbf{v} = 0 \qquad (1.3)$$
$$[\partial v_i / \partial x_i = 0]$$

is used.

The momentum conservation equation is conveniently written in the form

$$\rho \frac{D\mathbf{v}}{Dt} = \rho\mathbf{g} + \nabla . \mathbf{T} \qquad (1.4)$$

$$[\rho \, Dv_i/Dt = \rho g_i + \partial t_{ik}/\partial x_k]$$

where $\mathbf{g}$, the gravitational acceleration is taken as the only important body force. $\mathbf{T}$, the total stress tensor, is often decomposed into a rheological component $\mathbf{T}^E$, due to deformation, and a separate isotropic component $-p\mathbf{I}$, where p is the pressure and $\mathbf{I}$ the identity tensor $[\delta_{ij}]$. Because of the incompressibility condition, eqn (1.3), p is an arbitrary dynamical quantity determined only by eqn (1.4), since the normal equation of state

$$\rho = \rho(p, T) \qquad (1.5)$$

where T is temperature, no longer provides a relation between ρ and p.

Specification of $\mathbf{T}^E$ in terms of strain and velocity-gradient tensors (and of the temperature in many cases) is one of the most important issues in the mechanics of elastico-viscous fluids, of which polymer melts are an important example, and will be discussed in some detail in the next section and in later chapters.

For most confined flows, i.e. those without a significant interface with the ambient air, the gravity term can be combined with the pressure in terms of a reduced pressure

$$p_R = p - \rho\mathbf{g} . \mathbf{x} \qquad (1.6)$$

The momentum conservation equation then becomes

$$\rho \frac{D\mathbf{v}}{Dt} = -\nabla p_R + \nabla . \mathbf{T}^E \qquad (1.7)$$

Also, for many processing flows the inertia term $\frac{1}{2}\rho\mathbf{v} . \mathbf{v}$ is much smaller than either p or $\sqrt{\text{tr}(\mathbf{T}^{E2})}$ and so the left-hand side of eqn (1.7) can be neglected. This approximation, sometimes called the creeping-flow approximation, will be made in Chapters 3, 4, 5 and 7 (dealing with extrudate swell, extrusion, moulding, and film blowing, respectively). The momentum conservation equation then reduces to the stress equilibrium relation

$$\rho\mathbf{g} + \nabla . \mathbf{T} = 0 \qquad (1.8)$$

with the gravity term only explicitly included in Chapters 6–8. Inertia effects are included as a significant part of the flow description in Chapters 2 (general computational techniques), 6 (fibre spinning) and 8 (coating flows).

Most polymer processes of importance involve the softening (melting) and subsequent hardening (freezing) of the thermoplastics involved and so the temperature of the material during flow is an important factor in the process. Furthermore, the highly viscous nature of the softened polymer leads to large mechanical generation of heat; this combined with low internal heat transfer results in significantly large temperature gradients, leading to temperature differences within the flowing material large enough to affect its rheological behaviour. The full energy conservation equation has therefore to be included in the modelling of the process. It may be written as

$$\rho \frac{\mathrm{D}U}{\mathrm{D}t} = \mathbf{V} \cdot (\alpha \mathbf{V}T) + \mathbf{T}{:}\mathbf{V}\mathbf{v} \tag{1.9}$$

$$[\rho\, \mathrm{D}U/\mathrm{D}t = \partial(\alpha\, \partial T/\partial x_k)/\partial x_k + t_{ik}\, \partial v_i/\partial x_k]$$

where U is the internal energy, and α is the thermal conductivity defined from the constitutive equation for heat transfer

$$\mathbf{q} = -\alpha \mathbf{V}T \tag{1.10}$$

$$[q_i = -\alpha\, \partial T/\partial x_i]$$

where $\mathbf{q}$ is the heat flux vector. U is often written in terms of a specific heat γ where

$$\mathrm{d}U = \gamma\, \mathrm{d}T \tag{1.11}$$

(Alternative forms for the energy equation, including cases where ρ is not treated as constant, can be found in many text books including Bird et al. (1960) or Bird, Armstrong and Hassager (1977). The question is discussed in Chapter 5, Section 5.3.1.4.)

For reasons of simplicity, the energy equation is left out completely in Chapters 2 and 8 in the sense that the density and rheological equation of state are taken to be independent of temperature, so that the flow field is determined without reference to the temperature field. It is only introduced at the end of Chapter 3, where it is shown that expected temperature variations of viscosity could have a large effect on extrudate swell. The energy equation plays a central role in all the other chapters, except Chapter 8, from the point of view of both an effective phase change at the crystallisation or glass transition temperature and a significant decrease in viscosity with temperatures above the phase-change temperature.

At interfaces between the flowing polymer and the rigid boundaries of the processing equipment, it is usually assumed that the no-slip boundary condition holds, i.e. that **v** is continuous; similarly that the temperature T is continuous. In some cases, the wall temperature cannot be prescribed because of the nature of thermal control within the processing equipment; an effective heat flux or mixed temperature/heat flux condition is often therefore presumed to hold at the interface (see eqn (5.14) for example). Where melting or softening is assumed to take place at a sharp interface, with a jump in U, i.e. a non-zero latent heat, then an alternative interfacial condition of heat flux is required, as given by eqn (5.19) and implied by eqn (4.96). When a density change is associated with the phase change, a discontinuity in velocity occurs: for the steady-state solutions used in later chapters, this is never introduced explicitly.

At a 'free' surface, as with ambient air, it is assumed either that the interface is stress free (as in Chapter 7 on film blowing, Chapter 8 on coating flows and much of Chapters 3 and 6)

$$\mathbf{T} \cdot \hat{\mathbf{n}} = 0 \qquad (1.12)$$

where $\hat{\mathbf{n}}$ is a unit vector pointing outward from the flowing polymer and ambient pressure is taken as zero, or that small surface tension and air drag terms act on the polymer flow across the surface. The form used to obtain eqn (6.5) and implied by eqn (3.4) is

$$\mathbf{T} \cdot \hat{\mathbf{n}} = \sigma H \hat{\mathbf{n}} + \rho_a |\mathbf{v}| C_f \mathbf{v} \qquad (1.13)$$

$$[t_{ik} \hat{n}_k = \sigma H \hat{n}_i + \rho_a (v_k v_k)^{1/2} C_f v_i]$$

where σ is surface tension, H is the total curvature of the surface, ρ_a is the density of the ambient air and C_f a drag coefficient. A similar free-boundary condition is provided by eqn (8.2). Similarly, because of large temperature gradients in the ambient air, a heat transfer boundary condition of the form

$$\mathbf{q} \cdot \hat{\mathbf{n}} = h(T - T_a) \qquad (1.14)$$

is used, where h is a heat transfer coefficient and T_a the ambient air temperature, leading to the Nusselt number defined in Chapter 2, and the heat transfer term in eqns (6.6) or (7.14). In eqn (1.14) a radiation contribution is separately included. Lastly the interface must remain a material surface. This leads to very obvious boundary conditions in most of the succeeding chapters: for example, in steady plane or axisymmetric flows interfaces are composed of symmetric stream lines.

Certain difficulties arise at the intersection of fixed and free surfaces. Such contact lines are discussed fully in Chapter 8.

1.2 RHEOLOGICAL EQUATIONS OF STATE

The starting point for most of the complex finite-difference or finite-element calculations to be discussed later will be the linear viscous, or Newtonian, fluid model for which

$$\mathbf{T}^{E} = 2\eta_0 \mathbf{D} \qquad (1.15)$$

$$[t_{ik}^{E} = \eta_0(\partial v_i/\partial x_k + \partial v_k/\partial x_i)]$$

where

$$\mathbf{D} = \tfrac{1}{2}\{\nabla\mathbf{v} + (\nabla\mathbf{v})^{T}\} \qquad (1.16)$$

When substituted into eqn (1.4), this yields the Navier–Stokes equations

$$\rho\frac{D\mathbf{v}}{Dt} = -\nabla p_R + \eta_0\nabla^2\mathbf{v} \qquad (1.17)$$

$$[\rho\, Dv_i/Dt = -\partial p_R/\partial x_i + \eta_0\, \partial^2 v_i/\partial x_k\, \partial x_k]$$

from which p_R can be eliminated by the vector operation $\nabla \wedge$. Equation (1.17) together with eqn (1.1) leads to a classic set of partial differential equations exhaustively analysed in the literature, although there is no uniformity of opinion as to how they should best be solved numerically.

The Newtonian model is, however, not satisfactory for most of the polymer processing situations to be discussed in later chapters. Temperature dependence is usually introduced empirically in the form

$$\eta_0(T) = \eta_0(T_0)\exp\{-\zeta(T - T_0)\} \qquad (1.18)$$

with ζ^{-1} a characteristic rheological temperature difference (this form is used in Chapters 3, 5, 6 and 7) or in the Arrhenius form

$$\eta_0(T) = \eta_{0A}\exp(E/RT) \qquad (1.19)$$

with E an activation energy for viscous flow.

The most obvious and important characteristic of polymer melts, however, is that their simple shear viscosity—as calculated in steady-state from a Couette viscometer or from a long capillary viscometer—is a decreasing function of shear rate $\dot{\gamma}$. A satisfactory empirical statement of this property is given by

$$\eta = K(\dot{\gamma})^{\nu - 1} \qquad \dot{\gamma} \geq 0 \qquad (1.20)$$

where K is the consistency and ν the power-law index. K can obey a relation exactly equivalent to eqn (1.18) to allow for temperature dependence. A suitably invariant tensor representation of this observation provides the purely viscous constitutive relation for a power-law fluid,

$$\mathbf{T}^E = 2K\exp\{-\zeta(T - T_0)\}I_2^{1/2(\nu-1)}\mathbf{D} \qquad (1.21)$$

where the second invariant of $\mathbf{D}$ is given by

$$I_2 = 2\operatorname{tr}\mathbf{D}^2 \qquad (1.22)$$
$$[I_2 = 2d_{ik}\,d_{ki}]$$

This model is used extensively in later chapters, e.g. eqns (3.29), (4.4), (5.1) and (6.11). It is the basis of most of the calculations carried out in Chapters 4 and 5, as will be discussed in Section 1.3, in connection with the lubrication approximation.

An equally important characteristic, where the flow field experiences rapid changes in cross-section in the flow direction, is that of elasticity, whereby part of the deformation of the fluid is reversible, and the work done by the deforming stress is stored as elastic (recoverable) energy. Polymer rheology is concerned with an exhaustive analysis and description of this behaviour. For our present purposes, we seek no more than general approximate representations suitable for the flow fields of engineering interest.

If, as is sometimes the case, the elastic effects lead to no more than small perturbations in $\mathbf{D}$ to Newtonian behaviour, then a regular perturbation expansion about the Newtonian model can be used. This leads to the second-order fluid model

$$\mathbf{T}^E = \eta_0\mathbf{A}^{(1)} + (\nu_1 + \nu_2)\mathbf{A}^{(1)2} - \tfrac{1}{2}\nu_1\mathbf{A}^{(2)} \qquad (1.23)$$

where†

$$\mathbf{A}^{(1)} = 2\mathbf{D} \qquad (1.24)$$

$$\mathbf{A}^{(2)} = \frac{D\mathbf{A}^{(1)}}{Dt} + (\mathbf{W}\cdot\mathbf{A}^{(1)} - \mathbf{A}^{(1)}\cdot\mathbf{W}) + \mathbf{A}^{(1)2} \qquad (1.25)$$

$$\mathbf{W} = \tfrac{1}{2}(\nabla\mathbf{v} - (\nabla\mathbf{v})^T) \qquad (1.26)$$

and ν_1 and ν_2 are constants. This model has special relevance for steady simple shear flow, as defined in eqn (1.31), which is the basis of the

† $[A_{ik}^{(1)} = \partial v_i/\partial x_k + \partial v_k/\partial x_i]$
$[A_{ik}^{(2)} = \partial A_{ik}^{(1)}/\partial t + v_j\,\partial A_{ik}^{(1)}/\partial x_j + w_{ij}A_{jk}^{(1)} - A_{ij}^{(1)}w_{jk} + A_{ij}^{(1)}A_{jk}^{(1)}]$
$[w_{ik} = \tfrac{1}{2}(\partial v_k/\partial x_i - \partial v_i/\partial x_k)]$

8   J. R. A. PEARSON

lubrication approximation; indeed, provided v_1 and v_2 are regarded as functions of I_2 as defined by eqn (1.22) then eqn (1.23) is exact for such flow, and v_1 and v_2 are the first and second normal-stress coefficients.

However, the model, eqn (1.23), does not allow of instantaneous, or even rapid, elastic deformation, and so for many purposes a radical departure from the Newtonian model is required. This is provided in the most simple fashion by the upper-convected Maxwell model[†]

$$\mathbf{T}^E + \lambda_1 \overset{\triangledown}{\mathbf{T}}{}^E = 2\eta_0 \mathbf{D} \tag{1.27}$$

where λ_1 is a constant relaxation time, and the non-linear differential operator ($^\triangledown$) is given by[‡]

$$\overset{\triangledown}{\mathbf{J}} = \frac{D\mathbf{J}}{Dt} + (\mathbf{W}.\mathbf{J} - \mathbf{J}.\mathbf{W}) - (\mathbf{D}.\mathbf{J} + \mathbf{J}.\mathbf{D}) \tag{1.28}$$

This model is used as an archetypal form in Chapters 2, 3 and 6, i.e. eqns (2.2), (3.8) and (6.3). It can equally well be written in an integral form, eqns (2.5) or (2.7). In both the differential and the integral form, it can be regarded as the starting point for the development of more realistic models, such as the 'Oldroyd B' fluid model

$$\mathbf{T}^E + \lambda_1 \overset{\triangledown}{\mathbf{T}}{}^E = 2\eta_0(\mathbf{D} + \lambda_2 \overset{\triangledown}{\mathbf{D}}) \tag{1.29}$$

i.e. eqn (2.4), referred to in Chapter 3 at the end of Section 3.5.6, or the Phan-Thien and Tanner model (which employs a differential operator that is a linear combination of the upper and lower convected differential operators mentioned above) referred to in Chapters 2 and 6. The introduction of additional parameters into the rheological models allows of better agreement with rheological measurement using different deformation histories. The obvious objective in such models is to obtain as generally applicable a model as possible with as few parameters as possible: thus eqn (1.29) introduces a single additional constant λ_2, a retardation time, and the original Phan-Thien/Tanner model two additional constants.

It is obvious from small amplitude (dynamic) oscillatory shear measurements that a spectrum of relaxation times λ, rather than a single relaxation

† $[t_{ik}^E + \lambda_1(\partial t_{ik}^E/\partial t + v_j \partial t_{ik}^E/\partial x_j - \partial v_i/\partial x_j t_{jk}^E - \partial v_k/\partial x_j t_{ij}^E) = \eta_0(\partial v_i/\partial x_k + \partial v_k/\partial x_i)]$

‡ It is perhaps worth noting that the second-order fluid model can be written

$$\mathbf{T}^E = 2\eta_0\mathbf{D} - v_1\overset{\triangledown}{\mathbf{D}} + 4v_2\mathbf{D}^2$$

and that the tensor $\mathbf{A}^{(2)}$ is obtained from $\mathbf{A}^{(1)}$ by means of the lower convected differential operator ($^\triangle$). See Astarita and Marrucci, 1974, p.93 et seq., for definitions and use of convected derivatives.

time λ_1, is necessary to describe the behaviour of polymeric fluids, and so an obvious extension to the Maxwell model, eqn (1.27), is obtained by writing

$$T^E = \sum_K T_K^E \qquad (1.30)$$

where a λ_K is associated with each T_K^E, which obeys eqn (1.27), or indeed one of its extensions like the Phan-Thien/Tanner model, as used in Fig. 6.5 for example, and discussed in Chapter 6.

The viscoelastic parameters η_0, λ_1 and λ_2 and, to a much lesser extent G, are all temperature-dependent, and reasonable comparison with experiment often requires the dependence to be made explicit.

Further information about models for polymeric fluids is given in Bird, Armstrong and Hassager (1977) and in Bird, Hassager *et al.* (1977); helpful background information may be found in Astarita and Marrucci (1974), in Walters (1975), or in Lodge (1964).

1.3 KINEMATIC AND DYNAMICAL APPROXIMATIONS

Even with the relatively simple rheological equations of state introduced in Section 1.2 above, solution of the full conservation equations in any but geometrically simplified situations is difficult although not impossible.

Thus Chapters 2 and 3 are restricted to plane or axisymmetric flow with steady-state boundary conditions. Even so, significant computational complexity arises in the confined flows of Chapter 2 because of sharp changes in cross-section and corner boundaries and in the free-surface flow of Chapter 3 because of additional non-linearities introduced by the free surface and stress concentrations at the exit edges.

Chapters 4 and 5 report work that invokes the lubrication approximation. This is based on the requirement that one channel dimension (the depth in wide channel flow and the cross-sectional radius in axisymmetric pipe flow) be small compared with the others (the width and length in channel flow and the length in axisymmetric flow) and that the rate of change of this small dimension with distance downstream be small everywhere. A fuller description of this approximation as applied in polymer melt mechanics is given in Pearson (1983, Section 8.1). Its application to injection moulding will be discussed later in Section 5.3.1.3.

Briefly the approximation states that the flow everywhere within the channels in question is such as would arise in steady flow in a long wide uniform channel of constant depth or radius equal to the local value in the channel in question. Thus for a slowly varying channel depth, e.g. $H(x)$ in Fig. 5.3b, where $dH/dx \ll 1$ and $H/L \ll 1$, it may be assumed that in the neighbourhood of the plane $x = x_0$, the velocity will be such as would arise in a channel of length $\gg H_0$ of fixed width $W_0 \gg H_0$, with $H \equiv H(x_0) \equiv H_0$. Streamlines are all parallel to the x-axis, the only component of the rate-of-strain tensor $\mathbf{D}$ is dv_x/dy and this is a function of y only; $\mathbf{D}$ is constant along streamlines, and the flow is an example of viscometric flow with

$$\mathbf{v} = \{v_x(y), 0, 0\} \tag{1.31}$$

(e.g. Astarita and Marrucci, 1974, Section 5-2). It thus involves only three independent stress components,

$$\left.\begin{aligned} t_{xy} &= \eta_0 \, dv_x/dy \equiv \tau(I_2) \\ t_{xx} - t_{yy} &= v_1 (dv_x/dy)^2 \equiv \sigma_1(I_2) \\ t_{yy} - t_{zz} &= v_2 (dv_x/dy)^2 \equiv \sigma_2(I_2) \end{aligned}\right\} \tag{1.32}$$

where η, v_1 and v_2 are in general functions of dv_x/dy, all tending to constant values as dv_x/dy tends to zero. The second-order fluid model, eqn (1.23), thus becomes generally valid for lubrication flows provided η_0, v_1 and v_2 are regarded as functions of the shear rate.†

The continuity equation, eqn (1.3), is identically satisfied by eqn (1.31), and the no-slip boundary conditions at the walls yield

$$v_x(\pm\tfrac{1}{2}H_0) = 0 \tag{1.33}$$

The momentum conservation equation, eqn (1.7), becomes eqn (1.8), and if the reduced pressure is used as in eqn (1.7), only the x and y components of $\nabla \cdot \mathbf{T} = 0$ are relevant. From these it can readily be deduced that

$$p_R = \bar{p}_R(y) + \left(\frac{dp}{dx}\right)_0 x \tag{1.34}$$

where $(dp/dx)_0$ is a constant pressure gradient. The terms depending on v_1 affect $\bar{p}_R(y)$, but have no real significance in the lubrication approximation provided

$$\frac{L}{H} \gg \frac{\sigma_1}{\tau} \tag{1.35}$$

where τ and σ_1 are evaluated at the wall shear rate $\dot{\gamma}_w = (dv_x/dy)_w$. The dominant dynamical quantity then becomes the pressure gradient dp/dx.

† We write η for η_0 when it is shear-rate dependent.

The overall pressure drop is then obtained by integrating the local value dp/dx, now regarded as a slowly varying function of x, over the length L. It is in this way that the more precise concept of stress (a tensor) becomes replaced in many processing operations by the simple concept of pressure (a scalar). A further consequence of the lubrication approximation is that only the viscosity function $\eta(I_2)$ need be specified for computations of velocity field and pressure drop, and hence the widespread use of the power-law fluid model, eqn (1.21).

The classical lubrication approximation applies when ρ, η, v_1 and v_2 are effectively independent of temperature, so that there is no coupling back from the energy to the stress of flow fields. Minor modifications can be made to the theory if T can be regarded as a slowly varying function of x, whether T is also regarded as varying significantly with y or not. In practice, development lengths for T are much longer than those for $\mathbf{v}$ in a channel of constant depth, and so the fully developed approximation, which applies if T is determined only by the local flow field and the local wall boundary conditions on temperature, rarely applies. Physically, this means that the fluid temperature is significantly affected everywhere in the flow field by the entry temperature: even when $L \gg H$ the flow can be a developing flow.

A quite different type of flow field arises in Chapters 6 and 7, where the flow domain is again narrow in at least one coordinate direction. Filament flow is analogous to pipe flow in tubes of slowly varying radius in that it is axisymmetric with slowly varying radius, i.e. $dR/dz \ll 1$, where $\pi R^2 = A$, the cross-sectional area of the filament (Fig. 6.1). The stress and temperature fields are taken to be independent of r, because of the absence of stress and the poor heat transfer across the boundary $r = R$. The field variables v_z, $\mathbf{T}$ and T are thus functions of a single space variable z in the steady case. Furthermore, the flow is irrotational, so $\nabla \mathbf{v} = (\nabla \mathbf{v})^T = 2\mathbf{D}$, $\mathbf{W} = 0$, and $\mathbf{T}$ is diagonal in the (r, θ, z) coordinate system. The extensional flow field involved means that elastic effects are more likely to be relevant than in the simple shear flows of lubrication theory at the same value of I_2. For example, the Maxwell model, eqn (1.27), leads to differences in normal stress $t_{zz} - t_{rr}$ that can become arbitrarily large at extension rates of order λ^{-1} in filament flow, corresponding to large elastic stored energy.

Similarly, in steady film-blowing flows (see Fig. 7.1), the sheet thickness h is a slowly varying function of axial distance z, and the stress and temperature fields are independent of the coordinates x_3 (normal to the film surface) and x_2 (tangential to the film surface and normal to the flow direction x_1). This again leads to an irrotational ($\mathbf{W} = 0$) flow with $\mathbf{T}$

diagonal in the (x_1, x_2, x_3) coordinate system. These irrotational extensional flows are 'strong' in the sense of Tanner and Huilgol (1975) as opposed to the 'weak' viscometric flows, thus leading to enhanced elastic effects. This accounts for the use of elastic and elastico-viscous models for film-blowing flow.

The freedom associated with the unconfined filament and film flows makes them more sensitive to transient disturbances; it proves important as well as possible to investigate unsteady flows in Chapter 6. By contrast, most of the analyses described in Chapter 5 for injection moulding, which is basically a periodic process, are quasi-steady in the sense that the $\partial/\partial t$ terms in the conservation equations are largely neglected, while use of the power-law model, eqn (1.21), removes them from the constitutive equation.

1.4   DIMENSIONLESS FORMULATION

It is often both convenient and helpful to express the sets of equations modelling polymer flows in dimensionless form. For this, scale lengths, times and temperature differences are required.

The geometry of the flow field usually provides at least two lengths. It is usual to take the smallest of these, namely the channel depth or tube radius, as the basic length scale in confined flows; the lubrication approximation is then seen to require that the dimensionless length (and width where applicable) of the flow channel be very large compared to unity. The slowly varying depth requirement $|\nabla h| \ll 1$ is unaltered by non-dimensionalisation. The small dimension R in isothermal filament flow and h in isothermal sheet flow proves to be irrelevant dynamically and so filament length L or die width in film blowing is usually taken as the relevant length scale. In terms of this, it follows that the dimensionless filament radius or the dimensionless film thickness be very small compared to unity.

There is usually an imposed characteristic velocity V (the volume flow rate/unit cross-sectional area along channel, tube, filament or film), in terms of which the velocity field can be made dimensionless. If there is a characteristic relaxation time λ associated with the fluid rheology, as in eqn (1.27), then a length scale $V\lambda$ can immediately be defined. The dimensionless ratio

$$V\lambda/L \equiv De \qquad (1.36)$$

where L is a length along a streamline over which the velocity or deformation-rate field changes by order one, is often called the Deborah number. This is particularly relevant in filament and film blowing flows, as, for example, in Fig. 6.12. The Deborah number is often regarded as a measure of the importance of elastic effects. If $De \ll 1$, then a Newtonian model may well be relevant. In channel flow, the alternative dimensionless group $V\lambda/H$ is usually much larger; and is an indication that shear-dependence of viscosity will usually be relevant in polymer flows. It has already been argued that the generalised second-order fluid model, eqn (1.23), is relevant in confined flows, so the dimensionless group

$$Ws = v_1 \dot{\gamma}^*/\eta \equiv \sigma_1(\dot{\gamma}^*)/\tau(\dot{\gamma}^*) \qquad \dot{\gamma}^* = V/H \qquad (1.37)$$

often called the Weissenberg number, takes the place of the Deborah number. This is used in Chapters 2 and 3, and the aim of many workers is to discover stable computing schemes when Ws becomes large compared with unity. $Ws \ll 1$ effectively implies Newtonian flow, and computational difficulties, as discussed in Chapter 2, often arise when Ws becomes of the order of 1 to 10.

It will be noted that L/V and H/V provide time scales, and their ratios with relaxation time λ or v_1/η are what we have called Deborah and Weissenberg numbers respectively. They are also, of course, the inverse of characteristic extension and shear rates, respectively.

Inertial effects can be measured in terms of the Reynolds number

$$Re = \rho V l/\eta \qquad (1.38)$$

where l is the characteristic length scale. For confined channel flows this is given by H, but for fibre spinning by L. If $Re \ll 1$, as in Chapters 3, 4, 5 and 7, then inertial terms can be neglected and relation (1.8) is obtained, as explained earlier.

A further class of dimensionless groups is obtained when the energy conservation equation, eqn (1.9), is made dimensionless. The Péclet number

$$Pe = VH\rho\gamma/\alpha \qquad (1.39)$$

measures the relative importance of the convection term ($\rho DU/Dt$) to the conduction term ($\nabla . (\alpha \nabla T)$) and is analogous to the Reynolds number. However, because the length scale for diffusion, H, is much smaller than for convection, L, the important dimensionless group in channel flows is the Graetz number

$$Gz = PeH/L \equiv VH^2 \rho\gamma/\alpha L \qquad (1.40)$$

When $Gz \ll 1$, the thermal field is everywhere fully developed, as described in Section 1.3 above. When $Gz \gg 1$, conduction is important only in a boundary layer close to the walls and most of the flow field can be treated as adiabatic. This group is used to define various flow regimes in Chapter 5.

A characteristic temperature difference can be chosen in many ways. It may be provided by an externally imposed wall-temperature difference, e.g. by $T_w - T_m$ where T_w is the wall temperature and T_m is the softening or melting temperature; or by the rheological behaviour of the fluid to temperature variations, e.g. by ζ^{-1} in eqn (1.8); or in terms of an imposed pressure drop, e.g. by $P/\rho\gamma$. The relative importance of the generation term $\mathbf{T} : \nabla \mathbf{v}$ depends on which scale ΔT^* is used: relevant generation numbers are the Brinkman number

$$Br = \eta V^2 / \alpha (T_w - T_m) \qquad (1.41)$$

and the Nahme–Griffith number

$$Na = \eta V^2 \zeta / \alpha \qquad (1.42)$$

The former is a measure of the ratio of generated temperature differences to imposed wall-temperature differences, while the latter is a measure of the relative viscosity variations caused by heat generation, when Gz is not $\gg 1$. For large Gz, the ratio Na/Gz is more appropriate. These groups are used extensively in Chapters 4 and 5 because all can be significantly larger than unity in extrusion and injection moulding. Large Na or Na/Gz indicate strong coupling between the flow and energy equations. These groups refer primarily to confined flows, as in Chapters 3, 4 and 5.

In free flows, the controlling factor is often the external heat transfer coefficient h as in eqn (1.14). The relative importance of external to internal resistance is given by the Nusselt number

$$Nu = hR / \alpha \qquad (1.43)$$

which for fibre spinning flows is small, but not necessarily so in extrudate swell.

1·5 STABILITY AND SENSITIVITY

Most polymer processes are intended to be continuous and steady. Of those considered below, only injection moulding is a periodic batch-type process and most analyses treat even its flow mechanics as though it was only slowly varying in time. Thus if $t_{m.f.}$ is the time taken to fill the mould, then

$$H/Vt_{m.f.} \ll 1 \qquad (1.44)$$

H/V being the characteristic time associated with the channel width. By noting that $t_{m.f.} \sim L/V$, it can be readily seen that eqn (1.44) is effectively one of the criteria given in Section 1.3 for use of the lubrication approximation, namely, $H/L \ll 1$.

However, undesirable fluctuations do arise in most processing operations, and much effort is devoted to controlling and reducing these fluctuations. Their source may be wholly external (namely variations in the feed material, fluctuations in the driving motors, hunting in the temperature controllers, random variations in ambient air conditions) or largely internal (in the sense that self-sustained oscillations may arise within the flow field irrespective of external forcing). The time scale of these fluctuations can vary from the very long, in which case the instantaneous situation can be treated as effectively steady, to very short, in which case their effect is often no more than an imperceptible ripple. The most serious are usually those whose time scale is of the same order, even close to, one of the obvious time scales of the flow itself. Some of the fluctuations can, however, be primarily spatial, in that the actual flow departs from the geometrical symmetry intended and imposed, as far as possible, by the equipment geometry.

Mathematical analysis of these fluctuations proves to be much more difficult in the end than analysis of the primary, unperturbed, steady, symmetric flow and so is not emphasised in most later chapters. Mention is made in Chapter 3 of 'melt fracture' (Petrie and Denn, 1976) only to remind the reader that it is a genuine physical phenomenon. No reference is made in Chapter 4 to instabilities in screw extrusion, such as surging, simply because no available theory or calculation routine has explained them. Some successful results are reported in Section 8.6 in connection with coating flows. Non-uniformities in die flow are covered briefly in Chapter 9. Chapter 6 on fibre spinning examines the subject most thoroughly, largely because the spatially one-dimensional nature of the problem is retained even when the flow becomes unsteady. There the distinction between stability and sensitivity is made clear: instability is manifest in a phenomenon (known as draw resonance) whereby large self-sustained oscillations in fibre cross-section can arise without continuous external stimulus; sensitivity refers to the extent of the filament-flow response to various externally applied disturbances. Both Fourier and Laplace transformation in the time domain are employed, as well as direct forward step-by-step calculation in time.

This last-named direct method raises the very interesting issue of how to distinguish between true mechanical instability being correctly reflected by numerical calculations and solution-dependent numerical instability. When seeking numerical solutions to sets of non-linear partial-differential equations involving only spatial independent variables, iterative methods are almost invariably used; as explained in most of the later chapters on confined flow, methods which are successful when the non-linearities are weak often fail at some critical value of the parameter (Reynolds or Deborah/Weissenberg number) measuring non-linearity even when steady solutions are known analytically to exist. One way round this difficulty that has been attempted, is to make the original equations time-dependent, and to solve them as initial value problems, parabolic in time. Similar difficulties in convergence have been experienced in the marching problem form, which may or may not reflect true instability.

1.6   SIMULATION PACKAGES: COMPUTER-AIDED DESIGN AND CONTROL

The greatest advantage of using computers in process analysis comes when entire processes are being considered, and when a wide range of possible numerical values for material properties, machine dimensions and operating conditions are being assessed. At the lowest level, computers can be regarded as large, efficient, data-handling devices; at the highest, as virtually autonomous (by virtue of the wide range of computing algorithms permanently stored in their software systems) devices for numerical analyses of arbitrarily elaborate mathematical models. They can act as simulators for industrial processes, and so provide, in many ways, a cheap and rapid alternative to physical experiments on laboratory, pilot-plant or full scale.

This aspect of their use is strongly emphasised in several places in this text, particularly in Chapters 5, 6 and 8. Considerable progress is reported in fibre spinning, for which various simulation packages are described: as explained earlier, the advantage in this case is that the basic mechanics are relatively simple to model and so a wide range of external effects can be examined; there is the added interest that a variety of experimental results on sensitivity and stability can be used for comparison and that 'denier variability' (i.e. cross-sectional fluctuations) and filament breakage are commercially important issues.

Simulation of the injection moulding process is at present the most actively pursued of such schemes, largely because the procedure of mould design, machine selection and choice of operating cycle represents a lengthy and expensive part of what is often a relatively small-scale industrial operation. Despite a great deal of previous experience, many mould systems fail to operate satisfactorily when first installed, and so there is a compelling reason to improve on the present trial-and-error methods used to reach acceptable operation. Although most effort is at present devoted to modelling the melt flow and melt cooling process in the mould channels, the interaction of these with the machine characteristics—drive hydraulics, temperature control, coolant flow, platten elasticity and material preplasticisation—must be seen as an intrinsic part of the modelling objective.

Simulation packages can be used in various ways. One of the most widely canvassed is as part of computer-aided design packages. These are most valuable when they are used interactively by expert engineers, and are still a long way from providing automatic and unique solutions to design problems; they can be used for feasibility analyses of proposed designs, for optimisation of feasible, acceptable designs and for improvement in the operation of existing equipment. The information provided to a simulation package can be general and basic, as is implied by the fundamental mechanical analysis forming the core of most of this text, or empirical and specific, as a result of previous experience or experiment. Some of this specific numerical information fits neatly into a general and complete simulation program, providing, for example, the rheological parameters K, v and ζ in eqn (1.21), the heat transfer coefficient h in eqns (1.14) or (6.6), or the various lengths characterising the flow channels. Some, however, will be in the form of empirical correlation coefficients between various operating variables, which provide an alternative transfer-function-type representation of the flow processes (otherwise described in detail by the full program) for a particular process.

The latter are much relied upon in Chapter 9, which is concerned with control and optimisation of a given operating process. A very different approach and technique is then involved. Although a basic understanding of the flow mechanics and heat transfer helps in the choice of control strategy, the role of the computer is to build up and use in adaptive fashion a large amount of numerical information about an established process while it is running. The process of data collection is carried on simultaneously with those of statistical analysis and control. The computer is closely linked with the processing equipment and plays an intrinsic part in its operation. In newer designs of injection moulding equipment, for example, operating conditions are set by the operator through the computer console, which becomes a single electronic interface between machine and user.

One of the most difficult problems that faces those producing software packages, whether it be for analysis, design or control, is to provide a sufficient degree of flexibility in the overall system to simulate a wide enough range of equipment, material and product without making the package so complex that it can barely be understood by potential users. It is obviously advantageous to use short-cut methods when these are applicable, and to avoid elaborate calculations when these are unnecessary. There is no point in providing a larger amount of information from the program than is needed in any particular application. The only practical solution to this problem is to base its use on a series of graded options, limited on the one hand by the extent of the function to be carried out, and on the other by the depth or accuracy with which that function will be carried out. This implies that the software package has to be highly modular, with interchangeability between modules of different complexity for carrying out the same function, and flexibility in selection and sequencing of which functions to carry out. The secret of success, therefore, lies in the means for transferring pre-digested information between modules, so that information obtained through analysis can be used for the dual process of design, and then be employed to establish control of the operation. This is a very challenging task, and one that has yet to be completed satisfactorily for any single process, except at a rudimentary level. The problems involved are those of computing generally, and so the rest of this text is devoted largely to those aspects that are particular to polymer processing.

REFERENCES

Astarita, G. and Marrucci, G. (1974). *Principles of Non-Newtonian Fluid Mechanics*, McGraw-Hill, London.

Bird, R. B., Armstrong, R. C. and Hassager, O. (1977). *Dynamics of Polymeric Liquids: I. Fluid Mechanics*, Wiley, New York.

Bird, R. B., Hassager, O., Armstrong, R. C. and Curtiss, C. F. (1977). *Dynamics of Polymeric Liquids: II. Kinetic Theory*, Wiley, New York.

Bird, R. B., Stewart, W. E. and Lightfoot, E. N. (1960). *Transport Phenomena*, Wiley, New York.

Lodge, A. S. (1964). *Elastic Liquids*, Academic Press, London.

Pearson, J. R. A. (1983). *Mechanics of Polymer Processing*, McGraw-Hill (Hemisphere), Washington.

Petrie, C. J. S. and Denn, M. M. (1976). Instabilities in polymer processing, *AIChEJ.*, **22**, 209–36.

Tanner, R. I. and Huilgol, R. R. (1975). On a classification scheme for flow fields, *Rheol. Acta.*, **14**, 959–62.

Walters, K. (1975). *Rheometry*, Chapman and Hall, London.

CHAPTER 2

Computational Techniques for Viscoelastic Fluid Flow

M. J. CROCHET

*Unité de Mécanique Appliquée,
Université Catholique de Louvain, Belgium*

and

K. WALTERS

*Department of Applied Mathematics,
University College of Wales, Aberystwyth, UK*

2.1 INTRODUCTION

The materials used in polymer processing can usually be classified as non-Newtonian elastic liquids (see, for example, White, 1980). Such liquids are often processed in complex geometries. There is therefore a practical motivation for the current interest in the numerical simulation of non-Newtonian flow in complex geometries.

Under some conditions, the peculiar memory effects associated with elastic liquids do not have a strong influence on the flow, an obvious example being those situations where the lubrication approximation can be employed (see, for example, Pearson, 1967). Further, there are conditions where, although elastic memory effects may have a strong influence on the *stress* levels, they nevertheless have a negligible effect on the determination of the *velocity* field. We regard such situations as degenerate examples of the basic problem to be discussed in this chapter and to be best tackled by less sophisticated techniques. Our main concern is therefore with the numerical simulation of those flows where there is a strong interaction between fluid memory effects and complex flow geometries. Such a study necessarily involves a consideration of *highly* elastic liquids which possess a

21

long-range memory of past deformation. Further, the flow geometries are far more complex than simple channel or capillary flow and much of our concern will be concentrated on flows with sharp re-entrant corners. The combination of long-range memory and abrupt changes in geometry essentially rules out the possibility of employing simple rheological equations of state in the analyses *if the intention is to seek general validity.* Certainly the popular hierarchy equations of Coleman and Noll (1961) which are generally valid only for slow (and slowly varying) flows would be inappropriate in the present context (see, for example, Davies *et al.*, 1979; Perera and Walters, 1977a, b; Walters, 1979). At the same time, the flow problems we have in mind are far too complex to allow even the contemplation of the use of functional rheological equations of state which are required for general validity and, accordingly, pragmatism dictates that approximations must be used. Generality certainly cannot be claimed for any resulting numerical simulations and it would be naïve to expect future developments to remedy this deficiency. However, the simulations do have a facility for predicting, at least qualitatively, flow field characteristics and stress levels.

At the present time, existing numerical analyses are limited to simple approximate models of the Oldroyd–Maxwell type and related molecular-motivated models such as that due to Phan-Thien and Tanner (1977). In the present chapter, we shall concentrate on the upper-convected Maxwell model characterised by a constant viscosity coefficient η_0 and a constant relaxation time λ_1. This has equations of state in a rectangular Cartesian coordinate system x_i given by

$$p_{ik} = -p\delta_{ik} + p_{ik}^E \qquad (2.1)$$

$$p_{ik}^E + \lambda_1 \overset{\triangledown E}{p}_{ik} = 2\eta_0 e_{ik} \qquad (2.2)$$

where p_{ik} is the stress tensor, p_{ik}^E the extra-stress tensor, p an arbitrary isotropic pressure (in the case of incompressible fluids), δ_{ik} is the Kronecker delta, e_{ik} the rate-of-strain tensor and ∇ denotes an upper-convected derivative as given by Oldroyd (1950). If v_i is the velocity vector, we have

$$\overset{\triangledown E}{p}_{ik} = \frac{\partial p_{ik}^E}{\partial t} + v_m \frac{\partial p_{ik}^E}{\partial x_m} - \frac{\partial v_i}{\partial x_m} p_{mk}^E - \frac{\partial v_k}{\partial x_m} p_{im}^E \qquad (2.3)$$

We note for future reference that the model popularly known as 'Oldroyd B' is given by

$$p_{ik}^E + \lambda_1 \overset{\triangledown E}{p}_{ik} = 2\eta_0 [e_{ik} + \lambda_2 \overset{\triangledown}{e}_{ik}] \qquad (2.4)$$

where λ_2 is a constant retardation time.

To all intents and purposes, the developments we shall describe for the Maxwell model can be easily generalised to include more complicated models of the Oldroyd and Phan-Thien/Tanner type. A possible exception to any generalisation may occur in discussions concerning conditions in the immediate vicinity of any abrupt change in geometry where subtle changes in the make-up of the model can lead to significant differences in response. The corresponding *integral* form for the Maxwell model is given by

$$p_{ik}^{E}(\mathbf{x}, t) = \frac{2\eta_0}{\lambda_1} \int_{-\infty}^{t} \exp\left[-(t-t')/\lambda_1\right] \frac{\partial x_i}{\partial x_m'} \frac{\partial x_k}{\partial x_r'} e_{mr}(\mathbf{x}', t') \, dt' \quad (2.5)$$

where x_i' is the position at time t' of the element that is instantaneously at the point x_i at time t. These 'displacement functions' can be determined from the equation (see Oldroyd, 1950).

$$\frac{\partial x_i'}{\partial t} + v_m \frac{\partial x_i'}{\partial x_m} = 0 \quad (2.6)$$

where t' takes any constant value.

An alternative form to eqn (2.5) can be obtained by a simple integration by parts which yields (see Court *et al.*, 1981)

$$p_{ik}^{E}(\mathbf{x}, t) = \frac{\eta_0}{\lambda_1^2} \left[\int_0^{\infty} \exp\left[-\tau/\lambda_1\right][F_{ik} - \delta_{ik}] \, d\tau \right] \quad (2.7)$$

where $\tau \, (= t - t')$ is the time lapse and the Finger deformation tensor F_{ik} is given by

$$F_{ik} = \frac{\partial x_i}{\partial x_m'} \frac{\partial x_k}{\partial x_m'} \quad (2.8)$$

We note that the covariant analogue to F_{ik} is the Cauchy–Green tensor G_{ik}, where

$$G_{ik} = \frac{\partial x_m'}{\partial x_i} \frac{\partial x_m'}{\partial x_k} \quad (2.9)$$

and that F_{ik} and G_{ik} are related (in the obvious matrix notation) by

$$FG = I \quad (2.10)$$

I being the unit matrix.

If we neglect terms of order $\lambda_1 \overset{\triangledown E}{p}_{ik}$ in eqns (2.2), (2.5) or (2.7), we obtain the so-called Second-Order Equivalent (SOE) given by (see Davies *et al.*, 1979)

$$p_{ik}^{E} = 2\eta_0 e_{ik} - 2\eta_0 \lambda_1 \overset{\triangledown}{e}_{ik} \quad (2.11)$$

which is a special case of the second-order model of Coleman and Noll (1961). We introduce the SOE at this stage to make the point that although explicit differential models of this sort are conceptually inappropriate for the flow situations under discussion, it would be wrong to deprecate all numerical simulations for such models. Existing work (e.g. Crochet and Pilate, 1976) can be used as a helpful guideline for the more complex numerical simulations for the implicit differential model, eqn (2.2), or the explicit integral models, eqns (2.5) and (2.7). Further, there is some evidence (see, for example, Tiefenbruck and Leal, 1982) to suggest that simulations for the SOE have a range of applicability far beyond that which we might have anticipated. At the same time, there is no doubt that, in the long term, the use of models like eqns (2.2) or (2.7) must be encouraged and that of eqn (2.11) discouraged in the type of complex flow situations under discussion.

At this point, it is helpful to focus attention on the two-dimensional flow geometry shown in Fig. 2.1, where AB and HG are tacitly assumed to be far

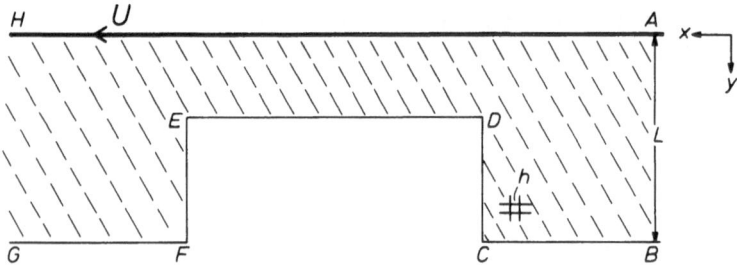

Fig. 2.1.

enough away from the protuberance for the flow there to be regarded as 'fully developed'. The flow is generated by movement of the plate AH in the direction indicated and the fully developed flows are therefore of the plane Couette type.

Taking Cartesian axes as shown, we denote the velocity components in the x and y directions by $u(x, y)$ and $v(x, y)$, respectively, and immediately introduce the non-dimensional variables

$$\left.\begin{array}{llll} x^* = \dfrac{x}{L} & y^* = \dfrac{y}{L} & u^* = \dfrac{u}{U} & v^* = \dfrac{v}{U} \\[2ex] p_{ik}^{E*} = p_{ik}^{E} \dfrac{L}{\eta_0 U} & p^* = \dfrac{pL}{\eta_0 U} & Re = \dfrac{\rho UL}{\eta_0} & Ws = \dfrac{\lambda_1 U}{L} \end{array}\right\} \quad (2.12)$$

where U is a characteristic velocity (which has already been specified in Fig. 2.1) and L is a characteristic length (e.g. the length AB in Fig. 2.1). The non-dimensional equations which govern the flow of the upper-convected Maxwell model, eqn (2.2), are then

$$\frac{\partial u}{\partial x} + \frac{\partial v}{\partial y} = 0 \tag{2.13}$$

$$f_x - \frac{\partial p}{\partial x} + \frac{\partial p_{xx}^{E}}{\partial x} + \frac{\partial p_{xy}^{E}}{\partial y} = Re \left[u \frac{\partial u}{\partial x} + v \frac{\partial u}{\partial y} \right] \tag{2.14}$$

$$f_y - \frac{\partial p}{\partial y} + \frac{\partial p_{xy}^{E}}{\partial x} + \frac{\partial p_{yy}^{E}}{\partial y} = Re \left[u \frac{\partial v}{\partial x} + v \frac{\partial v}{\partial y} \right] \tag{2.15}$$

$$p_{xx}^{E} \left(1 - 2Ws \frac{\partial u}{\partial x} \right) + Ws \left(u \frac{\partial p_{xx}^{E}}{\partial x} + v \frac{\partial p_{xx}^{E}}{\partial y} \right) - 2Wsp_{xy}^{E} \frac{\partial u}{\partial y} = 2 \frac{\partial u}{\partial x} \tag{2.16}$$

$$p_{yy}^{E} \left(1 - 2Ws \frac{\partial v}{\partial y} \right) + Ws \left(u \frac{\partial p_{yy}^{E}}{\partial x} + v \frac{\partial p_{yy}^{E}}{\partial y} \right) - 2Wsp_{xy}^{E} \frac{\partial v}{\partial x} = 2 \frac{\partial v}{\partial y} \tag{2.17}$$

$$- Wsp_{xx}^{E} \frac{\partial v}{\partial x} - Wsp_{yy}^{E} \frac{\partial u}{\partial y} + Ws \left(u \frac{\partial p_{xy}^{E}}{\partial x} + v \frac{\partial p_{xy}^{E}}{\partial y} \right) + p_{xy}^{E} = \frac{\partial u}{\partial y} + \frac{\partial v}{\partial x} \tag{2.18}$$

where f_x and f_y are the non-dimensional body forces in the x and y directions, respectively, and we have immediately dropped the star notation for convenience of presentation. Equation (2.13) is a statement of the equation of continuity, eqns (2.14) and (2.15) are the stress equations of motion and eqns (2.16)–(2.18) are determined from the rheological equations, eqn (2.2). The non-dimensional parameter Re is a Reynolds number and Ws is a non-dimensional elasticity parameter sometimes called the Weissenberg number.

From eqn (2.13) we can introduce a stream function ϕ, defined by

$$u = \frac{\partial \phi}{\partial y} \qquad v = -\frac{\partial \phi}{\partial x} \tag{2.19}$$

and we define the vorticity ω by

$$\omega = \frac{\partial v}{\partial x} - \frac{\partial u}{\partial y} \tag{2.20}$$

so that

$$\omega = -\nabla^2 \phi \tag{2.21}$$

∇^2 representing the Laplacian operator.

Eliminating the pressure p between eqns (2.14) and (2.15) yields

$$Re\left[\frac{\partial\phi}{\partial x}\frac{\partial\omega}{\partial y} - \frac{\partial\phi}{\partial y}\frac{\partial\omega}{\partial x}\right] = \frac{\partial^2 p^E_{xx}}{\partial x\,\partial y} + \frac{\partial^2 p^E_{xy}}{\partial y^2} - \frac{\partial^2 p^E_{xy}}{\partial x^2} - \frac{\partial^2 p^E_{yy}}{\partial x\,\partial y} \qquad (2.22)$$

The governing equations can now be taken to be eqns (2.16)–(2.18), (2.21) and (2.22). Inspection of these equations immediately reveals the distinctive difference between the present development and that corresponding to the Newtonian case for which $Ws = 0$, allowing the stress components to be substituted out of the governing equations yielding eqn (2.21) and

$$Re\left[\frac{\partial\phi}{\partial x}\frac{\partial\omega}{\partial y} - \frac{\partial\phi}{\partial y}\frac{\partial\omega}{\partial x}\right] = -\nabla^2\omega \qquad (2.23)$$

These equations must be solved for the *two* variables ϕ and ω. The stress components cannot be substituted out of the equations when $Ws \neq 0$ and we are therefore left with five equations in the five unknowns ϕ, ω, p^E_{xx}, p^E_{xy} and p^E_{yy}.†

Equations (2.10)–(2.18), (2.21) and (2.22) must be solved subject to certain prescribed boundary conditions and the general boundary-condition situation is one which merits some consideration.

Consider the solution to a general flow problem within a domain D as shown in Fig. 2.1. In the Newtonian case, we require knowledge of the velocity components over the boundary of D and the pressure at one point. In the case of highly elastic liquids, this is not in general sufficient on account of fluid memory, and we shall require kinematic information outside D, at least in the upstream direction. This requirement can present problems in actual flow situations, but it is often inadvertently overcome by assuming 'fully developed' flow conditions at the entry to a given domain (as in Fig. 2.1) which essentially implies knowledge of the flow outside the domain of interest.

For the geometry and flow indicated in Fig. 2.1, the boundary conditions on the stream function ϕ are supplied by the no-slip condition, which on the stationary boundary BCDEFG implies $u = v = 0$ while on AH we have $u = 1$, $v = 0$. Boundary conditions on the vorticity can also be approximated from the no-slip condition (cf. Section 2.2).

Conditions over AB are given by the known fully developed flow

† In the corresponding two-dimensional *axisymmetric* case, there are *six* equations with six unknowns comprising ϕ, ω and *four* stress components (see Crochet and Keunings, 1980).

behaviour there, which is easily obtained by substituting $u = u(y)$ into the governing equations.

Analytic solutions to eqns (2.16)–(2.18), (2.21) and (2.22) are out of the question and to make progress it is therefore necessary to resort to numerical techniques. Two basic choices are available.† The first involves the application of (conventional) finite-difference (FD) techniques, which are well documented in the Newtonian literature (see, for example, Roache, 1972). The second alternative involves the use of finite-element (FE) methods which are relatively new and are still under active development even in the Newtonian case. One can make out a case for employing both techniques. Finite-difference methods are certainly easier to implement than the equivalent FE techniques, but an FE algorithm, once constructed, allows the flow geometry to be changed with relative ease. Free boundaries have also succumbed more readily to FE techniques, although there is some evidence to suggest that FD methods can also be applied for this purpose (see, for example, Thames *et al.*, 1977; Thompson *et al.*, 1977). However, all existing non-Newtonian free boundary simulations have been obtained using FE techniques, and it must be acknowledged that there is a strong bias in favour of FE methods in present studies in non-Newtonian flow simulation. We shall nevertheless discuss both FD and FE techniques in the present chapter before ending with a discussion of the distinctive problems which are common to both.

2.2   FINITE-DIFFERENCE TECHNIQUES

The governing equations in Newtonian fluid mechanics are given by eqns (2.21) and (2.23) and it has been customary to use a (ϕ, ω) formalism to solve flow problems using finite differences rather than employ u, v and p as primitive variables.‡ Not surprisingly, developments in the non-Newtonian case have also tended to retain ϕ and ω as variables with the necessary addition of the stress components as pointed out in Section 2.1. Some attempts have been made to employ FD techniques to eqns (2.16)–

---

† Boundary integral, Spectral and Multigrid methods have also been employed with success in Newtonian calculations.

‡ The (ϕ, ω) formalism lends itself to a *conventional* finite-difference grid, whereas a (u, v, p) formalism demands the use of a *staggered* grid. Early calculations in Newtonian fluid mechanics favoured the former approach, but more recently the latter has been increasingly adopted, particularly when knowledge of the pressure field is required (for example, in free surface problems).

(2.18), (2.21) and (2.22) as they stand (see, for example, Gatski and Lumley, 1978a, b) but inspection of eqns (2.22) and (2.23) reveals that the basic structure in the vorticity variable has been lost in the non-Newtonian case. Accordingly, the most popular method of proceeding has been to introduce the substitutions (see Leal, 1979; Perera and Walters, 1977a)

$$
\left.
\begin{aligned}
p_{xx}^{\mathrm{E}} &= \bar{p}_{xx} + 2\,\frac{\partial u}{\partial x} \\[2mm]
p_{xy}^{\mathrm{E}} &= \bar{p}_{xy} + \left(\frac{\partial u}{\partial y} + \frac{\partial v}{\partial x}\right) \\[2mm]
p_{yy}^{\mathrm{E}} &= \bar{p}_{yy} + 2\,\frac{\partial v}{\partial y}
\end{aligned}
\right\}
\qquad (2.24)
$$

in which case eqn (2.22) becomes

$$
\nabla^2 \omega + Re\left[\frac{\partial \phi}{\partial x}\frac{\partial \omega}{\partial y} - \frac{\partial \phi}{\partial y}\frac{\partial \omega}{\partial x}\right] = \frac{\partial^2 \bar{p}_{xx}}{\partial x\,\partial y} + \frac{\partial^2 \bar{p}_{xy}}{\partial y^2} - \frac{\partial^2 \bar{p}_{xy}}{\partial x^2} - \frac{\partial^2 \bar{p}_{yy}}{\partial x\,\partial y} \qquad (2.25)
$$

and the elliptic nature of the Newtonian equation, eqn (2.23), is recovered. Similar substitutions have been made by Holstein (1981) and Townsend (1980b) with the same essential end in view.

Equations (2.16)–(2.18), (2.21), (2.24) and (2.25) represent a set of non-linear coupled differential equations in the variables ϕ, ω, $\bar{p}_{xx}$, $\bar{p}_{xy}$ and $\bar{p}_{yy}$.

The finite-difference method entails the replacement of *differential* operators by *difference* operators and results in a system of non-linear coupled algebraic equations in the dependent variables at a number of discrete points on an appropriate coordinate grid. It is possible to utilise a variable mesh but we shall limit attention in the present discussion to a square mesh defined by a grid length h.

With reference to Fig. 2.2, *central* differences such as

$$
\frac{\partial f_0}{\partial x} = \frac{f_3 - f_1}{2h} \qquad \frac{\partial^2 f_0}{\partial x^2} = \frac{f_3 - 2f_0 + f_1}{h^2} \qquad (2.26)
$$

can be defined (in the obvious notation) with truncation errors of order h^2 and *one-sided* differences such as

$$
\text{(i)}\ \frac{\partial f_0}{\partial x} = \frac{f_3 - f_0}{h} \qquad \text{or} \qquad \text{(ii)}\ \frac{\partial f_0}{\partial x} = \frac{f_0 - f_1}{h} \qquad (2.27)
$$

where (i) is a 'forward' difference and (ii) is a 'backward' difference. These have truncation errors of order h, so that implementing forward or

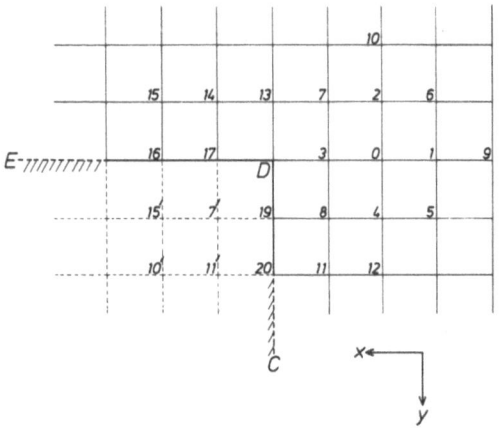

Fig. 2.2.

backward differences leads to errors that are more severe than in the central-differencing alternative.

The traditional numerical approach is to linearise the governing equations by a decoupling approach. This means, for example, that in solving the 'vorticity' equation, eqn (2.25), the stress components $\bar{p}_{xx}$, $\bar{p}_{xy}$ and $\bar{p}_{yy}$ and the stream function ϕ are considered as known, while in the solution of the rheological equations for the stress components, the velocity field is considered as known.

Applying difference formulae to the decoupled equations, each results in a matrix equation of the form

$$Ax = b \tag{2.28}$$

where A is the coefficient matrix, x is the unknown vector and b is a vector containing given information. A is invariably a sparse matrix and there are sound reasons for choosing the difference operators to ensure that A is diagonally dominant.

Many methods exist for the solution of eqn (2.28) and these fall into separate categories, i.e. direct or iterative schemes. Much work has been done in recent years on direct schemes applied to sparse matrix systems and a comprehensive survey has been undertaken by Reid (1977).

The alternative iterative methods begin with an initial approximation and produce a sequence of values or iterates, which, if convergent, reach an approximate solution after a finite number of steps, the precise number depending on such factors as the form of the coefficient matrix A, the initial approximation and the criterion for convergence. Classical iterative schemes are Successive Over Relaxation (SOR), Gauss–Seidel (GS), Alternating Direction Implicit (ADI) and Conjugate Gradients (CG). These are extensively described (Meijerink and van der Vorst, 1977; Reid, 1971; Varga, 1962; Young, 1971); Kershaw (1978) discussed the four methods using a model problem. Up to the present time, studies in non-Newtonian fluid mechanics have concentrated almost exclusively on SOR techniques although there have been some recent attempts to employ preconditioned CG methods (see Court et al., 1981).

The SOR technique has been a natural choice in numerical simulation in non-Newtonian fluid mechanics for a number of reasons, including the following:

(i)   it is well established in computational aspects of *Newtonian* fluid mechanics;
(ii) it is relatively easy to program;
(iii) a rigorous theory of the method exists and is well documented (see Varga, 1962; Young, 1971).

We shall summarise the simplest version of the method, namely Point SOR, by reference to eqn (2.21), which is of course common to incompressible Newtonian and non-Newtonian fluid mechanics.

Assuming the vorticity ω is known at all nodes on the interior of the domain of interest, we can apply at any point (i, j) a standard five point central difference formula to eqn (2.21) giving

$$4\phi_{i,j} - \phi_{i-1,j} - \phi_{i,j-1} - \phi_{i+1,j} - \phi_{i,j+1} = h^2 \omega_{i,j} \qquad (2.29)$$

In order to apply the method of SOR, we now rewrite eqn (2.29) in the form (see Roache, 1972)

$$\phi_{i,j}^{k+1} = \frac{f_\phi}{4} [\phi_{i+1,j}^k + \phi_{i-1,j}^{k+1} + \phi_{i,j+1}^k + \phi_{i,j-1}^{k+1} + h^2 \omega_{i,j}] + (1 - f_\phi)\phi_{i,j}^k \qquad (2.30)$$

where $k + 1$ is the current iteration number and f_ϕ is the so-called relaxation parameter for the SOR method which satisfies

$$0 < f_\phi < 2 \qquad (2.31)$$

For the complex flows under discussion, f_ϕ has to be determined by numerical experiment by solving eqn (2.30) for different values of f_ϕ and monitoring convergence for large k. If $1 < f_\phi < 2$ we have *over-relaxation*, if $0 < f_\phi < 1$ *under-relaxation* and $f_\phi = 1$ corresponds to *Gauss–Seidel* iteration.

An iteration such as eqn (2.30) is termed an *inner* iteration and ϕ^k the kth inner iterate. A sequence of such iterates is taken as converged if

$$\|\phi^k - \phi^{k-1}\| < \varepsilon_\phi \|\phi^{k-1}\| \tag{2.32}$$

where $\| . \|$ denotes some convenient norm (e.g. the supremum norm) and ε_ϕ is a relative tolerance factor. Once eqn (2.32) is satisfied, ϕ^k is called an *outer* iterate and is denoted by $\bar{\phi}^{(k)}$. In proceeding from one outer iteration to the next, the current solution is smoothed by taking its weighted average with the previous corresponding iterate as follows:

$$\tilde{\phi}^{(k)} = \xi_\phi \bar{\phi}^{(k)} + (1 - \xi_\phi)\tilde{\phi}^{(k-1)} \tag{2.33}$$

where ξ_ϕ $(0 < \xi_\phi < 1)$ is called a smoothing parameter. A convergence criterion is now applied to the outer iterates:

$$\|\tilde{\phi}^{(k)} - \tilde{\phi}^{(k-1)}\| < \tilde{\varepsilon}_\phi \|\tilde{\phi}^{(k-1)}\| \tag{2.34}$$

where $\tilde{\varepsilon}_\phi$ is another relative tolerance factor.

With careful ordering of the elements of the relevant x in eqn (2.28), the coefficient matrix A can be arranged to have desirable properties which will guarantee convergence of SOR. One property is diagonal dominance which is a sufficient condition for convergence of Gauss–Seidel and under-relaxation. However, in order to accelerate convergence it is well known that *over*-relaxation is useful, but here diagonal dominance is not in general sufficient, the desirable property of the matrix in this case being consistent ordering (see Varga, 1962).

The solution of the vorticity equation, eqn (2.25), and the rheological equations, eqns (2.16)–(2.18), require more detailed discussion. First, when second-order (central) discretisation formulae are employed, the resulting coefficient matrix is not diagonally dominant except for small *Re*. Accordingly, it is necessary to apply an upwind/downwind differencing scheme of the sort suggested by Greenspan (1968) to retrieve the diagonal dominance property. This reduces the formal order of accuracy of the numerical scheme as we have indicated, but this is a cost which has to be paid to retain the desired property of diagonal dominance, thus ensuring the convergence of Gauss–Seidel and under-relaxation schemes.

The boundary conditions on the vorticity and the stress components have to be determined carefully from the definition of ω and the rheological equations, respectively. So far as the vorticity is concerned, one can either apply first-order formulae of the sort described by Roache (1972) or second-order formulae associated with the names of Woods (1954) and Jensen (1959). The need for upwind/downwind differencing and other factors leads to the reluctant conclusion that no improvement can be expected in *non*-Newtonian fluid mechanics using schemes of higher order than that of Roache.

The corner points C, D, E and F in Fig. 2.1 need special attention since they are part of the boundary and it is necessary to specify the vorticity and the stress components there to develop a numerical solution. The salient corners C and F present no problems since it is easily argued that the vorticity and the extra stress components must be zero at such points. This is certainly not the case at the re-entrant corners D and E and indeed Moffatt (1964) has shown that the vorticity is infinite at a re-entrant corner even in the Newtonian case. To say the least, it is clearly essential to model the flow as accurately as possible near re-entrant corners and there is no doubt that the most appropriate line of attack is to seek an analytic asymptotic solution near the corner which can be matched to the finite-difference scheme being developed. This has already been successfully undertaken for the Newtonian case (see Holstein and Paddon, 1981) and is being actively pursued when the model is of the Oldroyd type with a non-zero retardation time λ_2. The Maxwell model has so far been reluctant to admit to an asymptotic corner analysis and this has led to a healthy research interest in the asymptotic behaviour of simple elastico-viscous models under conditions where changes in velocity and stress are severe. The problems are not yet fully resolved, but there is no doubt that subtle changes in the model structure can lead to dramatically different behaviour under the severe conditions encountered near abrupt changes in geometry. For example, inspection of the appropriate models reveals that when the time rates of change of p_{ik}^E and e_{ik} dominate the behaviour, the asymptotic form is given by (see Cochrane *et al.*, 1982)

$$p_{ik}^E = 2\eta_0 \frac{\lambda_2}{\lambda_1} e_{ik} \qquad (2.35)$$

for the Oldroyd model, eqn (2.4), but by

$$p_{ik}^E = \frac{2\eta_0}{\lambda_1} \varepsilon_{ik} \qquad (2.36)$$

for the Maxwell model, eqn (2.2), where ε_{ik} is a strain tensor. Here we have an important issue to decide. Do we wish our fluids to behave as a viscous fluid (i.e. like eqn (2.35)) or as an elastic solid (i.e. like eqn (2.36)) near an abrupt change in geometry? At the present time, the answer to this important question has not been completely resolved.

Until such time as a consistent asymptotic corner strategy is available, it is simply a case of choosing the best of the many available *ad hoc* methods.

A number of these are cited by Roache (1972) and the most popular within the non-Newtonian application has been that due to Kawaguti (1965) which involves the introduction of fictitious wall values of the variables near the re-entrant corner. The method treats ϕ as if it is symmetric about the corner point D in Fig. 2.2, creating fictitious values $\bar{\phi}$ along the walls (multi-valued points), i.e. $\bar{\phi}_{17} = \phi_3$ and $\bar{\phi}_{19} = \phi_{13}$ and the corner derivatives in the Poisson equation are evaluated at D based on these $\bar{\phi}$. Effectively this method evaluates ω_D as though the corner were an interior point and ω_D is given ultimately by

$$\omega_D = -\frac{2}{h^2} (\phi_3 + \phi_{13} - 2\phi_D) \qquad (2.37)$$

The asymptotic analysis work of Holstein and Paddon (1981) for the Newtonian problem gives some indirect support for the Kawaguti approach.

Some recent experimental results (Walters and Webster, 1982) have indicated that small changes in corner geometry, which lead to only small changes in flow characteristics in the Newtonian case except in the immediate vicinity of the corner, can, in the elastico-viscous case, have a dramatic effect on the flow over a substantial region which stretches significantly beyond the corner neighbourhood. This means that the corner problem needs to be carefully studied and satisfactorily resolved in the non-Newtonian case before complete confidence can be attached to any numerical simulations.

The final set of equations involved in the numerical simulation are the rheological equations of state, eqns (2.16)–(2.18), which are considered as linearly coupled equations in the dependent variables p_{xx}^E, p_{xy}^E and p_{yy}^E. In these equations, forward or backward finite-difference operators for the stress derivatives are employed, thereby ensuring diagonal dominance of the coefficient matrices and ensuring convergence of Gauss–Seidel and under-relaxation schemes.

We are now in a position to discuss the implementation of a global algorithm to solve eqns (2.16)–(2.18), (2.21), (2.24) and (2.25).

The basic approach is to proceed iteratively from an initial guess for all the unknown function values. Starting from these initial guesses, the main steps are as follows:

Step 1.   Solve the rheological equations, eqns (2.16)–(2.18) and (2.24) iteratively for the stress components, considering these equations to be linearly coupled equations in p_{xx}^E, p_{xy}^E and p_{yy}^E within each inner iteration.

Step 2.   Use the initial ϕ guess and the computed stress components from Step 1 to solve the vorticity equation, eqn (2.25).

Step 3.   Solve the stream function equation, eqn (2.21), for ϕ using the ω computed in Step 2.

Step 4.   Test for convergence. If this has not been achieved, return to Step 1 with the new ϕ field and proceed until convergence is reached.

Acceptable relative tolerance specifications for iterative convergence in the variables $[\phi$, stress, $\omega]$ can be taken to be $[10^{-4}, 10^{-2}, 10^{-1}]$. This choice is felt to be consistent with the *discretisation errors* of the formulae used to approximate the governing equations (see, for example, Cochrane *et al.*, 1981).

In implementing the SOR technique, numerous additional strategies arise through numerical experimentation which optimise the process. Rarely do these strategies find their way into published papers on the subject, but they are contained in numerous doctoral theses (Citroen, 1980; Court, 1980; Holstein, 1981; Manero, 1980; Paddon, 1979; Perera, 1976; Tiefenbruck, 1979; Webster, 1979) which provide a useful source of practical advice on numerical simulation in what is after all a relatively new area of research. We list some of the more important issues:

(i)   The 'Newtonian' solution provides a convenient starting guess for flow problems involving slightly elastic liquids, and a process of 'continuation' in the parameter Ws provides a convenient method of reaching relatively high values of Ws.

(ii)   To reach the final (and often severe) tolerance levels, it is often convenient to work to less severe tolerances in the first instance and to use the associated solution as the initial guess in the full problem.

(iii) To avoid oscillatory behaviour in the fully developed flow regions in the case of highly elastic liquids, it is advisable to 'anchor' the solution to a depth of 3 grid spacings in these regions.

Other additional strategies can be found in the theses already cited, but it is a salutary thought that any amount of numerical experimentation has not led to solutions beyond a fairly modest Ws level! Since this is a general problem which is not confined to fine-difference techniques or differential models, we shall leave a discussion of it to the concluding section of this chapter.

Since the integral equations, eqns (2.5) and (2.7), represent the same response as the differential equation, eqn (2.2), it is of interest to discuss the alternative possibilities of using FD techniques to handle explicit integral equations.

Inspection of the integral model is revealing since it highlights one of the problems of studying fluids with long-range memory in complex flow situations. Before one is able to solve the flow problem through a determination of u and v, it is necessary to obtain the displacement functions (x' and y') which are of course unknown until u and v are known. Some iterative technique is therefore essential if progress is to be made.

It is found to be more convenient to consider the integral model in the form of eqn (2.7) rather than of eqn (2.5). This avoids the necessity of employing an interpolation procedure to determine $e_{mr}(\mathbf{x}', t')$. Furthermore, after determining x' and y' from the hyperbolic equation, eqn (2.6), the most convenient method of calculating the components of F_{ik} is to determine the components of G_{ik} from eqn (2.9) and use eqn (2.10) through a simple matrix-inversion technique. In this way, the physical variables are determined numerically at grid points and no interpolation procedure is required (see Court et al., 1981).

The basic equations in the case of the integral model are eqns (2.6), (2.7), (2.21) or (2.24) and (2.25) and the dependent variables are ϕ, ω and the displacement functions x' and y'. A similar global strategy to that employed in the differential model case is possible with Step 1 now replaced by:

Step 1 (alternative).   Determine the displacement functions by solving eqn (2.6) corresponding to the initial guess for the ϕ field. Compute G_{ik}, F_{ik} and p_{ik}^{E}.

For steady flow, the equations for x' and y' can be written in the form

$$\left.\begin{array}{l} \dfrac{\partial x'}{\partial \tau} + u\, \dfrac{\partial x'}{\partial x} + v\, \dfrac{\partial x'}{\partial y} = 0 \\[3mm] \dfrac{\partial y'}{\partial \tau} + u\, \dfrac{\partial y'}{\partial x} + v\, \dfrac{\partial y'}{\partial y} = 0 \end{array}\right\} \tag{2.38}$$

where τ is now the non-dimensional time lapse given by $(U/L)(t - t')$.

The boundary conditions on ϕ and ω have already been discussed. Those on x' and y' for the flow geometry in Fig. 2.1 are given by

$$\left.\begin{array}{ll} x' = x, \; y' = y & \text{for all } \tau, \text{ on BDCEFG} \\[2mm] x' = x - (1-y)\tau, \; y' = y & \text{on AB and GH} \\[2mm] x' = x - \tau, \; y'; \; y & \text{on AH} \end{array}\right\} \tag{2.39}$$

Corresponding boundary conditions can also be found on the components of F_{ik} and p_{ik}^E (see, for example, Court et al., 1981 and Tiefenbruck, 1979).

Numerous methods are available to solve the first-order hyperbolic equations, eqn (2.38), for the displacement functions x' and y'. However, existing FD work has concentrated on the Lax Wendroff scheme as modified by Mitchell and Griffiths (1980). A convenient check on the accuracy of the computed displacement functions is provided by the condition

$$\det F_{ik} = 1 \qquad (2.40)$$

which is an alternative mathematical statement of the equation of continuity, eqn (2.13).

One of the difficulties associated with the numerical integration involved in eqn (2.7) is the choice of the finite time values at which the semi-infinite time interval of integration should be terminated. Too severe a truncation would result in important past-history information being neglected, whereas too long a time interval would not only prove expensive, but also numerical inaccuracies in the integral could emerge at long times. This is of course a problem which is common to FD and FE techniques and the present consensus of opinion favours the use of Gauss–Laguerre quadrature (see Court et al., 1981; Viriyayuthakorn and Caswell, 1980).

The proposed FD scheme for integral models has been successfully implemented for relatively small values of λ_1 (i.e. Ws), but the lack of convergence at high values of Ws already noted in connection with differential models is also apparent in the integral programs, failure taking place at comparable values of Ws.

The conventional FD method supplies ϕ and ω in the Newtonian problem and ϕ, ω and the stress components in the non-Newtonian case—but not the pressure, p. This has to be determined by returning to eqns (2.14) and (2.15). Simple integration of these equations (with known stress data) is one obvious way of determining p, but it is generally admitted that the method is open to numerical error (made manifest as multi-valued pressures). An alternative is to form a Poisson-type equation by differentiating eqns (2.14) and (2.15) to yield an equation of the form

$$\nabla^2 p = 2Re \left[\frac{\partial^2 \phi}{\partial x^2} \frac{\partial^2 \phi}{\partial y^2} - \left(\frac{\partial^2 \phi}{\partial x \, \partial y} \right)^2 \right] + \left[\frac{\partial^2 p_{xx}^E}{\partial x^2} + \frac{\partial^2 p_{xy}^E}{\partial x \, \partial y} + \frac{\partial^2 p_{yy}^E}{\partial y^2} \right] \qquad (2.41)$$

which, of course, has Neumann boundary conditions (except on free surfaces). The solution of eqn (2.41) is far from a straightforward matter, but the basic problem is not restricted to non-Newtonian situations and the Newtonian literature is available as a guide (see, for example, Briley, 1978).

The FD techniques discussed in this section have been successfully employed in non-Newtonian fluid mechanics to investigate the following flows:

(i)   flow past a stationary and a rotating cylinder (Townsend, 1980a);

(ii) flow past a sphere and a bubble (Tiefenbruck and Leal, 1982);

(iii) two-dimensional planar and axisymmetric pressure-driven flow past a protuberance, over a hole and in L-shaped and T-shaped geometries (Citroen, 1980; Cochrane *et al.*, 1981; Cochrane *et al.*, 1982; Court *et al.*, 1981; Davies *et al.*, 1979; Holstein, 1981; Perera and Strauss, 1979; Perera and Walters, 1977a; Townsend, 1980b; Walters and Webster, 1982);

(iv) flow through a planar contraction with and without abrupt changes in geometry (Cochrane *et al.*, 1981; Gatski and Lumley, 1978a, 1978b; Perera and Walters, 1977b).

2.3 FINITE-ELEMENT TECHNIQUES

The development of finite elements for solving non-Newtonian flow problems does not benefit from the long history of finite-difference techniques; the first application of finite elements for solving the Navier–Stokes equations is indeed attributed to Oden (1970). However, a continuous stream of scientific papers reveals intensive research in the field over the last ten years. A good overview may be obtained by inspecting the books edited by Gallagher (1975a,b; 1978), but the selection of the best method for solving a given type of problem remains a question of debate; a good example is given in the proceedings of a recent symposium devoted to the sole problem of the advective terms in the Navier–Stokes equations (Hughes, 1979), where papers alternate in favouring or dissuading the use of some new techniques for handling the non-linear terms. Finite elements have been developed by engineers within the context of linear elasticity theory where, fortunately, minimum principles provide means of evaluating the error, and uniqueness theorems give the necessary confidence; later, mathematicians were able to show that the engineers were right in their use of the finite-element concept, and built the theoretical background for linear elliptic problems (see, for example, Strang and Fix, 1973). In view of the analogy between the equations of elasticity and Stokes flow (i.e. creeping Newtonian flow), one may say that finite-element techniques for solving Stokes flow are built on solid foundations (see Girault and Raviart, 1981).

It may be shown that, under some conditions pertaining to the geometry of the domain and the smoothness of boundary conditions, some elements are valid while others may generate meaningless numbers; moreover, results have also been obtained for the full Navier–Stokes equations (Girault and Raviart, 1981; Témam, 1977). The situation is much more obscure in non-Newtonian fluid mechanics. The system, eqns (2.13)–(2.18), shows that the non-linear inertia terms seem almost harmless when they are compared to the memory terms identified by the factor Ws; uniqueness and existence theorems are not available. While today's technology in polymer processing is fond of numerical simulation which is altogether illuminating and, at least potentially, cheaper than experimentation, the numerical analyst is faced with two main problems: (i) generating readable numbers on the computer rather than overflow situations, (ii) verifying that the readable numbers, when they are obtained, are also meaningful. Unfortunately, the manpower and energy required by the former problem may deplete the attention devoted to the latter; a good example of the necessity of a careful assessment of available results is given by Mendelson et al. (1982).

Before explaining recent developments of numerical algorithms for solving non-Newtonian flow, we will first review briefly the most widely used techniques for solving the Navier–Stokes equations by means of finite elements, since they always constitute the starting point of more complex algorithms. Thus, let us consider the system formed by the field equations, eqns (2.13) to (2.15), together with the constitutive relations

$$p_{xx}^{E} = 2\,\frac{\partial u}{\partial x} \qquad p_{yy}^{E} = 2\,\frac{\partial v}{\partial y} \qquad p_{xy}^{E} = \frac{\partial u}{\partial y} + \frac{\partial v}{\partial x} \qquad (2.42)$$

where we recall that all variables are presently non-dimensional. The equations are written here for plane flow; we will see later how axisymmetric flow may be calculated without additional difficulty. While the use of the vorticity ω and stream function ϕ is widespread in finite-difference work, the opposite is true for finite-element algorithms; this observation is justified by the presence of high-order derivatives of ϕ and difficulties encountered with boundary conditions, which deprecate in particular the adoption of a $\phi - \omega$ finite-element formulation for solving non-Newtonian flow problems. The *mixed formulation* in which u, v and p are the primitive variables is certainly the main tool of engineering calculations; recent progress has been made on the elimination of the pressure by means of a

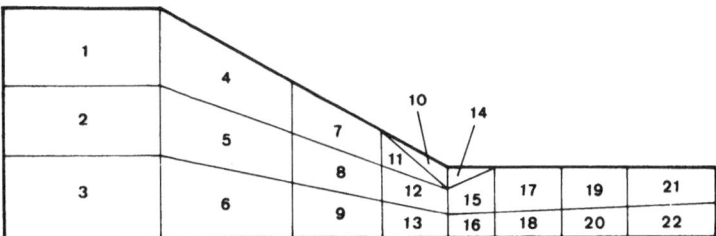

Fig. 2.3.   Finite element mesh.

penalty formulation (Heinrich and Marshall, 1981; Hughes *et al.*, 1979) but
the question will not be reviewed here since, with one exception (Bernstein
et al., 1981) the technique has not been used as yet in non-Newtonian fluid
mechanics.

In the finite-element formulation, the domain of integration D is covered
by a mesh of E *finite elements* Δ^n, $1 \leq n \leq E$ (Fig. 2.3); on each element, we
identify *nodes* and associated *nodal values* of the primitive variables. The
number of nodal values may differ from one primitive variable to another;
we will see that the number of nodal values for the pressure is lower than
the number of nodal values for u and v. It may also happen that a 'nodal
value' is not related directly to a node in the mesh; this is the case in
particular when the pressure is taken as a constant over an element. By
means of the nodal values, we wish to define an *interpolation* of the
unknown fields u, v and p, which will be denoted $\tilde{u}$, $\tilde{v}$ and $\tilde{p}$. Let N and M be
the number of nodal values for the velocity field and the pressure,
respectively; the interpolation is given by†

$$\tilde{u} = \sum_{i=1}^{N} U_i \psi_i(x, y) \qquad \tilde{v} = \sum_{i=1}^{N} V_i \psi_i(x, y) \qquad \tilde{p} = \sum_{i=1}^{M} P_i \phi_i(x, y) \quad (2.43)$$

The functions ψ_i and ϕ_i are called shape functions, and they are subject to
some conditions which are indispensable for the success of the finite
element method. Let x_j, y_j be the coordinates of node j associated with
nodal values U_j, V_j; we will require that

$$\psi_i(x_j, y_j) = \delta_{ij} \qquad 1 \leq i, j \leq N \qquad (2.44)$$

Moreover, let $\bar{\Delta}^i$ denote the union of elements to which node i belongs; the

† The shape functions ϕ_i should not be confused with the stream function ϕ
introduced earlier.

shape function ψ_i is of compact support, that is it must vanish outside $\bar{\Delta}^i$, i.e.

$$\psi_i(x, y) = 0 \qquad (x, y) \notin \bar{\Delta}^i \tag{2.45}$$

Similar conditions may be written for the shape functions ϕ_i and the associated pressure nodes. Condition (2.45) is one of the cornerstones of the finite-element procedure as opposed, for example, to fast Fourier transforms. When discussing the selection of a shape function, it will be sufficient to consider a single element belonging to $\bar{\Delta}^i$; the same symbol $\psi_i(x, y)$ will be used when (x, y) belongs to $\bar{\Delta}^i$ or $\Delta^k \subset \bar{\Delta}^i$ (Fig. 2.4).

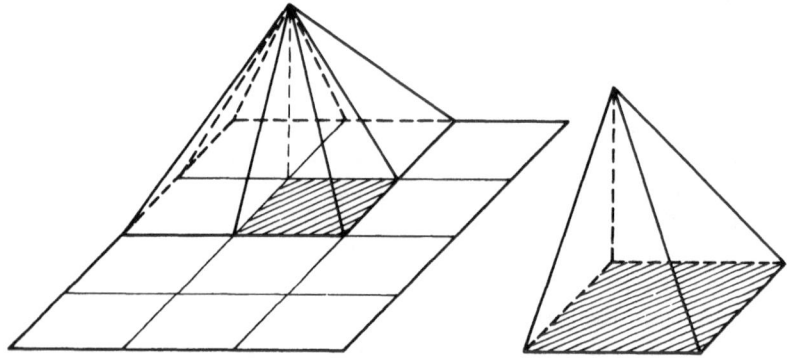

Fig. 2.4. Typical shape function over the domain of integration and at the element level.

The central problem of finite-element calculations is the selection of the shape functions $\psi_i(x, y)$ and $\phi_i(x, y)$ which will not be carried out however until we examine how they will be used. Let us therefore assume for the time being that the selection has been made, pursue the analysis, and return later to their identification. Let us introduce the approximation, eqn (2.43), in the constitutive equations, eqn (2.42), and obtain the stress values

$$p_{xx}^* = 2\frac{\partial \tilde{u}}{\partial x} \qquad p_{yy}^* = 2\frac{\partial \tilde{v}}{\partial y} \qquad p_{xy}^* = \frac{\partial \tilde{u}}{\partial y} + \frac{\partial \tilde{v}}{\partial x} \tag{2.46}$$

where the star superscript is used in opposition to the tilde since, at the present stage, the stresses *are not* associated with nodal values; we will later introduce symbols $\tilde{p}_{xx}$, $\tilde{p}_{yy}$ and $\tilde{p}_{xy}$ when expansions similar to eqn (2.43) are introduced for the extra-stress components. The approximations, eqns (2.43) and (2.46), will not in general satisfy exactly eqns (2.13), (2.14) and (2.15); their insertion in the field equations will give rise to *residuals* which are defined as follows:

$$\mathbf{R}_p(x, y) = \frac{\partial \tilde{u}}{\partial x} + \frac{\partial \tilde{v}}{\partial y}$$

$$\mathbf{R}_u(x, y) = \frac{\partial \tilde{p}}{\partial x} + \frac{\partial p_{xx}^*}{\partial x} + \frac{\partial p_{xy}^*}{\partial y} + f_x - Re\left(\tilde{u}\frac{\partial \tilde{u}}{\partial x} + \tilde{v}\frac{\partial \tilde{u}}{\partial y}\right)$$

$$\mathbf{R}_v(x, y) = -\frac{\partial \tilde{p}}{\partial y} + \frac{\partial p_{xy}^*}{\partial x} + \frac{\partial p_{yy}^*}{\partial y} + f_y - Re\left(\tilde{u}\frac{\partial \tilde{v}}{\partial x} + \tilde{v}\frac{\partial \tilde{v}}{\partial y}\right) \qquad (2.47)$$

A first method for evaluating the nodal velocity components and pressures is to require that the residual functions $\mathbf{R}_p$, $\mathbf{R}_u$ and $\mathbf{R}_v$ vanish at a discrete set of points. Let (x_i^v, y_i^v), (x_i^p, y_i^p) be the *collocation points* where the equations of motion and of mass conservation must be satisfied, respectively. We have

$$\mathbf{R}_p(x_i^p, y_i^p) = 0 \qquad 1 \le i \le M$$

$$\mathbf{R}_u(x_i^v, y_i^v) = 0 \qquad \mathbf{R}_v(x_i^v, y_i^v) = 0 \qquad 1 \le i \le N \qquad (2.48)$$

It will be found later that some of eqns (2.48) must be replaced by boundary conditions. Equations (2.48) form the *collocation method*; its foundation lies in the satisfaction of the field and constitutive equations at a finite number of nodes; the system contains as many equations as unknown nodal values. Despite its theoretical simplicity, the collocation method is difficult to apply; the reason is made evident by inspection of the right-hand sides of eqn (2.47). Indeed, the choice of shape functions in eqn (2.43) must be such that the residuals exist; since the gradients of the stresses p_{ij}^* contain *second-order derivatives* of the approximate velocity field, the shape functions ψ_i must be C^1-continuous (i.e. second-order spatial derivatives must be piecewise continuous), while the pressure gradients in $\mathbf{R}_u$ and $\mathbf{R}_v$ impose that ϕ_i be C^0-continuous. C^1-continuity requires elaborate elements and difficult programming. The only example of the use of collocation in solving non-Newtonian flow problems may be found in the work of Chang *et al.* (1979), and the matter will not be pursued here.

The continuity requirements on the shape functions will be relaxed with the use of the so-called *weighted residuals* approach; a detailed treatment of the method together with many applications to engineering problems may be found in the work of Finlayson (1972). Instead of requiring that the residuals vanish at a discrete set of nodes, let us select *M weighting functions*

$\bar{w}_i(x, y)$ and N weighting functions $w_i(x, y)$, and require that the residuals vanish *in the mean*, with the following rule:

$$\int_D \mathbf{R}_p \bar{w}_i \, dD = 0 \qquad 1 \le i \le M$$

$$\int_D \mathbf{R}_u w_i \, dD = 0 \qquad \int_D \mathbf{R}_v w_i \, dD = 0 \qquad 1 \le i \le N \qquad (2.49)$$

Provided the shape functions w_i are C^0-continuous, we may integrate by parts some of the terms in the weighted momentum equations, and obtain the following set:

$$\int_D \left(\frac{\partial \tilde{u}}{\partial x} + \frac{\partial \tilde{v}}{\partial y} \right) \bar{w}_i \, dD = 0 \qquad 1 \le i \le M$$

$$\int_D \left[(-\tilde{p} + p_{xx}^*) \frac{\partial w_i}{\partial x} + p_{xy}^* \frac{\partial w_i}{\partial y} + Re \left(\tilde{u} \frac{\partial \tilde{u}}{\partial x} + \tilde{v} \frac{\partial \tilde{u}}{\partial y} \right) w_i \right] dD$$

$$= \int_D f_x w_i \, dD + \int_{\partial D} t_x^* w_i \, dl$$

$$\int_D \left[p_{xy}^* \frac{\partial w_i}{\partial x} + (-\tilde{p} + p_{yy}^*) \frac{\partial w_i}{\partial y} + Re \left(\tilde{u} \frac{\partial \tilde{v}}{\partial x} + \tilde{v} \frac{\partial \tilde{v}}{\partial y} \right) w_i \right] dD$$

$$= \int_D f_y w_i \, dD + \int_{\partial D} t_y^* w_i \, dl \qquad (2.50)$$

where t_x^*, t_y^* are the components of the surface force on ∂D obtained from eqn (2.46). The set of equations, eqn (2.50), constitute the *weak form* of the Navier–Stokes equations; a mathematical discussion of the weak form and its validity may be found for example in the work of Girault and Raviart (1981). If, instead of eqn (2.48), we select eqn (2.50) as our fundamental equations for calculating the unknowns U_i, V_i and P_i, we find that the spatial derivatives do not exceed the *first-order* in $\tilde{u}$ and $\tilde{v}$, while no derivatives of $\tilde{p}$ appear in the equations. The continuity requirements on the velocity field will be that $\tilde{u}$ and $\tilde{v}$ be C^0-continuous while $\tilde{p}$ may even be piecewise continuous (C^{-1}-continuous). In obtaining eqn (2.50), we have not required that the weighting functions w_i and $\bar{w}_i$ be in some way related to the shape functions ψ_i and ϕ_i; an important application of the method of weighted residuals is obtained when the weighting functions are set equal to the corresponding shape functions, i.e.

$$w_i = \psi_i \quad 1 \le i \le N \qquad \bar{w}_i = \phi_i \quad 1 \le i \le M \qquad (2.51)$$

The resulting equations constitute the *Galerkin form* of the Navier–Stokes equations, which is their most widely used discretised form. Recent research on the calculation of the advection terms favours the use of weighting functions which differ from shape functions with so-called *upwinding* techniques (Hughes, 1979); eqns (2.50) are then called the *Petrov–Galerkin* form of the Navier–Stokes equations. Here we will concentrate on the Galerkin formulation.

Before showing how eqns (2.50) and (2.51) lead to an algebraic system of equations in the nodal values, let us discuss briefly how to deal with boundary conditions. Let us limit ourselves to problems with known boundaries; problems with free surfaces of unknown shape are the subject of Chapter 8. At every point of the boundary in plane flow, two boundary conditions must be imposed, i.e. two velocity components, two surface force components, or one of each. Imposing a velocity component is an *essential boundary condition*, in the sense that the approximation, eqn (2.43), may not be selected arbitrarily; a set of nodal vaues U_i ($i = i_1, i_2, \ldots, i_m$) and V_j ($j = j_1, j_2, \ldots, j_m$) are prescribed at the outset and take the values $\bar{u}_i$, $\bar{v}_j$. For example, nodal velocity components are forced to vanish on a stationary wall to which the fluid adheres. Essential boundary conditions are easily implemented without modifying the dimensions of the system, eqn (2.50); if we wish to specify the nodal component U_k, we simply replace the kth momentum equation in the x-direction by $U_k = \bar{u}_k$. When a velocity component is not specified along a section of the boundary, the corresponding component t_x^* or t_y^* of the contact force must be imposed, and this is called a *natural boundary condition*. The names 'natural' and 'essential' boundary conditions find their origin in variational calculus where their significance is obvious (Courant and Hilbert, 1962). Natural boundary conditions are implemented by replacing t_x^* or t_y^* in eqns (2.50) by their value $\bar{t}_x$ or $\bar{t}_y$ whenever it is specified; they will not be satisfied exactly by the approximate solution, eqn (2.43), and the error will depend upon the refinement of the mesh. The right-hand sides of the momentum equations in eqn (2.50) will be denoted by X_i and Y_i, respectively; they will contain a contribution from the line integral on ∂D only when node i is on the boundary. It is known from the analysis of the Navier–Stokes equations that the pressure is not unique unless a normal contact force is specified on a finite portion of the boundary; in that case, the indeterminacy must be lifted by imposing *one nodal pressure*. It does not make sense, either analytically or numerically, to impose the *pressure* at more than one point; in spite of the willingness of a computer program to generate numbers for such conditions, they should not be trusted.

We are now able to obtain the discretised form of the Navier–Stokes equations while using the approximation, eqn (2.43), and the Galerkin formulation. Introducing the symbols

$$a_{ij}^{xx} = \langle \psi_{i,x}; \psi_{j,x} \rangle \qquad a_{ij}^{yy} = \langle \psi_{i,y}; \psi_{j,y} \rangle \qquad a_{ij}^{xy} = \langle \psi_{i,x}; \psi_{j,y} \rangle$$

$$b_{ij}^{x} = \langle \psi_{i,x}; \phi_{j} \rangle \qquad b_{ij}^{y} = \langle \psi_{i,y}; \phi_{j} \rangle$$

$$c_{ijk}^{x} = \langle \psi_{i}; \psi_{j}\psi_{k,x} \rangle \qquad c_{ijk}^{y} = \langle \psi_{i}; \psi_{j}\psi_{k,y} \rangle \qquad (2.52)$$

where a comma denotes partial derivatives and $\langle \; ; \; \rangle$ denotes the L^2 scalar product over D,

$$\langle f; g \rangle = \int_{D} fg \, dD \qquad (2.53)$$

we obtain from eqn (2.50)

$$(2a_{xx}^{ij} + a_{yy}^{ij})U_{j} + a_{xy}^{ji}V_{j} - b_{im}^{x}p_{m} + Re(c_{ijk}^{x}U_{j} + c_{ijk}^{y}V_{j})U_{k} = X_{i}$$

$$a_{xy}^{ij}U_{j} + (2a_{yy}^{ij} + a_{xx}^{ij})V_{j} - b_{im}^{y}p_{m} + Re(c_{ijk}^{x}U_{j} + c_{ijk}^{y}V_{j})V_{k} = Y_{i}$$

$$- b_{jm}^{x}U_{j} - b_{jm}^{y}V_{j} = 0 \qquad (2.54)$$

with $1 \le i, j, k \le N$, and $1 \le m \le M$.

The discretised equation of continuity has been transposed to the end of the system in order to follow the usual order of elimination, while its sign has been changed in order to preserve a symmetric matrix when Re vanishes. The nodal values are obtained by solving the non-linear algebraic system, eqn (2.54). Various methods of solution have been reviewed by Gartling *et al.* (1977).

It appears that the most economical and successful technique is a combination of the classical Newton–Raphson method together with successive incrementation of the non-linear parameter Re, called *incremental loading* (see Oden, 1972, Chapter 17).

Now that the problem of solving the Navier–Stokes equations has been solved at least formally, we wish to comment on the shape functions ψ_i and ϕ_i which have led to successful applications. Let us first consider an arbitrary function $g(x, y)$ and see how it can be interpolated on the basis of its nodal values on a triangular element E. If the function $g(x, y)$ may be approximated by a discontinuous representation, we might write

$$\tilde{g}(x, y) = \bar{g} \qquad (x, y) \in E \qquad (2.55)$$

where $\bar{g}$ is the mean value of $g(x, y)$ over E and the shape function is the unit function. For a continuous approximation, we may calculate a linear interpolation between the values g_i of $g(x, y)$ at the vertices of the triangle and write

$$\tilde{g}(x, y) = \sum_{i=1}^{3} g_i \pi_i^1(x, y) \qquad (2.56)$$

where π_i^1 are linear shape functions which take the value 1 at node i and 0 at the others. Finally, we may design an interpolation passing through the exact values at the vertices and at midside nodes by constructing complete second-order polynomials in terms of the six known values, and write

$$\tilde{g}(x, y) = \sum_{i=1}^{6} g_i \pi_i^2(x, y) \qquad (2.57)$$

Again, π_i^2 is a second-order polynomial which takes the value 1 at node i and vanishes at the other nodes. Figure 2.5 shows typical shape functions on a triangular element. In the sequel, these three types of approximation on a triangle will be denoted respectively by the labels $P^0 - C^{-1}$, $P^1 - C^0$, $P^2 - C^0$, where the first part denotes the degree of the polynomial and the second the order of continuity.

Before considering a general quadrilateral, we will first study shape functions on a *parent element* which is a square defined by *local coordinates* $-1 \leq \xi, \eta \leq 1$. We may of course adopt again the form of eqn (2.55) within the parent element and obtain a discontinuous representation, $P^0 - C^{-1}$. If the value of g is given at the four corners, we may write

$$g(\xi, \eta) = \sum_{i=1}^{4} g_i \bar{\pi}_i^1(\xi, \eta) \qquad (2.58)$$

The shape functions $\bar{\pi}_i^1$ will not however be representable by first-order polynomials since they must obey four conditions, i.e. assume the value 1 at

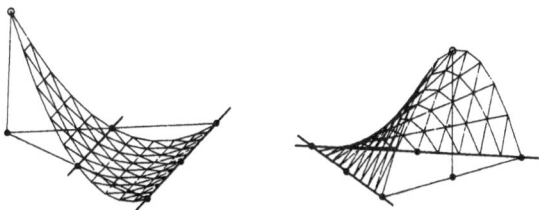

Fig. 2.5.   Shape functions on a triangular element.

one node and vanish at the other three. The functions π_i^1 are obtained by means of bilinear polynomials, and any complete first-order polynomial may be represented by eqn (2.58); the interpolation, eqn (2.58), is again of the $P^1 - C^0$ type. For obtaining polynomials of higher order, one may consider a nine-node quadrilateral; the nodes are located at the vertices, at mid-sides and at the centre of the quadrilateral and we write

$$\tilde{g}(\xi, \eta) = \sum_{i=1}^{9} g_i \bar{\pi}_i^2(\xi, \eta) \qquad (2.59)$$

The shape functions $\bar{\pi}_i^2(\xi, \eta)$ are biquadratic polynomials, and any second-order polynomial may be represented exactly by means of these shape functions. Another possibility is the Serendipity eight-node element, where the central node is omitted; both elements (nine nodes, eight nodes) are of the $P^2 - C^0$ type. Typical shape functions on a quadrilateral element are shown in Fig. 2.6.

We may now transform the quadrilateral element to an arbitrary quadrilateral by means of the transformation

$$x = \sum_{i=1}^{4} x_i \bar{\pi}_i^1(\xi, \eta) \qquad y = \sum_{i=1}^{4} y_i \bar{\pi}_i^1(\xi, \eta) \qquad (2.60)$$

where x_i, y_i are the coordinates of the vertices of the quadrilateral element. A nine (or eight)-node element may be transformed either into a

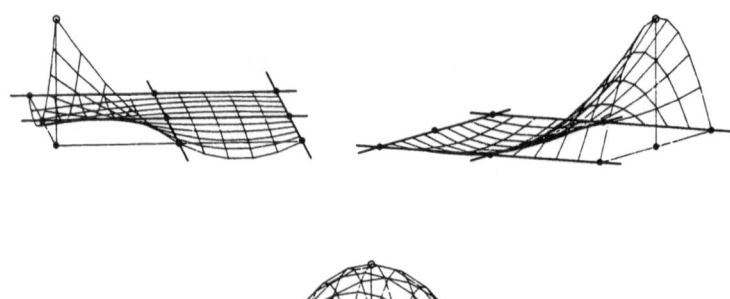

Fig. 2.6.   Shape functions on a quadrilateral element.

quadrilateral with straight edges by means of the transformation, eqn (2.60), or into a curvilinear quadrilateral by means of the transformation

$$x = \sum_{i=1}^{9} x_i \bar{\pi}_i^2(\xi, \eta) \qquad y = \sum_{i=1}^{9} y_i \bar{\pi}_i^2(\xi, \eta) \qquad (2.61)$$

The transformation from the parent to the actual element is called *isoparametric* when it is carried out with the help of the shape functions used for interpolating the unknown function on the element; it is *subparametric* when it is executed with shape functions of a lower order. For example, the transformation, eqn (2.60), is isoparametric for the four-node quadrilateral, and subparametric for the nine-node quadrilateral. We note that the isoparametric transformation may also be used with triangular elements for generating curved edges. Extensive data on various types of elements may be found in the work of Zienkiewicz (1977, Chapters 7 and 8). Subparametric and isoparametric transformations are shown in Fig. 2.7.

The problem now consists in selecting the appropriate shape functions ψ_i and ϕ_i for the velocity and pressure fields, respectively. From the early numerical experiments reported by Taylor and Hood (1973), it was found that the degree of the piecewise polynomials used for representing the pressure field should be at least one unit lower than the degree of the polynomials selected for the velocity field; this first rule has in fact been confirmed by the mathematical theory of finite elements for Stokes equations (see, for example, Girault and Raviart, 1981). The second rule stems from the form, eqn (2.50) of the Galerkin equations, and claims that the interpolation, eqn (2.43), should be at least C^0-continuous for the

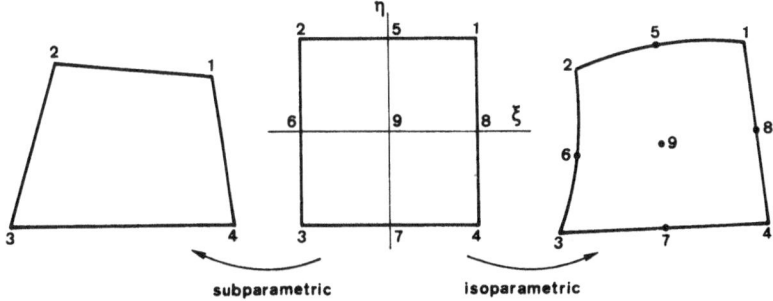

Fig. 2.7. Subparametric and isoparametric transformation.

velocity components, while C^{-1}-continuity is sufficient for the pressure field. The third rule for selecting the shape functions is the error of the finite element solution with respect to the exact solution; the error is usually given by an order magnitude of the type $O(h^m)$ where h is proportional to the size of the element, and the power m indicates how the numerical solution converges to the exact solution when the size of the elements decreases. The evaluation of the error is a difficult problem which will not be treated here (see Girault and Raviart, 1981; Témam, 1977).

Table 2.1 shows various combinations of shape functions. The first element is the simplest one might think of; it is easy to show that the incompressibility condition is so strong that no solution is possible.

TABLE 2.1
FINITE ELEMENTS FOR SOLVING STOKES FLOW

Shape	Velocity	Pressure	Remarks
1 Triangle	$P^1 - C^0$	$P^0 - C^{-1}$	Impossible element
2 Triangle	$P^2 - C^0$	$P^0 - C^{-1}$	
3 Triangle	$P^2 - C^0$	$P^1 - C^0$	Widely used
4 Quadrilateral	$P^1 - C^0$	$P^0 - C^{-1}$	Spurious pressure modes
5 Quadrilateral	$P^2 - C^0$	$P^1 - C^0$	Widely used
6 Quadrilateral	$P^2 - C^0$	$P^1 - C^{-1}$	

Elements 3 and 5 are the most widely used in engineering applications; for quadrilaterals, one may consider eight or nine nodes, but recent publications tend to prefer the nine-node element. Element 4 is often used in the literature, but the pressure field may under some circumstances be affected by spurious spatial oscillations of the *checkerboard* type (details on this may be found in the work of Sani *et al.*, 1981). For element 6, one selects a pressure field within the element which is a complete first degree polynomial with C^{-1} continuity at the interface of the elements; this element has not been widely used but is attractive in view of a good balance between nodal velocity components and pressures. Most of the literature on non-Newtonian flow uses elements 3 and 5, which are represented on Fig. 2.8. On the same figure we show the combination of triangles introduced by Nickell *et al.* (1974); however, they infer that this composite element might not be superior in precision to the nine-node quadrilateral, despite its larger number of nodal variables.

Our presentation has been limited until now to the analysis of plane flow; the theory is easily extended to the case of axisymmetric flow. The residuals,

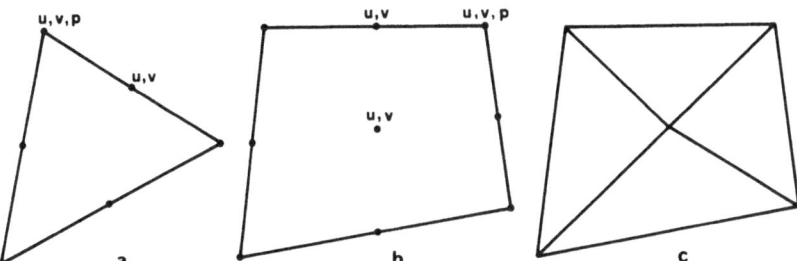

Fig. 2.8. Widely used elements: a, six-node triangle; b, nine-node quadrilateral; c, composite quadrilateral.

eqn (2.47), are then written in polar cylindrical coordinates; in order to apply the method of weighted residuals and obtain the Galerkin form of the Navier–Stokes equations, it is necessary to perform the integrations over a volume element of toroidal shape. The resulting system is again of the form of eqn (2.54), but the definitions, eqn (2.52), are different.

When a solution has been obtained in terms of velocity components, it is often desirable to exhibit the results in terms of a stream function, in order to visualise streamlines. Let the stream function $\tilde{\phi}(x, y)$ be defined by

$$\frac{\partial \tilde{\phi}}{\partial x} = -\tilde{v} \qquad \frac{\partial \tilde{\phi}}{\partial y} = \tilde{u} \qquad (2.62)$$

The problem of calculating $\tilde{\phi}$ is that $(-\tilde{v}\,dx + \tilde{u}\,dy)$ is not an exact differential for most of the usual elements, because conservation of mass is not satisfied everywhere. In elements 2, 4 and 6 of Table 2.1, it may be shown that the mass flux through the element boundary vanishes identically; under such conditions, $\tilde{\phi}$ may be calculated at each node by means of line integrals on the interfaces of elements. With elements 3 and 5 (the most widely used) continuity is only preserved in the mean, and we must resort to another technique for calculating $\tilde{\phi}$. From eqn (2.62), we obtain

$$\frac{\partial}{\partial x}\left(\frac{\partial \tilde{\phi}}{\partial x}\right) + \frac{\partial}{\partial y}\left(\frac{\partial \tilde{\phi}}{\partial y}\right) = \nabla^2 \tilde{\phi} = -\frac{\partial \tilde{v}}{\partial x} + \frac{\partial \tilde{u}}{\partial y} \qquad (2.63)$$

Calculating $\tilde{\phi}$ involves solving Poisson's equation; on impervious walls, Dirichlet boundary conditions are calculated easily while Neumann boundary conditions are used on entry and exit sections. Equation (2.63) is solved by means of finite elements; the same shape functions are used for $\tilde{\phi}$ and the velocity components.

The elements 3 and 5, and, to a lesser degree, element 4, are widely used for solving the Navier–Stokes equations; they are efficient for moderate Reynolds numbers. The upper limit depends upon the type of problem and the coarseness of the mesh and ranges approximately from 50 to a few hundred. The Reynolds number is usually quite low in polymer processing, and well below the limit of convergence of the iterative process for solving the non-linear algebraic system, eqn (2.54).

The finite-element techniques which have just been explained may be applied with much efficiency for calculating the *flow of shear-thinning fluids* whose behaviour is described by the model of the generalised Newtonian fluid. Let I_2 denote the second invariant of the rate of deformation tensor, written in such a way that $I_2^{1/2} = |\dot{\gamma}|$ in shear flow with a velocity gradient $\dot{\gamma}$,

$$I_2 = 2e_{ij}e_{ij} \tag{2.64}$$

Equations (2.42) are then replaced by

$$p_{xx}^E = 2\eta(I_2)\frac{\partial u}{\partial x} \qquad p_{yy}^E = 2\eta(I_2)\frac{\partial v}{\partial y} \qquad p_{xy}^E = \eta(I_2)\left(\frac{\partial u}{\partial y} + \frac{\partial v}{\partial x}\right) \tag{2.65}$$

We may still consider that the equations are written in non-dimensional form if η_0 in eqn (2.12) is now some fixed reference value for the shear viscosity. Once an interpolation, eqn (2.43), is selected for the velocity field, eqns (2.46) are replaced by

$$p_{xx}^* = 2\eta(\tilde{I}_2)\frac{\partial \tilde{u}}{\partial x} \qquad p_{yy}^* = 2\eta(\tilde{I}_2)\frac{\partial \tilde{v}}{\partial y} \qquad p_{xy}^* = \eta(\tilde{I}_2)\left(\frac{\partial \tilde{u}}{\partial y} + \frac{\partial \tilde{v}}{\partial x}\right) \tag{2.66}$$

where $\tilde{I}_2$ is given by eqn (2.64) in terms of $\tilde{u}$ and $\tilde{v}$. When the extra-stresses p_{xx}^*, p_{yy}^* and p_{xy}^*, are introduced in eqn (2.50), we obtain a non-linear algebraic system in U_i, V_i and P_i. The problem was first solved by Tanner *et al.* (1975) who used an elaborate relaxation technique for solving the system. In fact, it is sufficient to consider a sequence of solutions, $U_i^{(n)}$, $V_i^{(n)}$, $P_i^{(n)}$; the next solution $U_i^{(n+1)}$, $V_i^{(n+1)}$, $P_i^{(n+1)}$ is obtained by solving a Newtonian problem where the shear viscosity is calculated by means of the last solution. The sequence converges even for values of the power index as low as 0·1; the number of iterative steps for attaining a given level of accuracy increases when the power index decreases while remaining cost efficient. It should be noted that the technique is valid for *any viscosity law* $\eta(\tilde{I}_2)$ obtained from viscometric data. Applications of finite-element techniques to the flow of shear-thinning fluids in polymer processing may be found in the literature: the effect of shear-thinning upon die-swell is studied by Tanner *et al.* (1975); Crochet and Keunings (1981) extend the problem to dies of conical shape; entry and exit losses as a function of the power index are evaluated by Boger *et al.* (1978); and an analysis of the flow inside a coating die is given by Caswell and Tanner (1978).

While the flow of purely viscous shear-thinning fluids may be calculated easily by extending the method used for solving the Navier–Stokes equations (which we will henceforth designate as the u–v–p method since the velocity components and pressure are the primitive variables), the same is not true for the numerical solution of the flow of viscoelastic fluids. As a model problem for our discussion, we will study plane flow of a Maxwell fluid; the field and constitutive equations are given by the system, eqns (2.13)–(2.18), and we will also consider the case where the constitutive relations are given by their integral form, eqn (2.7).

The field equations, eqns (2.13) and (2.15) are identical for the flow of purely viscous or viscoelastic fluids. Provided we are able to supply an estimate of the extra-stress components, relations (2.47), (2.48) and their weighted residual form, eqn (2.50), remain entirely valid for our present purpose. The problem however is the evaluation of the extra-stresses. When the constitutive relations are given by eqns (2.16)–(2.18) in their differential form, the extra-stresses cannot be obtained as an explit expression in terms of the velocity components u and v, and the system, eqn (2.50) cannot be closed by substituting for p_{xx}^*, p_{yy}^*, p_{xy}^* in terms of $\tilde{u}$ and $\tilde{v}$. When the constitutive relations are given by their integral form, eqn (2.7), the extra-stress field is related to the velocity components through an intricate functional relation which will be considered later in this chapter.

When the constitutive relations are given in differential form and in view of the lack of an explicit expression, it seems natural to introduce for the extra-stress components a *finite-element interpolation* similar to eqn (2.43), i.e.

$$\tilde{p}_{xx} = \sum_{i=1}^{L} R_i \pi_i(x, y) \qquad \tilde{p}_{yy} = \sum_{i=1}^{L} S_i \pi_i(x, y) \qquad \tilde{p}_{xy} = \sum_{i=1}^{L} T_i \pi_i(x, y)$$

$$(2.67)$$

where the indexed symbols R, S and T have been used to denote the nodal values of the extra-stresses. We note that, at the present stage, the shape functions π_i and the number L of nodal stresses may differ from those used for the velocity components and pressure in eqn (2.43). We may now introduce the approximation, eqn (2.67), in the constitutive relations (2.16)–

(2.18); they will not be satisfied in general, and the residuals will be given as follows:

$$\mathbf{R}_R(x, y) = \tilde{p}_{xx}\left(1 - 2\dot{W}s\frac{\partial \tilde{u}}{\partial x}\right) + Ws\left(\tilde{u}\frac{\partial \tilde{p}_{xx}}{\partial x} + \tilde{v}\frac{\partial \tilde{p}_{xx}}{\partial y}\right) - 2Ws\tilde{p}_{xy}\frac{\partial \tilde{u}}{\partial y} - 2\frac{\partial \tilde{u}}{\partial x}$$

$$\mathbf{R}_S(x, y) = \tilde{p}_{yy}\left(1 - 2Ws\frac{\partial \tilde{v}}{\partial y}\right) + Ws\left(\tilde{u}\frac{\partial \tilde{p}_{yy}}{\partial x} + \tilde{v}\frac{\partial \tilde{p}_{yy}}{\partial y}\right) - 2Ws\tilde{p}_{xy}\frac{\partial \tilde{v}}{\partial x} - 2\frac{\partial \tilde{v}}{\partial y}$$

$$\mathbf{R}_T(x, y) = \tilde{p}_{xy} + Ws\left(\tilde{u}\frac{\partial \tilde{p}_{xy}}{\partial x} + \tilde{v}\frac{\partial \tilde{p}_{xy}}{\partial y}\right) - Ws\left(\tilde{p}_{xx}\frac{\partial \tilde{v}}{\partial x} + \tilde{p}_{yy}\frac{\partial \tilde{u}}{\partial y}\right) - \frac{\partial \tilde{u}}{\partial y} - \frac{\partial \tilde{v}}{\partial x}$$

$$(2.68)$$

Equations (2.47) are now replaced by similar expressions where, however, the extra-stresses are given by their interpolation, eqn (2.67)

$$\mathbf{R}_u(x, y) = -\frac{\partial \tilde{p}}{\partial x} + \frac{\partial \tilde{p}_{xx}}{\partial x} + \frac{\partial \tilde{p}_{xy}}{\partial y} + f_x - Re\left(\tilde{u}\frac{\partial \tilde{u}}{\partial x} + \tilde{v}\frac{\partial \tilde{u}}{\partial y}\right)$$

$$\mathbf{R}_v(x, y) = -\frac{\partial \tilde{p}}{\partial y} + \frac{\partial \tilde{p}_{xy}}{\partial x} + \frac{\partial \tilde{p}_{yy}}{\partial y} + f_y - Re\left(\tilde{u}\frac{\partial \tilde{v}}{\partial x} + \tilde{v}\frac{\partial \tilde{v}}{\partial y}\right)$$

$$\mathbf{R}_p(x, y) = \frac{\partial \tilde{u}}{\partial x} + \frac{\partial \tilde{v}}{\partial y}$$

$$(2.69)$$

The unknown variables U_i, V_i, P_i, R_i, S_i, T_i are obtained by using the method of weighted residuals. The Galerkin form of the constitutive and field relations, first obtained by Kawahara and Takeuchi (1977), is given by

$$\int_D \mathbf{R}_R \pi_i \, dD = 0 \qquad \int_D \mathbf{R}_S \pi_i \, dD = 0 \qquad \int_D \mathbf{R}_T^l \pi_i \, dD = 0 \qquad 1 \leq i \leq L$$

$$\int_D \mathbf{R}_u \psi_i \, dD = 0 \qquad \int_D \mathbf{R}_v \psi_i \, dD = 0 \qquad\qquad\qquad 1 \leq i \leq N$$

$$\int_D \mathbf{R}_p \phi_i \, dD = 0 \qquad\qquad\qquad\qquad\qquad\qquad\qquad 1 \leq i \leq M$$

$$(2.70)$$

An integration by parts is performed in the weighted momentum equations in order to recover their form given in eqn (2.50). The central problem at this stage is twofold: which shape functions should be selected in the interpolations, eqns (2.43) and (2.67), and how to solve the complex and large non-linear algebraic system resulting from eqn (2.70).

In view of our previous experience with the Navier–Stokes equations and the summary contained in Table 2.1, it seems natural to adopt triangular or quadrilateral elements of the $P^2 - C^0$ type for the velocity components and the $P^1 - C^0$ (or possibly $P^1 - C^{-1}$) type for the pressure. Since the extra-stresses are added to the pressure for generating the stress tensor of Cauchy, one might think of using the same order of interpolation for the extra-stresses as for the pressure, with the observation however that the extra-stresses must be continuous in view of the presence of spatial derivatives in eqn (2.68). It was shown however by Crochet (1982) that such a choice produces a singular stiffness matrix; the only valid remedy for solving the system, eqn (2.70), as it stands, is the adoption of the same shape functions for the velocity components and the extra-stresses, i.e. $\pi_i = \psi_i$ and $L = N$. Another remedy was found by Crochet and Bézy (1979) by adopting $P^1 - C^{-1}$ interpolation for the pressure and $P^1 - C^0$ interpolation for the stresses in a triangle; the element was however forsaken in view of the large number of constraints imposed on the velocity field by incompressibility. Table 2.2 gives a summary of elements which may be found in the literature. Elements 1 and 2 of Table 2.2 are shown on Fig. 2.9; they contain respectively 33 and 49 degrees of freedom. For axisymmetric flow, it is necessary to add nodal values of the extra-hoop stresses, since the extra-stress tensor of a Maxwell fluid is not traceless; the number of degrees of freedom of elements 1 and 2 become respectively 39 and 58.

The formal simplicity of the set, eqn (2.70), hides the fact that forming the algebraic system of equations is an elaborate and expensive task. Moreover, the resulting system is not well conditioned; when inertia effects can be neglected (this is often the case in polymer flow), the discretised momentum equations do not contain the velocity components, and the discretised equation for mass conservation does not contain the pressure.

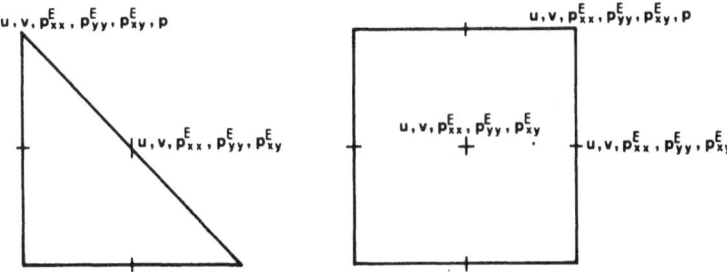

Fig. 2.9.   Elements for solving the flow of a Maxwell fluid.

TABLE 2.2

FINITE ELEMENTS FOR SOLVING THE FLOW OF A MAXWELL FLUID

Shape	Velocity	Extra-stress	Pressure	References
1 Triangle	$P^2 - C^0$	$P^2 - C^0$	$P^1 - C^0$	Kawahara and Takeuchi (1977) Crochet and Bézy (1979) Crochet and Keunings (1980)
2 Quadrilateral (9 nodes)	$P^2 - C^0$	$P^2 - C^0$	$P^1 - C^0$	Crochet and Keunings (1981) Crochet and Keunings (1982a) Crochet (1982)
3 Quadrilateral (8 nodes)	$P^2 - C^0$	$P^2 - C^0$	$P^1 - C^0$	Richards and Townsend (1981)
4 Triangle	$P^1 - C^0$	$P^1 - C^0$	$P^0 - C^{-1}$ over 2 triangles	Coleman (1980) Coleman (1981)

As a consequence, the large system of equations cannot be split into two smaller systems to be solved with an iterative procedure similar, for example, to a technique frequently used for solving non-isothermal Newtonian flow (see Gartling, 1977). The classical Newton–Raphson procedure for solving the global system is very efficient when the iterative procedure leads to a solution; 4 to 5 iterations will suffice in general for obtaining a relative change of the order of 10^{-4} between two iterations; it has been used with success for elements 1 and 2 of Table 2.2. A Picard iterative procedure is advocated by Richards and Townsend (1981) and Coleman (1980, 1981) under the premise that the radius of convergence of the Newton–Raphson method may be too small for reaching a solution when non-linear effects dominate the flow; results obtained so far are not conclusive in this respect.

These techniques have been used for calculating plane and axisymmetric entry flow (Coleman, 1980; Crochet and Bézy, 1979, 1980), for predicting the flow around a sphere (Crochet, 1982), for predicting die-swell (Coleman, 1981; Crochet and Keunings, 1980, 1981, 1982a), and for estimating hole pressure (Crochet and Bézy, 1979; Richards and Townsend, 1981). Results are obtained up to some value of the elasticity of the flow; a limitation which will be discussed later.

The desire to use lower degree polynomials for the extra-stresses and to split the global system into two smaller ones in order to obtain a cheaper algorithm, led to an alternate approach which is in fact based on the decomposition, eqn (2.24), used in finite-difference work. On the basis of eqns (2.24) and (2.16) to (2.18), we obtain

$$\bar{p}_{xx} = Ws \left(2p_{xx}^E \frac{\partial u}{\partial x} - u \frac{\partial p_{xx}^E}{\partial x} - v \frac{\partial p_{yy}^E}{\partial y} + 2p_{xy}^E \frac{\partial u}{\partial y} \right)$$

$$\bar{p}_{yy} = Ws \left(2p_{yy}^E \frac{\partial v}{\partial y} - u \frac{\partial p_{yy}^E}{\partial x} - v \frac{\partial p_{yy}^E}{\partial y} + 2p_{xy}^E \frac{\partial v}{\partial x} \right)$$

$$\bar{p}_{xy} = Ws \left(p_{xx}^E \frac{\partial v}{\partial x} + p_{yy}^E \frac{\partial u}{\partial y} - u \frac{\partial p_{xy}^E}{\partial x} - v \frac{\partial p_{xy}^E}{\partial y} \right) \qquad (2.71)$$

The decomposition, eqn (2.24), is then introduced in eqn (2.50), where we use $w_i = \psi_i$ and $\bar{w}_i = \phi_i$; one obtains the system

$$\int_D \left(\frac{\partial \tilde{u}}{\partial x} + \frac{\partial \tilde{v}}{\partial y} \right) \phi_i \, \mathrm{d}D = 0 \qquad 1 \le i \le M$$

$$\int_D \left[\left(-\tilde{p} + 2\frac{\partial \tilde{u}}{\partial x} \right) \frac{\partial \psi_i}{\partial x} + \left(\frac{\partial \tilde{u}}{\partial y} + \frac{\partial \tilde{v}}{\partial x} \right) \frac{\partial \psi_i}{\partial y} + Re \left(\tilde{u} \frac{\partial \tilde{u}}{\partial x} + \tilde{v} \frac{\partial \tilde{u}}{\partial y} \right) \psi_i \right.$$

$$\left. + \bar{p}_{xx} \frac{\partial \psi_i}{\partial x} + \bar{p}_{xy} \frac{\partial \psi_i}{\partial y} \right] \mathrm{d}D = \int_D f_x \psi_i \, \mathrm{d}D + \int_{\partial D} \bar{t}_x \psi_i \, \mathrm{d}l$$

$$\int_D \left[\left(\frac{\partial \tilde{u}}{\partial y} + \frac{\partial \tilde{v}}{\partial x} \right) \frac{\partial \psi_i}{\partial x} + \left(-\tilde{p} + \frac{\partial \tilde{v}}{\partial y} \right) \frac{\partial \psi_i}{\partial y} + Re \left(\tilde{u} \frac{\partial \tilde{v}}{\partial x} + \tilde{v} \frac{\partial \tilde{v}}{\partial y} \right) \psi_i \right.$$

$$\left. + \bar{p}_{xy} \frac{\partial \psi_i}{\partial x} + \bar{p}_{xy} \frac{\partial \psi_i}{\partial y} \right] \mathrm{d}D = \int_D f_y \psi_i \, \mathrm{d}D + \int_{\partial D} \bar{t}_y \psi_i \, \mathrm{d}l \qquad (2.72)$$

which must be solved jointly with

$$\int_D \mathbf{R}_R \pi_i \, \mathrm{d}D = 0 \qquad \int_D \mathbf{R}_S \pi_i \, \mathrm{d}D = 0 \qquad \int_D \mathbf{R}_T \pi_i \, \mathrm{d}D = 0 \qquad 1 \le i \le L$$

$$(2.73)$$

We note that, when Ws vanishes, $\bar{p}_{xx}, \bar{p}_{yy}$ and $\bar{p}_{xy}$ vanish identically, and the system, eqn (2.72) reduces to eqn (2.54); it may easily be shown (see Crochet, 1982) that the condition $\pi_i = \psi_i$ is not required for solving the new system. There are two ways of solving the system using eqns (2.72) and (2.73). In the first approach, $\bar{p}_{xx}, \bar{p}_{yy}$ and $\bar{p}_{xy}$ in eqn (2.72) are replaced by the right-hand sides of eqn (2.71), and the global system is solved in terms of the extra-stresses, velocity components and pressure. In the second approach, a trial velocity field is used for calculating the extra-stresses by means of the system, eqn (2.73); $\bar{p}_{xx}$, $\bar{p}_{yy}$ and $\bar{p}_{xy}$ are then calculated on the basis of eqn (2.71), and a new velocity field is obtained from eqn (2.72) where the old values of $\bar{p}_{xx}$, $\bar{p}_{yy}$ and $\bar{p}_{xy}$ are introduced. The decomposition, eqn (2.24), allows us to decouple the constitutive relations from the field equations, and to use polynomials of lower degree for interpolating the extra-stresses.

It must be observed that this new algorithm, as compared to eqn (2.70), is constructed like a perturbation of the u–v–p method for solving the Navier–Stokes equations. The decomposition, eqn (2.24), was first introduced by Chang et al. (1979), although the algorithm described in their paper led to some difficulties. It has been used by Crochet (1981) and Crochet and Keunings (1981, 1982a) with elements 1 and 2 of Table 2.2; Mendelson et al. (1982) and Jackson and Finlayson (1982) make use of the algorithm with $p^2 - C^0$ shape functions for the velocity components, and $p^1 - C^0$ shape functions for the pressure and the extra-stresses. Entry flow is studied by Mendelson et al. (1982), die swell by Chang et al. (1979) and Crochet and Keunings (1981, 1982a), the flow around a sphere by Crochet (1981), and the hole pressure problem by Finlayson (to appear).

Comparisons between the two algorithms, made by Crochet (1981), show that *smoother* results are obtained for low values of Ws with the second algorithm than with the first, while when Ws increases both algorithms behave essentially in the same fashion; in general the first algorithm allows one to obtain convergence of the iterative technique for values of Ws higher than with the second.

Both algorithms which have just been described may be extended without difficulty to other constitutive equations of the differential type like the Phan-Thien and Tanner (1977) model, or the White and Metzner (1963) model. They maintain in fact the same original form, eqn (2.2), of the upper-convected Maxwell fluid, where the material constants λ_1 and η_0 are now functions of the invariants of the rate of deformation tensor or of the extra-stress tensor. One may use an iterative technique where λ_1 and η_0 are calculated from the last iteration in order to obtain new values; an example may be found in the work of Crochet and Bézy (1980). The convergence of the iterative process is difficult to obtain when the relaxation time varies rapidly as a function of the rate of deformation tensor (for the White–Metzner fluid).

It has been shown in Section 2.2 that recent advances in finite-difference calculations have allowed calculation of the flow of a fluid of the integral type. Similar results have been obtained with finite elements (Bernstein *et al.*, 1981; Viriyayuthakorn and Caswell, 1980) and the method will be reviewed briefly. Further developments in that area may be expected over the next few years, since several attractive constitutive relations are based on integral models; moreover, non-isothermal constitutive relations for polymeric fluids may best be expressed in terms of integral models.

The main problem in developing an algorithm for calculating the flow of fluids given by an integral constitutive relation is the calculation of the extra-stress components for a given steady velocity field, given by a finite-element interpolation of the form of eqn (2.43). Since the integral in eqn (2.7) contains an exponential as a factor, one may calculate the integral up to any degree of accuracy, provided the strain tensor F_{ik} may be written as a polynomial in τ, by means of Gauss–Laguerre quadrature (tables of Gauss–Laguerre quadrature may be found in the work of Carnahan *et al.* 1969, p. 113). As with Gauss–Legendre quadrature, the integral is written as a sum of weights multiplied by the integrand at some fixed values of τ; the number of quadrature points depends upon the complexity of the flow. The difficult problem however is the calculation of the strain tensor. Available papers have not used until now the system, eqn (2.38), for finding at time $t - \tau$ the position (x', y') of a particle which occupies the position (x, y) at time t. Rather, they integrate in time the relations

$$\frac{dx'(t - \tau)}{d(t - \tau)} = \tilde{u}(x', y') \qquad \frac{dy'(t - \tau)}{d(t - \tau)} = \tilde{v}(x', t') \qquad (2.74)$$

and find the location of the nodes at time $t - \tau_i$, where τ_i is a Gauss–Laguerre point. A triangular finite element of type $P^2 - C^0$ for the velocity components is used by Viriyayuthakorn and Caswell (1980), and an elaborate tracking procedure is designed for solving eqn (2.74) with great accuracy. In the work of Bernstein *et al.* (1981), the use of quadrilaterals of type $P^1 - C^0$ allows one to approximate the velocity gradient by a constant tensor within the element (with some error), and the tracking procedure is greatly facilitated. Whereas Viriyayuthakorn and Caswell (1980) obtained the strains by comparing the deformed element at time $t - \tau_i$ to its shape at time t, Bernstein *et al.* (1981) calculated the relative deformation gradients along the paths of particles.

Let us now assume that an extra-stress field $\bar{p}_{xx}, \bar{p}_{yy}, \bar{p}_{xy}$ may be calculated on the basis of a velocity field $\tilde{u}, \tilde{v}$ associated with a pressure field $\tilde{p}$ which (in the mean) satisfies conservation of mass. Unless the exact solution has been obtained, the extra-stress field will not satisfy the momentum equations, eqns (2.50). The velocity components must be corrected by an amount $\delta\tilde{u}$,

$\delta \tilde{v}$ which affects the stress field. The corrected stress field is given (in non-dimensional form) by

$$p_{xx}^* \cong \bar{p}_{xx} + 2 \frac{\partial \delta \tilde{u}}{\partial x} \qquad p_{yy}^* \cong \bar{p}_{yy} + 2 \frac{\partial \delta \tilde{v}}{\partial y}$$

$$p_{xy}^* \cong \bar{p}_{xy} + \left(\frac{\partial \delta \tilde{u}}{\partial y} + \frac{\partial \delta \tilde{v}}{\partial x} \right) \tag{2.75}$$

The approximation, eqn (2.75), is obtained from a first-order asymptotic development of Finger's strain tensor in terms of τ. When the new stress values obtained in eqn (2.75) are inserted in the system, eqn (2.50), we recover precisely the system (2.72), where $\tilde{u}$ and $\tilde{v}$ are replaced by the increments $\delta \tilde{u}$ and $\delta \tilde{v}$ (when Re vanishes). Thus, correcting the velocity field amounts to solving the Stokes problem.

Results of high quality have been obtained by Viriyayuthakorn and Caswell (1980) for the calculation of axisymmetric die entry flow, while a tapered die has been studied by Bernstein et al. (1981).

Die swell is being considered by Viriyayuthakorn and Caswell (1982). A one-dimensional application to fibre spinning has been solved by Malkus (1981).

We have shown that finite-element techniques are now available for solving the flow of non-Newtonian fluids. In the absence of elasticity, excellent results are obtained whatever the form of the viscosity law. In the presence of elasticity, the problem is much more complex. Two basic techniques are available for solving the problem, i.e. the use of differential and integral models. They have in common a rather high computing cost; we believe however that this is of minor importance in view of the uninterrupted decrease in cost of computer time and memory storage. Moreover, the progressive installation in academic and industrial research centres of local computers with virtual memory allows long runs to be performed at no cost during nights and week-ends. Both techniques also show good results for low values of the Weissenberg number, when the solution is not too far removed from the Newtonian solution. The algorithms break down for higher values of Ws.

Several explanations may be found for the failure when Ws is high:

(i)　When the flow contains stress singularities, numerical experiments tend to show that they are enhanced by elasticity; local stress peaks destroy the accuracy of the finite-element solution. The implementation of finite elements containing a singularity might help, provided the nature of the singularity is known, which is not the case at the present time.

(ii)   Many available constitutive equations are not suited for numerical simulation. The upper-convected Maxwell model has an infinite elongational viscosity for a finite value of the elongation rate. The Phan-Thien/Tanner model has a viscosity which, beyond some value of the rate of deformation $\dot{\gamma}$, decreases faster than $\dot{\gamma}^{-1}$; this leads, in simple Poiseuille flow, to an infinite number of solutions, and an absolute maximum for the pressure gradient. However, the addition of a purely viscous component to the extra-stress tensor may remove this peculiar behaviour, and this has been recently used by Crochet and Keunings (1982b). Another equation, proposed by Bird *et al.* (1980), produces several branches for the elongational viscosity as a function of the rate of elongation. While it is possible in analytical work to reject unphysical branches, the computer is unable to do so, and we have actually been able to generate exact unphysical solutions.

(iii) First-order partial differential equations must be solved for the differential *and* for the integral model. For the former the first-order terms appear in the constitutive relations while for the latter they are present in the tracking procedure. It is known that much progress needs to be made in simulating (first-order) convective processes with finite elements (see, for example, Hughes, 1979). Crochet and Keunings (1982b) state that the addition of a viscous component to the extra-stress tensor leads indeed to a better conditioned system of algebraic equations, and the range of Weissenberg numbers where the calculations can be completed is greatly enlarged.

(iv)   No uniqueness, existence and stability proofs are available for the fluids and flows which have been considered so far.

2.4 CONCLUSION

We have seen that it is possible in principle to employ both FD and FE methods to handle implicit differential and explicit integral models in complex geometries. The techniques are now reasonably well established and non-controversial, *but they are all singly unable to break through the Ws barrier* and there is unfortunately an upper bound to the level of elasticity which can be accommodated before the programs cease to function. This level is exasperatingly low—so low in fact that many of the important and provocative experimental results which are available fall outside the present scope of numerical simulation. This conclusion is common to all workers in the field; it applies to FD and FE techniques, flow problems with and without abrupt changes in geometry, and to rheological models of the differential and integral type. There is no doubt at all that the problem of obtaining solutions for higher values of Ws is of critical importance as the present work in numerical simulation seeks to exert an influence on such practically important fields as polymer processing. This will undoubtedly require interactive work in rheology, experimentation, analytical work and numerical analysis.

REFERENCES

Bernstein, B., Kadivar, M. K. and Malkus, D. S. (1981). *J. Comp. Meth. Appl. Mech. Eng.*, **27**, 279.

Bird, R. B., Dotson, P. J. and Johnson, N. L. (1980). *J. Non-Newtonian Fluid Mechanics*, **7**, 213.

Boger, D. V., Gupta, R. and Tanner, R. I. (1978). *J. Non-Newtonian Fluid Mechanics*, **4**, 239.

Briley, W. R. (1978). *J. Comp. Phys.*, **14**, 8.

Carnahan, B., Luther, H. A. and Wilkes, J. O. (1969). *Applied Numerical Methods*, Wiley, New York.

Caswell, B. and Tanner, R. I. (1978). *Polym. Eng. Sci.*, **18**, 416.

Caswell, B. and Viriyayuthakorn, M. (1983). *J. Non-Newtonian Fluid Mechanics*, **12**, 13.

Chang, P. W., Patten, Th. W. and Finlayson, B. A. (1979). *Comp. and Fluids*, 7, 267.

Citroen, S. L. (1980). Ph.D. Thesis, University of Wales.

Cochrane, T., Walters, K. and Webster, M. F. (1981). *Phil. Trans. Roy. Soc.*, **301**, 163.

Cochrane, T., Walters, K. and Webster, M. F. (1982). *J. Non-Newtonian Fluid Mechanics*, **10**, 95.

Coleman, B. D. and Noll, W. (1961). *Ann. N.Y. Acad. Sci.*, **89**, 672.

Coleman, C. J. (1980). *J. Non-Newtonian Fluid Mechanics*, **7**, 289.

Coleman, C. J. (1981). *J. Non-Newtonian Fluid Mechanics*, **8**, 261.

Courant, R. and Hilbert, D. (1962). *Methods of Mathematical Physics*, Vol. 1, Interscience, New York.

Court, H. (1980). Ph.D. Thesis, University of Wales.

Court, H., Davies, A. R. and Walters, K. (1981). *J. Non-Newtonian Fluid Mechanics*, **8**, 95.

Crochet, M. J. (1982). In *Finite Elements in Fluids*, Vol. 4 (Ed. R. H. Gallagher), Wiley, New York, Chapter 26, pp. 573–97.

Crochet, M. J. and Bézy, M. (1979). *J. Non-Newtonian Fluid Mechanics*, **5**, 201.

Crochet, M. J. and Bézy, M. (1980). In *Rheology*, Vol. 2 (Ed. G. Astarita), Plenum, New York, p. 53.

Crochet, M. J. and Keunings, R. (1980). *J. Non-Newtonian Fluid Mechanics*, **7**, 199.

Crochet, M. J. and Keunings, R. (1981). Proceedings of 2nd World Congress of Chemical Engineering, Montreal, Vol. 6, pp. 285–9.

COMPUTATIONAL TECHNIQUES FOR VISCOELASTIC FLUID FLOW   61

Crochet, M. J. and Keunings, R. (1982a). *J. Non-Newtonian Fluid Mechanics*, **10**, 85.
Crochet, M. J. and Keunings, R. (1982b). *J. Non-Newtonian Fluid Mechanics*, **10**, 339.
Crochet, M. J. and Pilate, G. (1976). *J. Non-Newtonian Fluid Mechanics*, **1**, 247.
Davies, A. R., Walters, K. and Webster, M. F. (1979). *J. Non-Newtonian Fluid Mechanics*, **4**, 325.
Finlayson, B. A. (1972). *The Method of Weighted Residuals and Variational Principles*, Academic Press, New York, London.
Gallagher, R. H. (Ed.) (1975a). *Finite Elements in Fluids*, Vol. 1, Wiley, New York.
Gallagher, R. H. (Ed.) (1975b). *Finite Elements in Fluids*, Vol. 2, Wiley, New York.
Gallagher, R. H. (Ed.) (1978). *Finite Elements in Fluids*, Vol. 3, Wiley, New York.
Gartling, D. K. (1977). *Comp. Meth. Appl. Mech. Eng.*, **12**, 365.
Gartling, D. K., Nickell, R. E. and Tanner, R. I. (1977). *Int. J. Num. Meth. Eng.*, **11**, 1155.
Gatski, T. B. and Lumley, J. L. (1978a). *J. Comp. Phys.*, **27**, 42.
Gatski, T. B. and Lumley, J. L. (1978b). *J. Fluid Mech.*, **86**, 623.
Girault, V. and Raviart, P. A. (1981). *Finite Element Approximation of the Navier–Stokes Equations*, Springer-Verlag, Berlin.
Greenspan, D. (1968). *Lectures on Numerical Solutions of Linear, Singular and Non-Linear Differential Equations*, Prentice-Hall, New Jersey.
Heinrich, J. C. and Marshall, R. S. (1981). *Comp. and Fluids*, **9**, 73.
Holstein, H. (1981). Ph.D. Thesis, University of Wales.
Holstein, H. and Paddon, D. J. (1981). *J. Non-Newtonian Fluid Mechanics*, **8**, 81.
Hughes, T. J. R. (Ed.) (1979). *Finite Element Methods for Convection Dominated Flows*, Vol. 34, The American Society of Mechanical Engineers, AMD.
Hughes, T. J. R., Liu, W. K. and Brooks, A. (1979). *J. Comp. Phys.*, **30**, 1.
Jackson, N. R. and Finlayson, B. A. (1982). *J. Non-Newtonian Fluid Mechanics*, **10**, 55.
Jensen, V. G. (1959). *Proc. Roy. Soc.*, **A249**, 346.
Kawaguti, M. (1965). MRC Report No. 574, University of Wisconsin, Madison.
Kawahara, M. and Takeuchi, N. (1977). *Comp. and Fluids*, **5**, 33.
Kershaw, D. S. (1978). *J. Comp. Phys.*, **26**, 43.
Leal, L. G. (1979). *J. Non-Newtonian Fluid Mechanics*, **5**, 33.
Malkus, D. S. (1981). *J. Non-Newtonian Fluid Mechanics*, **8**, 223.
Manero, O. (1980). Ph.D. Thesis, University of Wales.
Meijerink, J. A. and Van Der Vorst, H. A. (1977). *Maths of Computation*, **31**, 148.
Mendelson, M. A., Yeh, P. W., Brown, R. A. and Armstrong, R. C. (1982). *J. Non-Newtonian Fluid Mechanics*, **10**, 31.
Mitchell, A. R. and Griffiths, D. F. (1980). *The Finite Difference Method in Partial Differential Equations*, Wiley, New York.
Moffatt, H. K. (1964). *J. Fluid Mech.*, **18**, 1.
Nickell, R. E., Tanner, R. I. and Caswell, B. (1974). *J. Fluid Mech.*, **65**, 189.
Oden, J. T. (1970). *Trans. ASCE, J. Eng. Mech. Div.*, **96**, 529.
Oden, J. T. (1972). *Finite Elements of Non-linear Continua*, McGraw-Hill, New York.
Oldroyd, J. G. (1950). *Proc. Roy. Soc.*, **A200**, 523.
Paddon, D. J. (1979). Ph.D. Thesis, University of Wales.

62   M. J. CROCHET AND K. WALTERS

Pearson, J. R. A. (1967). In *Non-linear Partial Differential Equations*, Academic Press, New York,,London, p. 73.
Perera, M. G. N. (1976). Ph.D. Thesis, University of Wales.
Perera, M. G. N. and Strauss, K. (1979). *J. Non-Newtonian Fluid Mechanics*, 5, 269.
Perera, M. G. N. and Walters, K. (1977a). *J. Non-Newtonian Fluid Mechanics*, 2, 49.
Perera, M. G. N. and Walters, K. (1977b). *J. Non-Newtonian Fluid Mechanics*, 2, 191.
Phan-Thien, N. and Tanner, R. I. (1977). *J. Non-Newtonian Fluid Mechanics*, 2, 353.
Reid, J. K. (1971). *Proceedings of Conference on Large Sparse Sets of Linear Equations*, Academic Press, New York, London, p. 231.
Reid, J. K. (1977). In *The State of the Art in Numerical Analysis* (Ed. D. A. H. Jacobs). Academic Press, New York, London, p. 85.
Richards, G. D. and Townsend, P. (1981). *Rheol. Acta*, 20, 261.
Roache, P. J. (1972). *Computational Fluid Dynamics*, Hermosa Publications, Albuquerque, New Mexico.
Sani, R. L., Gresho, P. M., Lee, R. L. and Griffiths, D. F. (1981). *Int. J. Num. Meth. Fluids*, 1, 17.
Strang, G. and Fix, G. J. (1973). *An Analysis of the Finite Element Method*, Prentice-Hall, New Jersey.
Tanner, R. I., Nickell, R. E. and Bilger, R. W. (1975). *Comp. Meth. Appl. Mech. Eng.*, 6, 155.
Taylor, C. and Hood, P. (1973). *Comp. and Fluids*, 1, 73.
Témam, R. (1977). *Navier–Stokes Equations (Theory and Numerical Analysis)*, North Holland, Amsterdam.
Thames, F. C., Thompson, J. F., Mastin, C. W. and Walker, R. L. (1977). *J. Comp. Phys.*, 24, 245.
Thompson, J. F., Thames, F. C. and Mastin, C. W. (1977). *J. Comp. Phys.*, 24, 274.
Tiefenbruck, G. (1979). Ph.D. Thesis, California Institute of Technology.
Tiefenbruck, G. and Leal, L. G. (1982). *J. Non-Newtonian Fluid Mechanics*, 10, 115.
Townsend, P. (1980a). *J. Non-Newtonian Fluid Mechanics*, 6, 219.
Townsend, P. (1980b). *Rheol. Acta*, 19, 1.
Varga, R. S. (1962). *Matrix Iterative Analysis*, Prentice-Hall, New Jersey.
Viriyayuthakorn, M. and Caswell, B. (1980). *J. Non-Newtonian Fluid Mechanics*, 6, 245.
Walters, K. (1979). *J. Non-Newtonian Fluid Mechanics*, 5, 113.
Walters, K. and Webster, M. F. (1982). *Phil. Trans. Roy. Soc.*, A308, 199.
Webster, M. F. (1979). Ph.D. Thesis, University of Wales.
White, J. L. (1980). *Rheometry: Industrial Applications* (Ed. K. Walters), Wiley, New York, Chapter 5.
White, J. L. and Metzner, A. B. (1963). *J. Appl. Polym. Sci.*, 7, 1867.
Woods, L. C. (1954). *Aeronautical Quarterly*, 5, 176.
Young, D. M. (1971). *Iterative Solution of Large Linear Systems*, Academic Press, New York, London.
Zienkiewicz, O. C. (1977). *The Finite Element Method*, 3rd Edn, McGraw-Hill, New York.

CHAPTER 3

Extrudate Swell

R. I. TANNER

*Department of Mechanical Engineering,
University of Sydney, Australia*

It is well known that a stream of fluid emerging from a plain circular tube (or die) does not usually have the same diameter as the tube. With polymeric fluids the extrudate often has a diameter several times that of the tube. This spectacular effect is shown in Fig. 3.1 (Lodge, 1964). If gravitational forces are significant compared with the viscous forces acting on the fluid, then the extrudate shape will partly be governed by gravity, and the details of the orientation of the tube are relevant. To confine the discussion, we suppose here that gravity is negligible or effectively absent; this is often the relevant case for polymer melts. Additionally, the flows to be considered will be steady unless it is explicitly noted otherwise. There is of course an intrinsic interest in finding the mechanism of swelling and, more practically, the production of precision extrusions also needs an accurate prediction of swelling.

Two difficulties arise in a study of extrudate swell:

(a) material non-linearity;
(b) the free surface of the jet.

The first problem is common to all non-Newtonian flows; the second poses a special interest, quite separate from the first, involving mixed boundary conditions on the fluid. By mixed boundary conditions we mean that part of the fluid surface S_v has velocity (usually no-slip) boundary conditions and a part S_t has traction (or stress) boundary conditions. In addition, there may be thermal boundary conditions. Figure 3.2 shows a typical case; it may be that an external medium or surface tension (to be mentioned later)

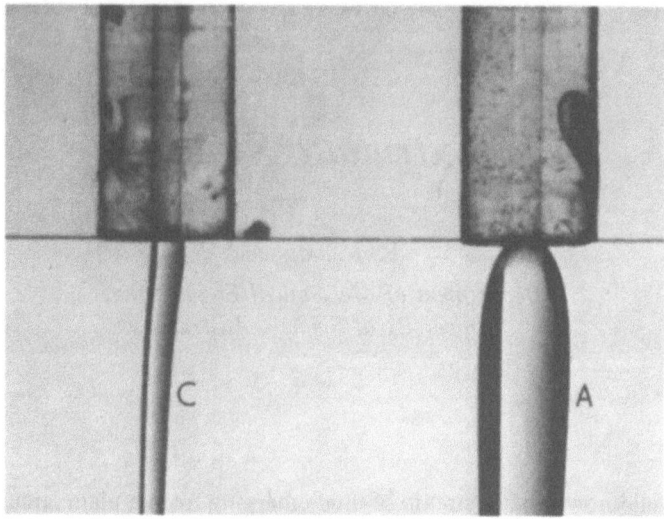

Fig. 3.1.   Extrudate swelling. At C is a Newtonian stream exiting from a capillary, showing little swell; at A a viscoelastic fluid swells to three times the capillary diameter. (Reproduced from A. S. Lodge (1964), *Elastic Liquids*, p. 242, with kind permission of the publishers, Academic Press, London.)

will cause normal and tangential stresses on the free boundary, or that slip at the wall may occur in the die, but in the main the conditions shown in Fig. 3.2 suffice to describe extrusion.

Of the two difficulties just mentioned, the boundary condition problem is the more basic as free boundaries present fundamental difficulties even for simple material properties. To see this, we may look at the problem of

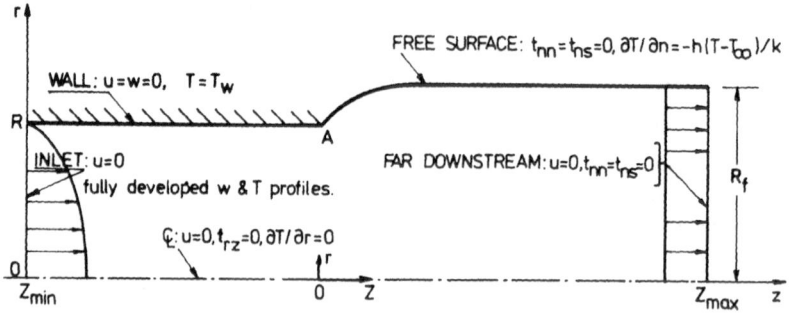

Fig. 3.2.   Boundary conditions in extrusion.

computing the flow of a free jet from a sharp-edged orifice. For the two-dimensional problem of a thin plate orifice (slit) emitting a plane jet of inviscid fluid the problem of locating the unknown free surface and computing the velocity field was solved elegantly using complex variable methods over a century ago. However, the axisymmetric potential jet cannot be solved in this manner, and numerical methods had to be used in the solution (Hunt, 1968). For viscous Newtonian fluids, certain boundary layer calculations in the jet itself can be made, but they do not solve the entire problem, and certain difficulties are also encountered using finite differences, one of which is the irregular shape of the boundary; only in 1972 was a creeping (inertialess) solution obtained for this problem. This solution (Nickell et al., 1974; Tanner, 1973) uses a finite-element method and shows that at the point of exit where there is a sudden change in boundary condition there is a stress singularity as predicted by the local analysis of Michael (1958). This singularity, compounded with the difficulty of locating the (unknown) free surface, argues against the possibility of finding a simple analytical method and suggests we commit ourselves to numerical methods for this class of problems. Available methods include finite differences, finite elements, and the so-called boundary-element methods which rely on integral equation methods. We now consider these in turn, restricting the discussion initially to the incompressible, isothermal case.

3.1 THE FINITE-DIFFERENCE METHOD

Rouse and Abul-Fetouh (1950) used finite-difference methods for axisymmetric potential flow free jet problems and the results for final jet size were not accurate (Hunt, 1968), probably due to difficulties in enforcing the normal force boundary condition on the free surface which, in general, does not fit the mesh neatly.

Horsfall (1973) used a finite-difference formulation (see Chapter 2) with stream function and vorticity as dependent variables for a viscous jet and produced a swelling result of about half of that observed. It seems likely that this failure was again due mainly to problems in trying to enforce the normal stress boundary condition where the mesh fits worst. Hill et al. (1981) used a marker-and-cell-type finite-difference program and produced results that show too little swelling (about 60 % of observed swelling). Apart from the difficulty in the free surface fit, they used too short a field to accommodate the proper boundary conditions shown in Fig. 3.2 and most likely this has affected their swelling ratios. Since this problem arises with all computer methods we shall discuss it now. Looking at Fig. 3.2 one can regard the transition from a Poiseuille flow inside the tube to the rigid body motion downstream in terms of a disturbance superposed on a Poiseuille flow or on a rigid cylinder motion. In both cases the disturbance flow has a zero net flux along the z-axis, and it is known (Tanner, 1963; Waldron, 1965) that in creeping flows these zero-flux disturbances in a tube or channel attenuate exponentially (this is a fluid mechanical analogue of St. Venant's principle) away from the source of disturbance (in this case, the exit plane). The rate of attenuation (upstream) inside the tube (the smallest eigenvalue in the problem) is about $\exp(+4\cdot466z/R)$ in axisymmetric creeping flow and for plane flow about $\exp(+3\cdot749x/R)$. (Note x and z are

TABLE 3.1

NEWTONIAN SWELLING RATIOS χ_0

Investigator	Method	Degrees of freedom	χ_0
(a) *Axisymmetric*			
Batchelor and Horsfall (1971)	Experiment	—	$1\cdot135 \pm 0\cdot01$
Nickell *et al.* (1974)	FEM	1 000	$1\cdot128$
Allan (1977)	FEM	988	$1\cdot132$
Tanner (1976)	FEM	254	$1\cdot136$
Trogdon and Joseph (1981)	Series matching	50	$1\cdot111$
Ryan and Dutta (1981)	FD	110	$1\cdot120$
Crochet and Keunings (1980)	FEM	1 178	$1\cdot126$
Chang *et al.* (1979)	FEM	306	$1\cdot139$
		378	$1\cdot126$
Extrapolated value		∞	$1\cdot127 \pm 0\cdot003$
(b) *Plane*			
Crochet and Keunings (1981a)	FEM	174	$1\cdot227$
		338	$1\cdot207$
		562	$1\cdot200$
		1 178	$1\cdot196$
Crochet and Keunings (1980)	FEM	1 178	$1\cdot188$
Chang *et al.* (1979)	FEM	306	$1\cdot206$
	Collocation	252	$1\cdot155$
Reddy and Tanner (1978b)	FEM	254	$1\cdot199$
J. F. Milthorpe (private communication, 1981)	FEM	2 928	$1\cdot189$
Extrapolated value		∞	$1\cdot190 \pm 0\cdot002$

negative here.) In the exterior free jet the rates are about $\exp(-2{\cdot}811z/R_f)$ and $\exp(-2{\cdot}106x/R_f)$ respectively. Thus for axisymmetric creeping flows an upstream and downstream field of about $2R$ is safe, giving a minimum attenuation of 250, which would be usually at the numerical noise level. For plane flows it is safer to use $2{\cdot}5$–$3R$ as a field length. With viscoelastic and/or inertial effects longer fields will often be needed, especially downstream, and failure to provide these will render the results questionable; the jet form is sensitive to small constraints. If analytical results, such as the ones quoted above for creeping flow, are available, they should be used, otherwise careful numerical experiments are necessary.

In another finite-difference study Ryan and Dutta (1981) avoided the lack of fit between the free surface and the mesh by mapping the jet and tube wall on to a rectangular space and performing difference calculations in the rectangular space. They also used a stream function/vorticity scheme in the mapped space and satisfied the normal stress boundary condition on the transformed free surface. Results for a circular die are better than the previous efforts described above (but somewhat low, only 12 % swelling for a circular tube instead of the nearly 13 % expected; see Table 3.1). However, the scheme failed to work properly or at all for annular dies when the ratio inner radius:outer radius lay outside the range $0{\cdot}3$–$0{\cdot}7$. Thus more development seems to be needed on the finite-difference approach before it is a reliable tool even for creeping flow: it seems likely that better results could be obtained with this method and its performance in viscoelastic swelling remains to be investigated.

3.2 THE FINITE-ELEMENT METHOD

One advantage of the finite element system is that it permits a closer 'mesh' near the singular exit points. Of course, no finite mesh can exactly capture the singular behaviour there and it would be possible, if one knew the order of the singularity, to insert a special semi-analytical element at the singular point to give a good representation near these points. This device has been used in fracture mechanics (Tracey, 1977); it has not so far been used in extrusion studies. The finite-element method's ability to model the singularity more or less well without special elements is helpful but probably not decisive in favouring finite-element methods over finite difference methods; more useful is the ability to have a mesh fit the extrudate shape as exactly as is needed.

Even within the finite-element field, a wide choice of computational methods is available. Restricting ourselves to plane or axisymmetric incompressible problems, we have the option of working with a stream function ψ or with the velocity components, u, w and the pressure p as primitive variables; or, we may retain the velocities, pressure and stresses (the 'mixed' method; see Chapter 2). The free boundary conditions are best handled by not using the stream function method (Nickell *et al.*, 1974). Briefly, if we consider a computation using a stream function in a Newtonian fluid, then the pressure is eliminated and is recovered afterwards by solving a separate Poisson-type equation. For problems in which velocity boundary conditions are given the scheme works satisfactorily, since boundary conditions are given in terms of ψ and its normal derivative $\partial\psi/\partial n$. For boundaries partially made up of free surfaces, without traction, we require on a part of the boundary (S_t) that the normal (t_{nn}) and tangential (t_{ns}) stress components vanish there. For the Newtonian fluid this means that

$$\frac{\partial v_n}{\partial s} + \frac{\partial v_s}{\partial n} = 0 \qquad -p + 2\eta\frac{\partial v_n}{\partial n} = 0 \qquad (3.1)$$

where v_n and v_s are the components of velocity normal and tangential to the boundary, respectively, and η is the (constant) viscosity. Of course on the free surface we must have the normal velocity component (v_n) zero, so that $\partial v_n/\partial s$ vanishes. To express eqn (3.1) in terms of the stream function is difficult; it is the presence of p in the normal stress term which is awkward, since p is not known. These problems are bypassed by not using ψ in the computation and electing to work instead with the velocity components u, w, the pressure p, and possibly the stresses, as computational variables. No completely successful examples of free-surface computations using the stream function are known; for the primitive variable method using u, w and p we proceed as follows.

In a Cartesian tensor representation we have the conservation laws

$$\frac{\partial t_{ij}}{\partial x_j} + \rho(f_i - a_i) = 0 \qquad (3.2)$$

and

$$\frac{\partial v_i}{\partial x_i} = 0 \qquad (3.3)$$

where v_i is the velocity component in the direction of x_i, f_i are the (known) body force components, and t_{ij} are stress components; a_i signifies, for steady flow, the acceleration component $v_j\partial v_i/\partial x_j$.

Let the boundary S be composed of two parts; $S = S_v + S_t$. Velocity boundary conditions are given on S_v, while traction (stress) boundary conditions are given on S_t. On S_t we shall assume

$$t_{ij}n_j = T_i \qquad (3.4)$$

where the traction vector components T_i are given and n_j is the outward-pointing normal unit vector on the surface.

We now discuss the choice of constitutive equation. Finite-element methods have been used for Reiner–Rivlin fluids, where extremum principles may be shown to exist (Johnson, 1960) but it is known that such a constitutive equation is not a realistic representation of many real non-Newtonian fluids and in fact there seems to be no experimental evidence that any real non-Newtonian fluid is accurately described by such a constitutive relation. For special flows (the class of viscometric flows) it may be shown that a very wide class of fluid is described by the following constitutive equation (Pipkin and Tanner, 1972):

$$t_{ij} = -p\delta_{ij} + \eta A_{ij}^{(1)} + (v_1 + v_2)A_{ik}^{(1)}A_{kj}^{(1)} - \tfrac{1}{2}v_1 A_{ij}^{(2)} \qquad (3.5)$$

Here v_1 and v_2 are material parameters, called the first and second normal stress coefficients, and η is the viscosity function. The $A_{ij}^{(1)}$ and $A_{ij}^{(2)}$ are the first and second Rivlin–Ericksen tensors respectively, defined by

$$A_{ij}^{(1)} = \frac{\partial v_i}{\partial x_j} + \frac{\partial v_j}{\partial x_i} \qquad (3.6)$$

and

$$A_{ij}^{(2)} = \frac{\partial A_{ij}^{(1)}}{\partial t} + v_k \frac{\partial}{\partial x_k} A_{ij}^{(1)} + A_{jk}^{(1)} \frac{\partial v_k}{\partial x_i} + A_{ik}^{(1)} \frac{\partial v_k}{\partial x_j} \qquad (3.7)$$

(Higher-order tensors may be defined by changing the superscripts (1) and (2) in eqn (3.7) to (n) and $(n + 1)$ respectively.) The material functions η_1, v_1 and v_2 are functions of $\tfrac{1}{2}\mathrm{tr}\,(A^{(1)})^2$ (which is the square of the 'shear rate' in a generalised sense), so that one obtains a realistic description of real fluids in the restricted class of viscometric flows. In the simplest case, when η, v_1 and v_2 are constants, we obtain the second-order fluid equation; it may be shown from a very general standpoint that this describes many fluids for 'slow' flows. To define a slow flow, note that the dimensions of the fraction v_1/η are time, and hence v_1/η is a characteristic time of the fluid; a slow flow may be defined as steady flow in which $v_1/\eta\sqrt{\mathrm{tr}(A^{(1)})^2}$ is less than unity.

Other equations that have been used are the convected Maxwell model and the Oldroyd model (Oldroyd, 1958). The convected Maxwell model has a constitutive equation of the form

$$\lambda \frac{\Delta t_{ij}^{E}}{\Delta t} + t_{ij}^{E} = \eta_0 A_{ij}^{(1)} \tag{3.8}$$

where the operator $\Delta/\Delta t$ is given by

$$\frac{\Delta t_{ij}^{E}}{\Delta t} = \frac{\partial t_{ij}^{E}}{\partial t} + v_k \frac{\partial t_{ij}^{E}}{\partial x_k} - \frac{\partial v_i}{\partial x_k} t_{kj}^{E} - \frac{\partial v_j}{\partial x_k} t_{ki}^{E} \tag{3.9}$$

and t_{ij}^{E} is the deviatoric or extra stress, defined as

$$t_{ij}^{E} = t_{ij} + p\delta_{ij} \tag{3.10}$$

More complex constitutive models may also be used; most of the difficulties that arise are already present in eqns (3.5) and (3.9). The finite-element method now reduces the above system of partial differential equations to a set of (usually non-linear) algebraic equations by one of several approaches. In the displacement method (strictly velocity in this case) all stress variables are expressed in terms of the velocities and the pressures, and the stresses are then back-calculated after finding the velocities u, w, and the pressure p. This is the approach followed by Nickell *et al.* (1974), and Viriyayuthakorn and Caswell (1980). The following steps are made to effect this process:

(i)   We assume that the nodal variables (a node usually occurs where two element bounding curves cross but it may also be elsewhere on the element boundary) are the unknowns.

(ii)   We assume an interpolation function for the variables inside each element. Often linear or quadratic interpolation is used. Thus the representation of the radial velocity component u inside the mth element will take the form

$$u^{(m)} = \sum_{i=1}^{r} N_i^{(m)}(x,y) u_i^{(m)} \tag{3.11}$$

where the superscript denotes that the field is for the mth element, and r is the number of nodes associated with this element; the shape or weighting functions $N_i^{(m)}$ are chosen in advance and the nodal point variables $u_i^{(m)}$ are the unknowns.

(iii) We substitute the global expression for the variables into the mass-conservation equation and the equation of motion, multiply by the relevant shape functions in turn for each element and integrate over the body. The process then sets the resulting expression equal to zero. This is the Galerkin method. It is usual to eliminate the stresses at this point and to make an integration by parts on the equation of motion, eqn (3.2); this reduces by one the number of times one has to differentiate the shape functions. Care must be taken in this step if the shape functions are of limited differentiability.

Taking the mass-conservation equation as an example of the process, we form

$$\int_{\text{body}} \frac{\partial v_j}{\partial x_j} N_i^{(m)} \, dV = 0 \qquad (m = 1, N) \qquad (3.12)$$

where $N_i^{(m)}$ are the known shape functions for the pressure. Clearly, eqn (3.12), after integration, gives a linear combination of the unknown nodal point velocities and forms part of a system of equations; the remainder comes from the equations of motion.

Thus the result of these integrations is a system of simultaneous equations, usually non-linear, for the nodal point velocities and pressures.

(iv) When integrating the weighted equation of motion by parts we obtain a surface integral which can be evaluated on the parts of the body (S_t) where traction boundary conditions are given; this takes care of these boundary conditions which contribute to the known right-hand side of the system of equations to be solved. The ease with which these traction boundary conditions can be satisfied constitutes one great advantage of the finite-element method. The remaining (velocity) boundary conditions are now set in; we recognise that some nodal velocities are known and these velocities are removed from the list of unknowns.

(v) Finally, we find a set of non-linear simultaneous equations for the remaining unknown nodal point velocity and pressure components arranged as a vector $\mathbf{V}$

$$(\mathbf{K} + \mathbf{C})\mathbf{V} = \mathbf{F} \qquad (3.13)$$

The matrix $\mathbf{C}$ depends on $\mathbf{V}$ and is often unsymmetric. Hence iteration has to be used to solve eqn (3.13). $\mathbf{K}$ is symmetric and constant and arises from the (linear) creeping flow part of the problem.

(vi) After solution for u, w and p, the stresses and other quantities (i.e. streamlines) are calculated.

72   R. I. TANNER

There are some constants on the shape functions. It has been found essential for accuracy that the pressure field be interpolated with a polynomial one order lower than the velocity terms. Two successful (u, w, p) elements are shown in Fig. 3.3; these have linear pressure fields over the element and quadratic velocity fields.

Crochet and Keunings (1981a) have used the mixed method where no attempt is made to eliminate the stresses; these, plus the pressure and the velocities are used as primitive variables and are interpolated on the elements. Thus one has larger matrices and longer solution vectors for a given number of elements but there are compensating advantages in that the constitutive equation can more easily be changed with this scheme and also the construction of a Newton–Raphson equation-solving scheme is facilitated. It remains to be seen which method gives the best results; for inelastic fluids the displacement method is superior in that it gives smoother stress fields; all methods, as we shall see, have problems at high

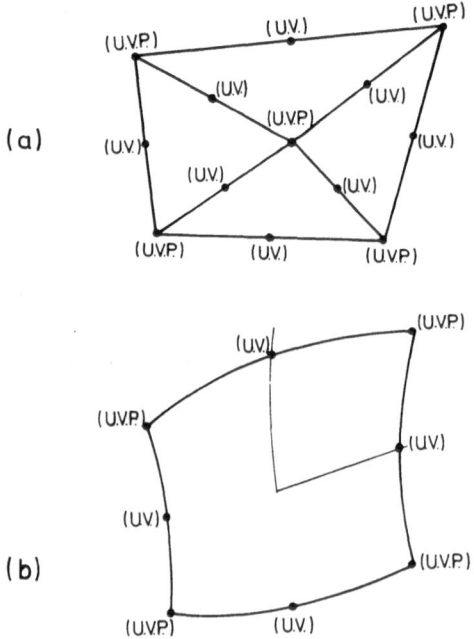

Fig. 3.3. Some successful elements for finite-element fluid mechanics. (a) Four combined biquadratic velocity and bilinear pressure area-coordinate triangular elements. (b) Combined biquadratic velocity and bilinear pressure isoparametric quadrilateral element.

Weissenberg numbers. In either case the solution of the non-linear equations, eqn (3.13) needs to be done. Many authors have simply used a Picard iteration where all non-linear terms are set in the right-hand side of eqn (3.13) and so a sequence of linear problems is solved; others have used Newton–Raphson methods or approximations to this scheme. Gartling *et al.* (1977) and Engelman *et al.* (1981) survey the various schemes available and their convergence rates; they show that the choice of solution scheme is partly problem-dependent. However, the Newton–Raphson scheme nearly always performs well, even if it is sometimes hard to implement.

3.3 THE BOUNDARY-ELEMENT METHOD

In the finite-element (and finite-difference) schemes, one finds the complete solution field for the problem whether or not this is of interest. In some problems only the swelling shape and/or the pressure losses may be needed, and it is therefore interesting to look at the boundary-element method and its derivatives which avoid evaluating all of the internal variables, at least for linear creeping flow problems. Essentially, the techniques are derived from reciprocal theorems familiar in linear elasticity, combined with a knowledge of the relevant Green's function. The effective adaptation of these methods for computer use in elasticity is due to Rizzo (1967) and Cruse (Cruse and Rizzo, 1968). Here we begin by discussing viscous incompressible flows; the problem follows closely the discussion for linear elasticity (Brebbia, 1978).

The equations of motion and mass conservation are given above in eqns (3.2) and (3.3).

Now consider an arbitrary set of fields v_i^*, p^*, t_{ij}^*. We multiply the equation of motion, eqn (3.2) and the mass conservation, eqn (3.3) by v_i^*, and p^*, respectively, and integrate over the body; this is the familiar Galerkin procedure (Nickell *et al.*, 1974) already used in the finite-element approach, but we also add on a surface term over the part S_v of the surface (Fig. 3.2) where the velocities are given. Thus we have (where Ω denotes the entire body)

$$0 = \int_\Omega \left[\frac{\partial t_{ij}}{\partial x_i} + \rho(f_i - a_i) \right] v_i^* \, dV + \int_\Omega p^* \frac{\partial v_i}{\partial x_i} \, dV + \int_{S_v} (v_i - \hat{v}_i) T_i^* \, dS$$

(3.14)

where T_k^* is the traction vector $t_{kj}^* n_j$ formed from the starred stress tensor and the outward unit normal vector **n**, and $\hat{v}$ are the given boundary conditions on S_v. Any solution (t_{ij}, p, v_i) that satisfies the equations of motion, eqn (3.2), the mass conservation equation, eqn (3.3) and the boundary condition $v_i = \hat{v}_i$ on S_v will make eqn (3.14) vanish, and hence eqn (3.14) is then satisfied for arbitrary (t_{ij}^*, p^*, v_i^*).

Now consider the following expression denoted by $I(*, 0)$ where

$$I(*, 0) = \int_\Omega \frac{\partial t_{ij}^*}{\partial x_j} v_i \, dV - \int_S v_k T_k^* \, dS - \int_\Omega p^* \frac{\partial v_i}{\partial x_i} \, dV \qquad (3.15)$$

Here S is the whole body surface.

By using Green's theorem we can show that for a Newtonian fluid there is a reciprocal theorem where starred and unstarred fields are interchanged

$$I(0, *) = I(*, 0) \qquad (3.16)$$

This reciprocal theorem can be used to replace some of the terms in eqn (3.14), finding, when both v_i^* and v_i are incompressible fields,

$$\int_\Omega \frac{\partial t_{ij}}{\partial x_j} v_i \, dV - \int_S v_i T_i^* \, dS + \int_S v_i^* T_i \, dS + \int_{S_v} (v_i - \hat{v}_i) T_i^* \, dS$$

$$+ \int_\Omega \rho(f_i - a_i) v_i^* \, dV = 0 \qquad (3.17)$$

We can combine the second and fourth terms to read $-\int_S v_i T_i^* \, dS$, understanding that $v_i = \hat{v}_i$ on S_v. We shall also suppose here that f_i and a_i are known; for the latter this implies an iteration scheme that does not converge for very large Reynolds numbers. (See also Section 3.6.) For the examples given here, both f_i and ρ are assumed zero, so we have creeping flow under no body forces. We will also restrict ourselves to plane and axisymmetric flows.

We now assume that the (*)-fields are produced by a concentrated unit force in the l-direction. The plane-flow solutions for the $T_k^{*(l)}$ and $v_k^{*(l)}$ in Cartesian coordinates x_i are well-known (Brebbia, 1978)

$$v_k^{*(l)} = \frac{1}{4\pi\eta} \left[(-\ln r)\delta_{lk} + \frac{\partial r}{\partial x_l} \frac{\partial r}{\partial x_k} \right] \qquad (3.18)$$

$$T_k^{*(l)} = -\frac{1}{\pi r} \left[\frac{\partial r}{\partial n} \frac{\partial r}{\partial x_k} \frac{\partial r}{\partial x_l} \right] \qquad (3.19)$$

where r is the distance from the point of application of the force (r) and the position at which v is evaluated (x). δ_{lk} is the unit tensor, equal to unity if l equals k, zero otherwise. Hence

$$r^2 = (x_k - r_k)(x_k - r_k) \tag{3.20}$$

and by differentiating eqn (3.20) we find

$$\frac{\partial r}{\partial x_i} = \frac{x_i - r_i}{r} \tag{3.21}$$

In eqn (3.19) n refers to the direction normal to the surface across which the traction is being computed. The corresponding formulas for axisymmetric flows contain Legendre functions and are given by Bush and Tanner (1981).

Consider the body of fluid in Fig. 3.4. Suppose the boundary is discretised into linear 'elements', ab, bc, cd, etc., and at the centre of ab we place node 1. (In this simplest of boundary element schemes no nodes occur where elements join.) We assume a (*)-field which consists of the response to a unit force in the x_1-direction applied at node 1. Then we have (Brebbia, 1978)

$$\frac{\partial t_{ij}}{\partial x_j} = - \delta_{1_i} \tag{3.22}$$

In the case of creeping flow with no body forces, substituting eqn (3.22) in eqn (3.17) we find

$$0 = -\tfrac{1}{2}v_1 - \int_S v_k T_k^* \, dS + \int_S v_k^* T_k \, dS \tag{3.23}$$

where v_k^* and T_k^* are known exactly from eqns (3.18) and (3.19). If we now assume that the field is uniform over each segment, so that we may speak of

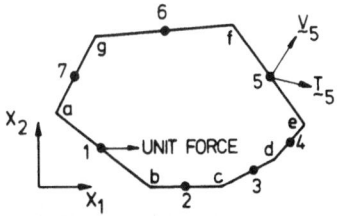

Fig. 3.4. Boundary element discretisation. The traction and velocity components are uniform on each segment (element).

$v_1^{(m)}$, $v_2^{(m)}$, $T_1^{(m)}$ and $T_2^{(m)}$ as the uniform components on the segment containing the mth node, then eqn (3.23) becomes a set of linear equations

$$0 = -\tfrac{1}{2}v_1 + \sum_{m=1}^{N} \left[\int_{S_m} T_1^* \, \mathrm{d}S \int_{S_m} T_2^* \, \mathrm{d}S \right] \begin{bmatrix} v_1^{(m)} \\ v_2^{(m)} \end{bmatrix}$$

$$+ \sum_{m=1}^{N} \left[\int_{S_m} v_1 \, \mathrm{d}S \int_{S_m} v_2 \, \mathrm{d}S \right] \begin{bmatrix} T_1^{(m)} \\ T_2^{(m)} \end{bmatrix} \qquad (3.24)$$

where N is the number of nodes. The $\tfrac{1}{2}$-factor in eqn (3.24) arises because the point force is applied on the boundary and not in the fluid interior, (Brebbia, 1978).

The integrals in eqn (3.24) may be computed, since $\mathbf{v}^*$ and $\mathbf{T}^*$ are known and eqn (3.24) represents a linear equation connecting $2N$ components of velocity and $2N$ components of traction. Similarly, one can apply a unit force in the x_2 direction at node 1, generating a second equation, and so on for all nodes, finally generating $2N$ equations. There are $2N$ velocity components and $2N$ traction components, but only a total of $2N$ unknowns, since at each node two out of the four unknowns are given as boundary conditions. After eliminating these known quantities, one can solve a set of linear equations for all $2N$ boundary unknowns. To obtain values in the interior, one places a point force where needed as the (*)-field, and uses eqn (3.17) again, thus producing the needed value of $\mathbf{v}$; a different starred field will produce the tractions and stresses. Thus results at all points of the body can be found, but only if needed.

In a test on a plane creeping (slit) extrusion (Milthorpe and Tanner, 1981) comparable accuracy (meaning that the extrudate profile settled to within 0·001 of the die radius in both cases) was obtained in a computer time of only between 5% and 10% of the corresponding finite-element time. In this test it was necessary to have about the same number of iterations (four) for convergence of the shape, and each iteration was 10–20 times faster using the boundary method. The difference in the methods is that whereas the finite-element method (FEM) used 353 unknowns the boundary element method (BEM) used only 38 unknowns. However, the FEM matrix is banded (but not very narrow in the present case; due to the incompressibility constraint, the half-bandwidth is 41) whereas the BEM generates a full matrix. By considering the number of operations needed to solve the equations, we do find that the BEM is quicker for this type of problem, even though it does not give rise to narrow banded matrices;

Bettess (1981) has come to less definite conclusions, and remarks about relative speed need to be treated with caution. The difficulty with the method is that non-linear problems have to be solved as a series of perturbations about the viscous solutions, and exploratory work is now proceeding in this direction; Bush and Tanner (1982) indicate that Reynolds numbers greater than 15 can be solved for Newtonian flow in the Hamel (or converging-wall) problem. Thus it is believed that the boundary method has considerable potential and should be further explored.

All the above techniques are applicable whether or not the flow has a free boundary; we now look at algorithms for finding the boundary.

3.4 LOCATING THE FREE BOUNDARY

On the free surface we have to satisfy three conditions simultaneously:

(i)   zero normal velocity;
(ii) zero (or prescribed) shear stress;
(iii) zero (or prescribed) normal stress.

In the most elementary cases without surface tension, so that the shear and normal stresses are zero, one approach is to ignore condition (i), set up conditions (ii) and (iii) on an assumed contour, and then calculate the normal velocities on the assumed contour. From these (Fig. 3.5(a)) a new

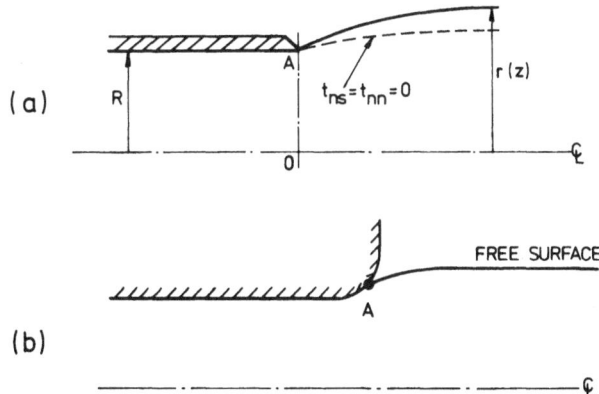

Fig. 3.5.   (a) Locating unknown boundary by iteration: dashed line, nth approximation to free streamline; full line, new streamline. Separation is always at A. (b) Rounded exit with indeterminate separation point A.

streamline can be constructed, which can serve as a new assumed contour, and the process can be continued until a satisfactory level of convergence is obtained, i.e. the normal velocity is small enough for the final contour to be regarded as satisfying all conditions. To construct the boundary streamline at each iteration we know that the streamline begins at point A (Fig. 3.5(a)), and we have an estimate of the velocity vector on the assumed boundary (broken line in Fig. 3.5(a)). If we assume that the velocity components u and w on the true boundary are the same as those on the broken line, then the streamline coordinates $r(z)$ are given by

$$r(z) = R + \int_{z=0}^{z} dz \frac{u}{w} \tag{3.25}$$

The integral was found using Simpson's rule (Tanner *et al.*, 1975). This defines another bounding curve and a new iteration can start with this curve. This simple approach has been used by most workers and is adequate for practical extrudate swell calculations where surface tension is not dominant. If the swelling were very large, it would be useful to employ a more complex scheme to find the free surface, perhaps employing an interpolation (or extrapolation) scheme to find the values of (u/w) in eqn (3.25); the simple scheme described uses the (u/w) values on the previous surface. It should be noted that the streamline springs from a sharp corner in all these problems; up to the present time no computations seem to have been done with a more realistic rounded corner exit (Fig. 3.5(b)); since the point of separation is now an unknown, the problem is more difficult.

For cases when surface tension is very large, the above scheme is unstable (Reddy and Tanner, 1978a). This means that when the dimensionless number $\sigma/\eta\bar{w}$, which gives the ratio of surface tension to viscous forces, becomes very large another scheme for finding the boundary shape is needed. Orr and Scriven (1978) have in this case set the normal velocity and the shear stress equal to zero and then made the normal stress equal the surface tension force by adjusting the boundary position. For small surface tension numbers, such as are found in polymer processing, the previous simpler scheme is superior.

In the next section we report some results and compare the various approaches.

3.5 FACTORS AFFECTING EXTRUDATE SWELL

We now consider some results. The object of our studies is to understand the pressure losses at exit, the shape of the extrudate and ultimately the stability of the flow in terms of basic parameters. Although we are discussing extrusion without tension in the extrudate here, the methods and some of the results have obvious relevance to other flow problems such as melt spinning, film blowing and wire coating.

Let us first concentrate on the simplest problem of a long (plane slit or axisymmetric circular) die. This will ensure that the flow is a fully developed viscometric flow far upstream of the exit plane. Let us now consider the swelling ratio χ. This ratio is defined (see Fig. 3.2) as

$$\chi = \frac{R_f}{R} \tag{3.26}$$

The swelling ratio is a function of several parameters. These are:

(1) Geometry—R is the only length parameter.
(2) Flow kinematics—only $\bar{w}$, the average entry speed, is needed if the die is long enough.
(3) Gravity (g). For many polymer melts, ignoring this factor is a good assumption which will be adopted here.
(4) Fluid properties. Clearly we need ρ, η for an isothermal incompressible Newtonian fluid (we only consider incompressible fluids). For other fluids, especially viscoelastic fluids, other parameters are needed, and these will be introduced as required. Surface tension (σ) is also a relevant parameter.
(5) If heat transfer is important, then the wall temperature (T_w), the surroundings temperature (T_∞), a surface heat transfer coefficient (h), the product of density and specific heat ($\rho\gamma$) and the conductivity (k) will need to be specified.
(6) Other factors such as slip at the wall can also be considered.

We can study the influence of the above factors by dimensional methods; once the relevant dimensionless quantities have been selected, then computer solutions can be made to explore the range of parameters required.

3.5.1 Newtonian fluids
In this case we have (ignoring gravity)

$$\chi = \chi(R, \bar{w}, \rho, \eta, \sigma) \tag{3.27}$$

Here R is the tube radius and the other parameters are defined above.

Formation of dimensionless groups shows that

$$\chi = \chi(2\rho\bar{w}R/\eta, \sigma/\eta\bar{w}) \tag{3.28}$$

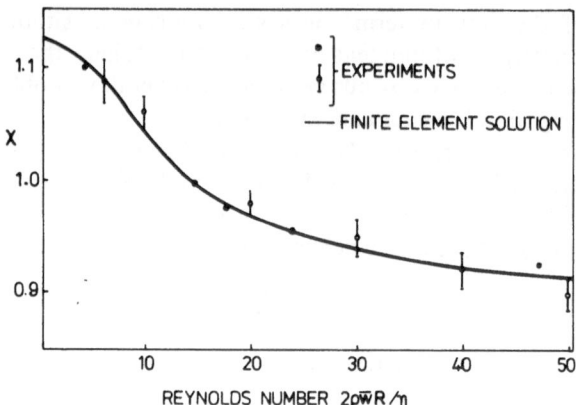

Fig. 3.6. Extrudate swell ratio χ as a function of Reynolds number Re for $\sigma/\bar{w}\eta = 0.3$ for axisymmetric flow.

The influence of the Reynolds number ($Re = 2\rho\bar{w}R/\eta$) and the surface tension number on χ have been investigated; both produce some change in jet diameter. In all cases, we used a finite-element program that has been previously described (Nickell *et al.*, 1974). The results are shown in Figs 3.6 and 3.7; the final diameters agree well with available experiments. The effect of large surface tension is to destabilise the calculation; small 'kinks' in the surface tend to grow, and no convergence was obtained for $\sigma/\eta\bar{w} >$ 4·0 (Reddy and Tanner, 1978a); as $\sigma/\eta\bar{w} \to \infty$ the swelling should vanish. We have not attempted to improve the algorithm but Orr and Scriven's (1978) method may be useful here.

The effect of increasing inertia is to slow down the rate of convergence; iteration for the non-linear effects and the free surface correction are done in each iteration cycle. Without surface tension calculations up to Reynolds numbers of 1000 have been converged (Reddy and Tanner, 1978a) without difficulty but at considerable expense due to the large number of iterations required (~ 50 at the highest Reynolds numbers). In Fig. 3.6 the computations were done to match the experiments of Goren and Wronski (1966) and higher Re calculations were not attempted with $\sigma/\bar{w}\eta = 0.3$.

At high Reynolds numbers one must permit a long jet region to exist; the analytical results of Goren and Wronski (1966) for the rates of decay in a jet are useful for setting the field length.

From the polymer processing point of view, the Reynolds number is usually negligible, and in the following we shall ignore it; we shall also ignore surface tension effects which are also usually small. In brief, inertia

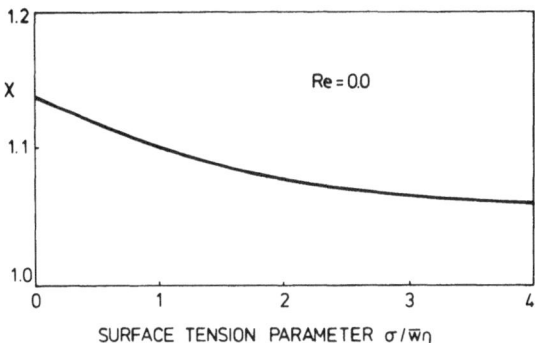

Fig. 3.7. χ as a function of $\sigma/\eta \bar{w}$ for $Re = 0$; axisymmetric flow.

tends to make jets smaller, and surface tension tends to pull the jet shape back towards a cylinder of radius R. When we ignore ρ and σ in our list of parameters in eqn (3.27), then no dimensionless group can be formed from R, $\bar{w}$ and η, and we have simply that χ is a constant χ_0.

A selection of χ_0 values from the literature for the basic Newtonian creeping flow case is shown in Table 3.1. It is noticeable that the FEM results with a greater number of degrees of freedom generally imply less swell. Several programs are involved in the results of Table 3.1. By plotting χ_0 against the reciprocal of the number of degrees of freedom for each family of results, the extrapolated values $\chi_0 = 1 \cdot 127 \pm 0 \cdot 003$ (axisymmetric) and $\chi_0 = 1 \cdot 190 \pm 0 \cdot 002$ (plane) have been estimated for an infinite number of degrees of freedom. The error ranges represent differences in calculation method and grid choice. It is noticeable that the FD method (Ryan and Dutta, 1981), the series matching technique of Trogdon and Joseph (1981), and the collocation method of Chang et al. (1979) yield results that are far from the experimental results, and are presumed to be inaccurate. In the remainder of this article we shall assume $\chi_0 = 1 \cdot 13$ and $1 \cdot 19$ for the base cases in axisymmetric and plane flow respectively.

3.5.2 Inelastic non-Newtonian fluids

The next level of complexity is to permit the viscosity to vary with shear rate. A realistic viscosity–shear rate curve will usually involve several parameters to describe the Newtonian behaviour at low shear rates ($\dot{\gamma}$) and the power-law behaviour at higher shear rates. The most important features can be investigated by using the simple power-law rule which gives (in simple shearing)

$$\eta = K|\dot{\gamma}|^{\nu - 1} \tag{3.29}$$

82   R. I. TANNER

where K is a consistency parameter and v is the flow index. (In a general flow, we replace $|\dot{\gamma}|$ by $\sqrt{I_2}$, where I_2 is the second invariant of the rate-of-deformation tensor. When $v = 1$, we have the Newtonian case and $K = \eta$; when $v < 1$, the flow is 'pseudoplastic' (shear-thinning) and when $v > 1$, it is shear-thickening. Stability considerations forbid a negative value of v, so that $v = 0$ represents the extreme lower limit for v. In that case, we have slug (or plug) flow in the tube and the extrudate expansion is zero so that $\chi = 1 \cdot 0$; the 'fluid' just slips at the wall.) Dimensional theory then shows us that (for zero gravity and surface tension)

$$\chi = \chi(v) \tag{3.30}$$

in the creeping-flow limit. Figure 3.8 shows computed results for this case.

Again, it is clear that no great change of expansion takes place due to change in viscosity with shear rate. In fact, the usual variation of viscosity with shear rate demands $1 > v$ (shear-thinning) and this yields somewhat less expansion than the Newtonian case. Thus power-law fluids are not very interesting, and we conclude that non-Newtonian inelastic behaviour is not the cause of large extrudate swell.

3.5.3 Viscoelastic effects

Introduction of memory effects may be accomplished in several ways. The Maxwell fluid (eqn (3.8)) has a relaxation time (λ), in addition to the viscosity and density, as a fluid parameter. Under creeping-flow conditions, dimensional theory shows that

$$\chi = \chi(\lambda \bar{w}/R_0) \tag{3.31}$$

The group $\lambda \bar{w}/R_0$ is the Weissenberg number (Ws) (the use of this terminology is preferred, reserving the Deborah number for Eulerian

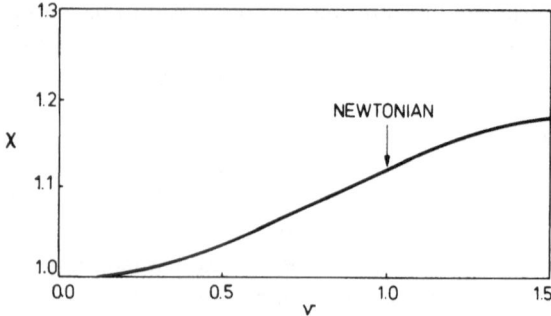

Fig. 3.8. χ as a function of power-law index v for zero inertia and surface tension; axisymmetric flow.

unsteady flows). As mentioned above, in the limit of slow flow, a viscoelastic fluid behaves as a second-order fluid (eqn (3.5)) with η, v_1, and v_2 constant. In this case dimensional analysis gives

$$\chi = \chi\left(\frac{v_2}{v_1}, \frac{v_1 \bar{w}}{\eta R_0}\right) \tag{3.32}$$

In plane flows, it can be shown that v_2 is irrelevant, and hence can be omitted from eqn (3.32) so that χ is only a function of $v_1 \bar{w}/\eta R$. The best available experimental evidence shows that v_2/v_1 is in any case roughly a constant (about -0.10) for most polymeric fluids and is probably not a very important factor in die-swell. If we note that v_1/η has the dimensions of time, then the equivalence of eqns (3.5) and (3.8) is apparent. In a simple shearing flow where the velocity field is given by $\mathbf{v} = \mathbf{i}y\dot{\gamma}$, we define the first normal stress difference N_1 by $\sigma_{xx} - \sigma_{yy}$, equal by definition to $v_1 \dot{\gamma}^2$, and the second normal stress difference N_2 by $\sigma_{yy} - \sigma_{zz}$, equal by definition to $v_2 \dot{\gamma}^2$. Hence the difference N_1 is proportional to $v_1(\bar{w}/R_0)^2$, and the group $v_1 \bar{w}/\eta R_0$ is proportional to $(N_1/\tau)_w$ evaluated at the tube wall; thus the equivalence of the present formulation to those based on 'recoverable shear' $(N_1/2\tau)_w$ is demonstrated. In fact, for equivalence, we set $\lambda = v_1/2\eta$. Results for a limited range of $\lambda\dot{\gamma}_w$ are shown in Fig. 3.9; these computations are based on the second-order fluid model and the Maxwell model and did not converge for higher Ws values. (No results are yet available for the axisymmetric case with the second-order model.) Note there is a slight difference between the u, v, p and the mixed methods (Crochet and Keunings, 1981b); both are shown in Fig. 3.9. In the plane cases there is a slight initial contraction, then an expansion as $\lambda\dot{\gamma}_w$ increases; this is an expected result (Tanner, 1980). Also shown are the results of the analytic formula of the elastic-recovery type (Tanner, 1970) modified for plane flow

$$\chi = 0.19 + \left(1 + \frac{1}{12}\left(\frac{N_1}{\tau}\right)^2\right)^{1/4} \tag{3.33}$$

and the corresponding result for axisymmetrical flow.

χ computed from eqn (3.33) increases more slowly with Ws than other computations or experiments indicate, although it often seems to fit experimental results well at higher Weissenberg numbers; for axisymmetrical flows the analytical result agrees better.

There is evidence that the expansion is beginning to increase rapidly for $\lambda\dot{\gamma}_w > 1$. Thus, as expected, we see viscoelasticity as a cause of enhanced swelling.

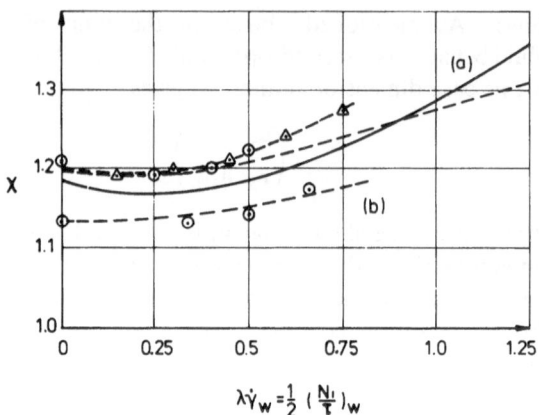

$$\lambda\dot\gamma_w = \tfrac{1}{2}\left(\tfrac{N_1}{\tau}\right)_w$$

Fig. 3.9. χ as a function of Weissenberg number $\lambda\dot\gamma_w$ for Maxwell and second-order fluids. Note slight reduction in χ for small $\lambda\dot\gamma_w$ in the plane case only. No convergence was obtained for higher $\lambda\dot\gamma_w$. (Note: $\lambda\dot\gamma_w = \tfrac{1}{2}(N_1/\tau)_w$ for these curves.) (a) Plane flow: ——, mixed methods, Maxwell model (Crochet and Keunings, 1981a); $-\triangle-$, u, v, p method, second-order fluid (Reddy and Tanner, 1978b); $\odot$, u, v, p method, Maxwell fluid (Crochet and Keunings, 1981a); $---$, eqn (3.33). (b) Axisymmetric flow: $\odot$, mixed method, Maxwell model (Crochet and Keunings, 1980); $---$, elastic eqn $= 0.13 + [1 + \tfrac{1}{8}(N_1/\tau)^2_w]^{1/6}$ (Tanner, 1970).

Crochet and Keunings (1982) have been able to compute large swellings ($\sim 100\%$) at a Weissenberg number of 4·5 by carefully choosing long entry and extrudate regions and by changing to an Oldroyd B fluid, which also contains a retardation time. These changes appear to result in extra numerical stability for reasons not yet known.

3.5.4 Thermal effects
In the case where the flow is not isothermal, several more dimensionless groups appear. In order to have any significant changes in swelling, we need to couple the thermal and mechanical fields by permitting the viscosity η to vary with temperature. Thus, we let

$$\eta = \eta_w \exp\{-\zeta(T - T_w)\} \qquad (3.34)$$

where T is the temperature, ζ is a positive constant, T_w is the die-wall temperature, and η_w is the viscosity at T_w. Note that this means the fluid is still Newtonian, since $\dot\gamma$ does not enter eqn (3.34).

Besides these factors, we also need the parameters mentioned above. We will take T_w as a reference temperature and then we find

$$\chi = \chi(hR/\alpha, \rho\gamma\bar{w}R/k, (T_w - T_\infty)\alpha/\eta_w\bar{w}^2, \eta_w\bar{w}^2\zeta/\alpha) \qquad (3.35)$$

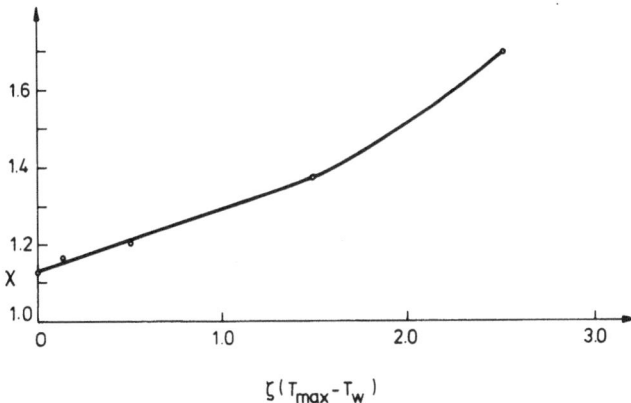

Fig. 3.10. Thermal swelling of a Newtonian fluid with variable viscosity as a function of temperature difference between centreline and die wall (made dimensionless with ζ).

where h is the overall heat transfer coefficient from the jet surface. Phuoc and Tanner (1980) have investigated the effect of varying the rate of extrusion $\bar{w}$ for a fluid that has properties similar to low-density polyethylene (except that it is Newtonian) and assuming it is extruded through a 2 mm diameter tube; thus the Nusselt number $\mathrm{Nu}(\equiv hR/\alpha)$ is fixed and the Péclet number $Pe(\equiv \rho\gamma\bar{w}R/\alpha)$ and the Nahme–Griffith number $Na(\equiv \eta_w\bar{w}^2\zeta/\alpha)$ are functions only of $\bar{w}$. To fix the remaining group, they took $T_w - T_\infty$ to be 125 °C; χ then is only a function of $\bar{w}$. Results are shown in Fig. 3.10, where the large swelling at higher extrusion rates is apparent; in this figure the more readily appreciated dimensionless function of $\bar{w}$, $\zeta(T_{\max} - T_w)$, has been used instead of $\bar{w}$ as dependent variable; here $T_{\max}$ is the temperature on the centreline in the fully developed (low Graetz number) tube flow.

3.5.5 Wall slip
In a recent paper, Silliman and Scriven (1980) have shown that permitting slip between the fluid and the wall near the tube lip (where the fluid is under severe shear stress) reduces swelling in Newtonian fluids in the plane case. This is not an unexpected result, since complete slip would produce no swelling at all. Hence, we conclude that wall slip is not important in explaining large swells.

3.5.6 Die geometry effects
So far only the results for long dies have been noted. Tanner (1976), Allan (1977) and Crochet and Keunings (1981b) have looked at other geometries

including the flow from a reservoir through a short die and flow from a converging or diverging tube die of variable length. The results are given for Newtonian fluids and for low Weissenberg numbers, but even so Allan (1977) reported some convergence problems for very short dies (length/ diameter < 0·05). Few other results are available for viscoelastic and non-Newtonian fluids in short dies as yet.

We have shown how to use finite-element methods to investigate the extrusion process; similar results are now being obtained using boundary elements. It is confirmed that viscoelasticity is important as a factor in extrudate swell, while non-Newtonian behaviour (alone) and wall slip are not. Surprisingly, we find that variations of temperature, through the agency of temperature-sensitive viscosity, can produce large swelling effects. In the investigation reported here, the temperature difference between inside and outside of the extrudate is produced by viscous dissipation, but we can easily infer that the results obtained for swelling due to *any* method of producing a similar temperature distribution near the exit will be both qualitatively and quantitatively similar to the present results. Thus we can expect that shooting hot fluid into a cold, fairly short die will produce similar effects to those computed; the results are quite sensitive to the exact temperature distributions.

3.6 STABILITY CONSIDERATIONS

We have discussed the methods so far used for the simulation of extrudate swell. There is much more to be attempted, particularly in the computation of the non-axisymmetric and unsteady phenomena often observed (variously called melt flow instability, elastic turbulence, melt fracture) and in more realistic three-dimensional calculations. The latter depend on developing fast reliable programs and have not been seriously attempted yet. Here we shall look at current stability problems. 'Melt fracture' is a physical instability in the flow system and is not related to the computational instabilities previously mentioned in connection with surface tension and viscoelastic effects. Similarly, there are inertial instabilities at high Reynolds numbers leading to turbulence. We shall not consider these physical instabilities; at present, current computational methods do not permit us to reach sufficiently high Weissenberg numbers to observe them in computations. The situation with inertial problems is better, and experiments by Gartling *et al.*, (1977) showed that use of a sufficiently fine grid plus the application of a Newton–Raphson method of iteration permitted Reynolds numbers in excess of 1000 to be reached. During the course of this investigation several methods of attacking the non-linear system, eqn (3.13), were tried; it was found that attempts to rearrange eqn (3.13) in the Picard iteration form

$$\mathbf{KV}^{(n)} = \mathbf{F}^{(n)} - (\mathbf{CV})^{n-1} \qquad (3.36)$$

(where n refers to the iteration step number) were unsuccessful except at very low Reynolds numbers (based on element size) of order unity. Similarly, attempts to split $\mathbf{C}$ into symmetric and unsymmetric parts, and keep only the symmetric part on the left-hand side of eqn (3.36) were also not very successful. Thus it is necessary to deal with the storage and solution of unsymmetric matrices. In general, the application of an unsymmetric solver and a Newton–Raphson scheme was successful. It should also be noted that mesh refinement aids convergence in the high Reynolds number case.

The difficulties at high Weissenberg numbers, which are particularly evident with free boundaries, have been mentioned. In this case it has been found by many workers that mesh refinement *decreases* stability, and that a Newton–Raphson scheme does not increase stability to the degree expected, although it does help somewhat. Thus one is led to ask whether there is a solution to the problem, and if so, is there a bifurcation present above certain critical Weissenberg numbers. Mendelson *et al.* (1982), have concluded, at least for some problems with second-order fluids, that a unique solution does exist, and we are not able to compute it due to instability. Thus they reject the hypothesis of bifurcation and suggest computational instability (a reasonable conclusion in the author's view). Due to the similarity of experiences with second-order and Maxwell fluids, it is also hard to believe that the constitutive equation is at fault, as has sometimes been suggested. Accepting this conclusion, then it is clear that a stability analysis of the various computational schemes is in order. This is very difficult in general, but the following analysis of a very simple scheme is given as an example of what needs to be done.

3.6.1 A result for plane creeping flow

If a stream function finite-difference formulation is adopted for the second-order fluid, eqn (3.5), we find an equation for the stream function ψ (Tanner, 1966)

$$\nabla^4 \psi - \left(\frac{v_1}{2\eta}\right) \mathbf{v} \cdot \nabla(\nabla^4 \psi) = 0 \qquad (3.37)$$

If we denote $\nabla^4 \psi$ by f, then a finite-difference scheme for f is

$$f_{i,j}^{n+1} = \left(\frac{v_1}{2\eta}\right)\left[\frac{u_{i,j}^n}{2h}\{f_{i+1,j}^n - f_{i-1,j}^n\} + \frac{v_{i,j}^n}{2h}\{f_{i,j+1}^n - f_{i,j-1}^n\}\right] \quad (3.38)$$

where $u_{i,j}^n$, $v_{i,j}^n$ are the x and y components of the velocity vector at the node (i,j) at the nth iteration step, and $f_{i,j}^n$ is the value of $\nabla^4 \psi$ at (i,j) at the nth iteration step; h is the grid spacing. We can linearise eqn (3.37), considering a perturbation of a stream of speed U flowing in the x-direction, so that $\mathbf{v} = U\mathbf{i} + u\mathbf{i} + v\mathbf{j}$.

Then eqn (3.38) can be linearised to read (ignoring small quantities of order $(u^2 + v^2)$)

$$f_{i,j}^{n+1} = \frac{v_1 U}{4\eta h}\{f_{i+1,j}^n - f_{i-1,j}^n\} \quad (3.39)$$

If we consider an unbounded space, and decompose f into Fourier modes, then we can show from a classic stability analysis (Richtmyer and Morton, 1967) that instability will set in if

$$\frac{v_1 U}{\eta h} > 2 \quad (3.40)$$

The quantity $\frac{1}{2}v_1 U/\eta h$ is a Weissenberg number, and the particular scheme given above is thus unstable if $Ws > 1$.

While most of the schemes proposed in the literature are not based on this simple streamfunction algorithm, the replacement by any other linear operator probably will not affect the lack of convergence of the scheme, which appears to be due to the higher order terms on the right-hand side. One of the most disquieting aspects of the stability criterion, eqn (3.40), is that any attempt to refine the solution by using a finer mesh will cause lack of convergence, and in the limit $h \to 0$, no solution is ever possible with these simple schemes. Of course, a Newton–Raphson approach will eliminate some of these problems, and such schemes are more stable.

There is evidence in the literature that the above criteria are relevant. Table 3.2 shows that many of the schemes that have been tried fail in the vicinity of $v_1 U/\eta h = 2$. The Newton–Raphson scheme of Crochet and Bézy (1980) is outstandingly stable. Because of the many variables in each problem, especially boundary conditions, these tests do not prove conclusively that non-convergence always occurs as suggested, and it would be useful to check on this point. Nevertheless, care will need to be taken to avoid the present types of instabilities in future schemes.

TABLE 3.2
CONVERGENCE OF COMPUTATIONS

Authors	Model	Problem	Scheme	Max $v_1 u/\eta h$	Reynolds no.
Reddy and Tanner (1978b)	Second-order	Free jet	FEM	2·5	Zero
Pilate and Crochet (1977)	Second-order	Cylinder	FD	2	Non-zero
Court et al. (1981)	Maxwell	Constriction	FD	2	Non-zero
Crochet and Bézy (1980)	Maxwell	Contraction	FEM	7	Zero
Milthorpe and Tanner (1981) (extended up to $\lambda\dot\gamma = 0·75$)	Maxwell	Extrusion	FEM	1·5	Zero
Viriyayuthakorn and Caswell (1980)	Maxwell	Contraction	FEM	3	Zero
Crochet and Pilate (1976)	Second-order	Contraction	FD	4	Non-zero

3.7 MECHANISMS OF SWELLING

The above results are of interest in their own right as computer solutions to complicated boundary value problems, but they can also be regarded as a set of controlled experiments in the search for a physical explanation of swelling. This is in line with the philosophy that one does computing for the sake of enlightenment, not just to obtain numbers. This philosophy is likely to be very valuable in polymer processing operations where actual experiments are hard to perform because of size, for example. In the present problem one now sees three contributions to swelling:

(i) the small 'Newtonian' swelling due to rearrangement of the velocity field;

(ii) the 'elastic recovery' swell described by equations such as eqn (3.33);

(iii) 'inelastic' swelling due to, for example, thermal effects (Fig. 3.10).

The latter was discovered by numerical experiments and this represents a useful byproduct of computing research.

Finally, we note that before viscoelastic computations can be directly useful for simulating industrial processes, methods of dealing with Weissenberg numbers in the range 10–100 need to be devised. One possible method, so far relatively untried, is to use a time-dependent code. This line of research holds considerable interest and needs to be developed further.

REFERENCES

Allan, W. (1977). *Int. J. Num. Meth. Eng.*, **11**, 1621.

Batchelor, J. and Horsfall, F. (1971). 'Die Swell in Elastic and Viscous Fluids', Rubber and Plastics Research Assoc. of Great Britain, Report No. 189.

Bettess, P. (1981). *Int. J. Num. Meth. Eng.*, **17**, 306.

Brebbia, C. A. (1978). *The Boundary Element Method for Engineers*, Wiley, New York.

Bush, M. B. and Tanner, R. I. (1981). In *Numerical Methods in Laminar and Turbulent Flow* (Ed. B. A. Schrefler), Pineridge Press, Swansea, p. 119.

Bush, M. B. and Tanner, R. I. (1982). *Int. J. Num. Meth. Fluids.* (in press).

Chang, P. W., Patten, T. W. and Finlayson, B. A. (1979). *Comp. and Fluids*, **7**, 285.

Court, H., Davies, A. R. and Walters, K. (1981). *J. Non-Newtonian Fluid Mechanics*, **8**, 95.

Crochet, M. J. and Bézy, M. (1980). *Proc. 8th Intl. Cong. Rheol. (Naples)*, **2**, 53.

Crochet, M. J. and Keunings, R. (1980). *J. Non-Newtonian Fluid Mechanics*, **7**, 199.

Crochet, M. J. and Keunings, R. (1981a). *J. Non-Newtonian Fluid Mechanics*, **10**, 85.

Crochet, M. J. and Keunings, R. (1981b). *Proc. 2nd World Congress on Chem. Engn.*, **6**, 285.

Crochet, M. J. and Keunings, R. (1982). *J. Non-Newtonian Fluid Mechanics*, **10**, 339.

Crochet, M. J. and Pilate, G. (1976). *J. Non-Newtonian Fluid Mechanics*, **1**, 247.

Cruse, T. A. and Rizzo, F. J. (1968). *J. Math. Anal. Appl.*, **22**, 244.

Engelman, M. A., Strang, G. and Bathe, K. J. (1981). *Int. J. Num. Meth. Eng.*, **17**, 707.

Gartling, D. K., Nickell, R. E. and Tanner, R. I. (1977). *Int. J. Num. Meth. Eng.*, **11**, 1155.

Goren, S. L. and Wronski, S. (1966). *J. Fluid Mech.*, **25**, 185.

Hill, G. A., Shook, C. A. and Esmail, M. N. (1981). *Canadian J. Chem. Eng.*, **59**, 100.

Horsfall, F. (1973). *Polymer*, **14**, 262.

Hunt, B. W. (1968). *J. Fluid Mech.*, **31**, 361.

Johnson, M. W. (1960). *Phys. Fluids*, **3**, 871.

Lodge, A. S. (1964). *Elastic Liquids*, Academic Press, London.

Mendelson, M. A., Yeh, P.-W., Brown, R. A. and Armstrong, R. C. (1982). *J. Non-Newtonian Fluid Mechanics*, **10**, 31.

Michael, D. H. (1958). *Mathematika*, **5**, 82.

Milthorpe, J. F. and Tanner, R. I. (1981). *Proc. 7th Australasian Conf. on Hydraulics and Fluid Mechanics, Brisbane*, p. 103.

Nickell, R. E., Tanner, R. I. and Caswell, B. (1974). *J. Fluid Mech.*, **65**, 189.

Oldroyd, J. G. (1958). *Proc. Roy. Soc. London*, **A245**, 278.

Orr, F. M. and Scriven, L. E. (1978). *J. Fluid Mech.*, **84**, 145.

Phuoc, H. B. and Tanner, R. I. (1980). *J. Fluid Mech.*, **98**, 253.

Pilate, G. and Crochet, M. J. (1977). *J. Non-Newtonian Fluid Mechanics*, **2**, 323.

Pipkin, A. C. and Tanner, R. I. (1972). *Mechanics Today*, **1**, 262.

Reddy, K. R. and Tanner, R. I. (1978a). *Comp. and Fluids*, **6**, 83.

Reddy, K. R. and Tanner, R. I. (1978b). *J. Rheology*, **22**, 661.

Richtmyer, R. D. and Morton, K. W. (1967). *Difference Methods for Initial-Value Problems* (2nd edn), Interscience, New York.

Rizzo, F. J. (1967). *Q. Appl. Math.*, **25**, 83.

Rouse, H. and Abul-Fetouh, A. (1950), *J. Appl. Mech.*, **17**, 421.

Ryan, M. E. and Dutta, A. (1981). *Proc. 2nd World Congress of Chem. Eng., Montreal*, **6**, 277.

Silliman, J. J. and Scriven, L. E. (1980). *J. Comp. Phys.*, **34**, 287.

Tanner, R. I. (1963). *J. Fluid Mech.*, **17**, 161.

Tanner, R. I. (1966). *Phys. Fluids*, **9**, 1246.

Tanner, R. I. (1970). *J. Polym. Sci. (A.2)*, **8**, 2067.

Tanner, R. I. (1973). *Applied Polymer Symposium*, **20**, 201.

Tanner, R. I. (1976). In *Proc. Seventh Int. Congress on Rheology, Göthenburg* (Ed. C. Klason and J. Kubat), Swedish Soc. of Rheol., Gothenburg, p. 140.

Tanner, R. I. (1980). *J. Non-Newtonian Fluid Mechanics*, **7**, 265.

Tanner, R. I., Nickell, R. E. and Bilger, R. W. (1975). *Comp. Meth. Appl. Mech. Eng.*, **6**, 155.

Tracey, D. M. (1977). *Int. J. Num. Meth. Eng.*, **11**, 1225.

Trogdon, S. A. and Joseph, D. D. (1981). *Rheol. Acta*, **20**, 1.

Viriyayuthakorn, M. and Caswell, B. (1980). *J. Non-Newtonian Fluid Mechanics*, **6**, 245.

Waldron, K. J. (1965). M. Eng. Sc. Thesis, University of Sydney.

CHAPTER 4

Extrusion (*Flow in Screw Extruders and Dies*)

R. T. FENNER

*Department of Mechanical Engineering,
Imperial College of Science and Technology, London, UK*

The extrusion process is one of shaping a molten polymeric material by forcing it through a die. By far the most common method of generating the required pressure, and usually of melting the material as well, is by means of one or more screws rotating inside a heated barrel. While the form of the die determines the initial shape of the extrudate, its dimensions may be further modified before final cooling and solidification takes place. Screw extrusion is the most important polymer process in that most thermoplastics and rubbers pass through an extruder at least once during their lives; if not during fabrication of the end product, then during the homogenisation stage. Also, in most modern injection-moulding machines, the preparation of melt for injection is carried out in what is essentially a screw extruder. Similarly, blow-moulding and calendering are often post-extrusion operations.

Figure 4.1 shows the diagrammatic cross-section of a typical single-screw plasticating extruder, which both melts and pumps polymer. Solid material in the form of either granules or powder is usually gravity-fed through the hopper. After entering a helical screw channel, the polymer passes in turn through the feed, compression and metering sections of the screw. The channel is relatively deep in the feed section, the main functions of which are to convey and compact the solids. The compression section owes its name to its progressively decreasing channel depth. Melting occurs there as a result of the supply of heat from the barrel and mechanical work from the rotation of the screw. The shallow metering section, which is of constant depth, is intended to control the output of the machine, generate the

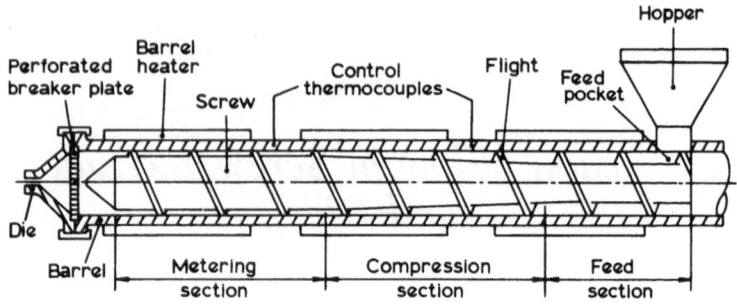

Fig. 4.1. Diagrammatic cross-section of a typical single-screw plasticating extruder.

necessary delivery pressure and mix the melt. On leaving the screw, the melt is usually forced through a perforated breaker plate before flowing to the die.

An extrusion die consists of a set of flow channels designed to convert the melt ouptut from an extruder into a continuous and uniform extrudate of the required cross-sectional shape. Most dies start with a wide distribution channel which feeds melt to the final narrow forming channels which help to meter the flow. The simplest dies are those used to make axisymmetric products such as lace and rod. Other types are used in the production of flat film, sheet, pipe and tubular film, and in covering wire and cable.

The purpose of this chapter is to describe some of the computational techniques which have been developed for analysing flow in screw extruders and dies. Only the internal flows occurring in screw and die channels are considered, the free surface flows associated with, for example, the production of tubular film being treated elsewhere.

4.1   MELT FLOW IN SINGLE-SCREW EXTRUDERS

Except in the case of melt-fed homogenising extruders, melt flow in the absence of solid polymer only occurs towards the delivery end of the machine. Conveying of solid material and melting are important stages of the extrusion process. Early attempts to analyse extruder performance were, however, concentrated on melt flow. It is for this reason, and also because the analysis of melting involves treatments of melt flow, that the latter is considered first here.

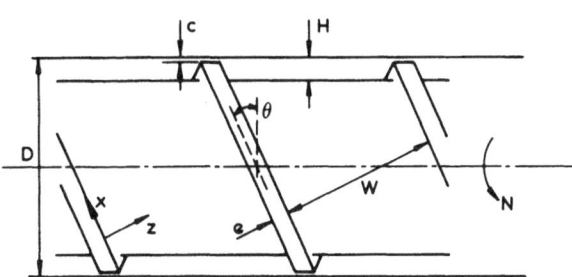

Fig. 4.2.   Geometry of an extruder screw.

Figure 4.2 shows an enlarged view of part of the screw and barrel shown in Fig. 4.1, and serves to define the geometry. D is the internal barrel diameter, H the screw channel depth, W and e the channel and flight widths, respectively, c the radial clearance between flight and barrel, and θ is the helix angle. The screw rotates as shown at N revolutions per unit time. Simplifying assumptions usually associated with the analysis have been discussed in detail by Fenner (1970, 1979). For example, assuming body and inertia forces to be negligible, the barrel may be treated as rotating about a stationary screw. A coordinate system fixed relative to the screw is also selected.

The most natural coordinate system to use is a helical one. Zamodits (1964) set up the relevant equations and obtained some solutions for non-Newtonian melts. Nebrensky *et al.* (1973) used a variational analysis and helical coordinates, and a limited number of solutions using a finite-element method have been presented by Choo *et al.* (1981). In the melt flow regions of most extruders, the screw channel is relatively shallow, and it is reasonable to assume that the channel may be unrolled and treated as rectilinear, and Cartesian coordinates used. Although the plane for unrolling onto is often taken as the one generated by the flight tip, Martin (1970) has suggested that the plane at the mean channel depth would be more appropriate. Taking the x coordinate to lie in the downstream direction (Fig. 4.2), the y and z coordinates may be chosen to be in the radial and transverse directions as shown in Fig. 4.3, which is a view along the screw channel in the downstream direction.

The components in the downstream and transverse directions of the barrel surface velocity relative to the screw are

$$V_x = \pi D N \cos \theta \qquad V_z = \pi D N \sin \theta \qquad (4.1)$$

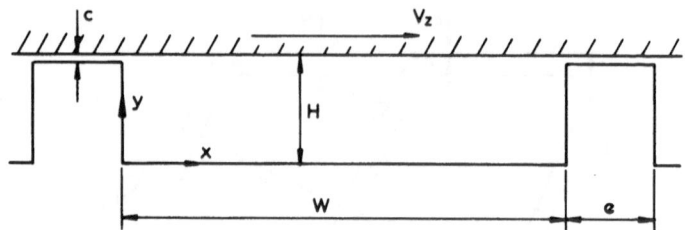

Fig. 4.3.   View down screw channel in the positive x direction.

For slow flows, the momentum conservation equations, eqns (1.4), are (see eqn (1.8))

$$\frac{\partial p}{\partial x_i} = \frac{\partial P^E_{ij}}{\partial x_j} \qquad (4.2)$$

and for steady flow of a melt with locally constant thermal properties, the energy conservation equation, eqn (1.9), reduces to

$$\rho \gamma v_i \frac{\partial T}{\partial x_i} = \alpha \nabla^2 T + P^E_{ij} e_{ij} \qquad (4.3)$$

Because melts are subjected to large rates of deformation for relatively long times while flowing in extruders, it is customary to treat them as inelastic viscous fluids. The empirical power-law constitutive equation, eqn (1.21), is frequently employed:

$$P_{ij} = 2\eta e_{ij} \qquad \eta = K\{\sqrt{I_2}\}^{\nu-1} \exp\left[-\zeta(T - T_0)\right] \qquad (4.4)$$

From Figs 4.1 and 4.2, the velocity boundary conditions are

$$
\begin{aligned}
v_x = v_y = v_z = 0 \qquad &\text{on} \qquad z = 0,\, z = W,\, y = 0 \\
v_x = V_x,\, v_y = 0,\, v_z = V_z \qquad &\text{on} \qquad y = H
\end{aligned}
\qquad (4.5)
$$

assuming no slip at the channel boundaries. At least for the purposes of solving the channel flow equations, it is reasonable to assume negligible leakage of melt over the flight tips:

$$\int_0^H v_z \, dy = 0 \qquad (4.6)$$

The thermal boundary condition at the barrel surface is generally taken to be one of prescribed temperature:

$$T = T_b(x) \qquad \text{on} \qquad y = H \qquad (4.7)$$

Similarly, at the screw surface

$$T = T_s(x) \qquad \text{on} \qquad y = 0, z = 0, z = W \qquad (4.8)$$

although the screw surface temperature, T_s, is not usually known. At least for an uncooled screw, it would appear to be reasonable to assume no conduction of heat to or from the screw surfaces. The best choice of temperature boundary condition for the screw is, however, intimately linked with the particular choice of mathematical model for melt flow, and is discussed further in Section 4.1.1. If thermal convection terms are retained in the energy equation, it is necessary to specify some initial temperature profile at the beginning of the region where melt flow is to be analysed.

Further simplifications to the channel flow equations may be introduced with the aid of the lubrication approximation, which was discussed in Section 1.3. Applying the approximation in the x direction to velocities, they may be assumed to be fully developed in the downstream direction: $v_x = v_x(y, z)$, $v_y = v_y(y, z)$ and $v_z = v_z(y, z)$. Continuity equation, eqn (1.3), becomes

$$\frac{\partial v_y}{\partial y} + \frac{\partial v_z}{\partial z} = 0 \qquad (4.9)$$

and equilibrium equations, eqns (4.2), become

$$\frac{\partial p}{\partial x} = G_x = \frac{\partial P_{xy}^E}{\partial y} + \frac{\partial P_{xz}^E}{\partial z} \qquad (4.10)$$

$$\frac{\partial p}{\partial y} = \frac{\partial P_{yy}^E}{\partial y} + \frac{\partial P_{yz}^E}{\partial z} \qquad (4.11)$$

$$\frac{\partial p}{\partial z} = \frac{\partial P_{zy}^E}{\partial y} + \frac{\partial P_{zz}^E}{\partial z} \qquad (4.12)$$

where the pressure gradient G_x is independent of y and z. Although it is much less reasonable to apply the lubrication approximation in the x direction to temperatures, if it is applied then the temperatures are fully developed in the downstream direction, $T = T(y, z)$, and energy equation, eqn (4.3), reduces to

$$\rho \gamma \left(v_y \frac{\partial T}{\partial y} + v_z \frac{\partial T}{\partial z} \right) = \alpha \left(\frac{\partial^2 T}{\partial y^2} + \frac{\partial^2 T}{\partial z^2} \right) + \eta I_2 \qquad (4.13)$$

The next possible stage of simplification is to apply the lubrication approximation in the z direction, which is reasonable provided $H \ll W$, and which implies $v_x = v_x(y)$, $v_y = 0$, $v_z = v_z(y)$ and $\partial p / \partial y = 0$. Equations (4.10) and (4.12) reduce to

$$G_x = \frac{\mathrm{d}P^{\mathrm{E}}_{xy}}{\mathrm{d}y} \qquad \frac{\partial p}{\partial z} = G_z = \frac{\mathrm{d}P^{\mathrm{E}}_{zy}}{\mathrm{d}y} \tag{4.14}$$

For temperatures, a very useful combination of assumptions has been found to be that of negligible temperature variations in the transverse direction, while retaining thermal convection in the downstream direction. The energy equation becomes

$$\rho\gamma v_x \frac{\partial T}{\partial x} = \alpha \frac{\partial^2 T}{\partial y^2} + \eta I_2 \tag{4.15}$$

where

$$I_2 = \left(\frac{\mathrm{d}v_x}{\mathrm{d}y}\right)^2 + \left(\frac{\mathrm{d}v_z}{\mathrm{d}y}\right)^2 \tag{4.16}$$

In order to define dimensionless parameters for screw channel melt flow, it is convenient to take the characteristic velocity as V_x, in the downstream direction of flow. The dimensionless pressure gradient, Péclet (eqn (1.39)) and Nahme (eqn (1.42)) numbers become

$$\pi_P = \frac{G_x H}{\bar{P}} \tag{4.17}$$

$$Pe = \frac{\rho\gamma V_x H}{\alpha} \tag{4.18}$$

$$Na = \frac{\bar{\eta} V_x^2 \zeta}{\alpha} \tag{4.19}$$

where the mean viscosity and shear stress, $\bar{\eta}$ and $\bar{P}$, are determined at the mean shear rate, V_x/H, and barrel temperature, T_{b}

$$\bar{\eta} = K\left(\frac{V_x}{H}\right)^{\nu - 1} \exp\left[-\zeta(T_{\mathrm{b}} - T_0)\right] \qquad \bar{P} = \bar{\eta}\frac{V_x}{H} \tag{4.20}$$

A further useful parameter is the dimensionless volumetric flowrate:

$$\pi_V = \frac{\dot{V}}{WHV_x} \qquad \dot{V} = \int_0^W \int_0^H v_x \, \mathrm{d}y \, \mathrm{d}z \tag{4.21}$$

4.1.1 Fully developed flow in an extruder screw channel

Most of the early attempts to analyse melt flow in screw extruders involved gross simplifications of the mathematical model, such as the assumptions of Newtonian material behaviour and isothermal flow conditions. Such solutions have been reviewed by Fenner (1970, 1977). On the other hand, prohibitive computing costs have so far ruled out fully three-dimensional solutions. Two forms of two-dimensional solution which have proved useful in practice restrict variations of flow conditions to either the x, y or y, z plane. While the former are of the developing flow type considered in the next section, the latter are fully developed. Although fully developed flow solutions provide a considerable amount of insight into the details of the velocity and temperature profiles in an extruder channel, they are of only limited use in process analysis for design purposes because thermally fully developed conditions are frequently not reached, even at the delivery end of the screw.

The governing equations for fully developed flow are eqns (4.9) to (4.13) subject to boundary conditions, eqns (4.5), (4.7) and, say, (4.8). The usual approach is to seek numerical solutions to a set of four simultaneous coupled second-order non-linear partial differential equations, two governing transverse flow (derived from eqns (4.9), (4.11) and (4.12)) and the other two governing downstream flow and heat transfer (eqns (4.10) and (4.13)). This problem has been studied extensively by Martin (1969) using finite-difference techniques. Finite-element methods are also applicable (Hami, 1977; Palit, 1972). Finite-difference and finite-element treatments for equations of this type are discussed in more detail in Section 4.4.

The work of Martin (1969) and Martin *et al.* (1969) provided a number of important results. It showed the dominant effect of thermal convection, at least in transverse flow, confirming that for the high values of Péclet number encountered in practice the isotherms tend to take up shapes which lie along streamlines. For typical screw channels having relatively small depth-to-width ratios, it is reasonable to apply the lubrication approximation in the z direction to both velocities and temperatures, provided a screw temperature boundary condition is employed which accounts for thermal convection in the recirculating transverse flow. Neglecting the small amount of melt that may flow over the tips of the screw flights, the screw surface forms the same streamline as the barrel surface, namely, the one completely enclosing the recirculating flow. It is therefore to be anticipated that the screw temperature is equal to the barrel temperature, a result confirmed by experiment (Cox and Fenner, 1978). A suitable form of screw temperature boundary condition, eqn (4.8), is therefore

$$T = T_b(x) \quad \text{on} \quad y = 0, z = 0, z = W \quad (4.22)$$

Martin also demonstrated the potentially important effects of leakage flow over the screw flight, assuming that such flow occurs, on melt temperatures in the channel. While the no leakage condition imposed by eqn (4.6) is reasonable for the purposes of computing velocity profiles in the channel, owing to the intense shear in the clearance between screw and barrel, a significant amount of heat may be convected into the channel rather than be conducted into the barrel or flight tip surfaces (Martin, 1970).

4.1.2 Developing flow in an extruder screw channel

Few attempts have been made to include the development of temperature profiles in the downstream direction. Yates (1968) worked with eqns (4.14) to (4.16) and (4.4) subject to boundary conditions, eqns (4.5) to (4.8), together with an assumed temperature profile at the beginning of melt flow, and obtained solutions for relatively small values of Nahme number. Martin (1970) subsequently reported that the limitation of the degree of coupling between temperatures and velocities had been overcome. Fenner (1975a) used the same mathematical model to study the design of large melt extruders. A computer solution technique for the generalised Yates model may be outlined as follows.

Figure 4.4 shows the cross-section of a typical unwrapped screw channel in the x, y plane, including feed, compression and metering sections of the screw, all of which must be treated in a melt-fed extruder. Note that the y coordinate will be assumed to have its origin always at the screw root. Because of the slowly varying channel depth, the resulting distortion of the truly Cartesian coordinate system implied by Fig. 4.4 is not significant. Also, the lubrication approximation is applicable in the x direction to velocities, though not to temperatures. Velocities may change slowly with downstream position as a result of the changing temperature profile.

Equations (4.14) to (4.16) and (4.4) subject to boundary conditions, eqns

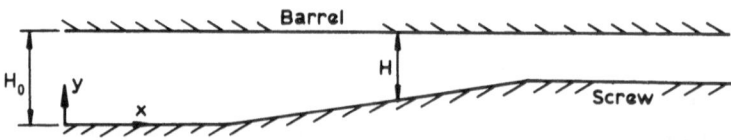

Fig. 4.4. Cross-section of an unwrapped screw channel in the downstream direction.

(4.5) to (4.7) and (4.22), govern the flow, and overall conservation of mass is preserved by requiring the volumetric flowrate

$$\dot{V} = W \int_0^H v_x \, dy \qquad (4.23)$$

to be independent of downstream position.

Introducing the following dimensionless variables and relative channel depth, S,

$$U^* = \frac{v_x}{V_x} \qquad W^* = \frac{v_z}{V_x} \qquad Y^* = \frac{y}{H}$$

$$X^* = \frac{x}{H_0} \qquad S = \frac{H}{H_0} \qquad T^* = \zeta[T - T_b(0)] \qquad (4.24)$$

where H_0 and $T_b(0)$ are, respectively, the channel depth and barrel temperature at $x = 0$, the equilibrium and energy equations become, using constitutive eqn (4.4)

$$\pi_P = \frac{d}{dY^*} \left[\frac{dU^*}{dY^*} (I_2^*)^{1/2(\nu - 1)} \exp(-T^*) \right] \qquad (4.25)$$

$$\pi_T = \frac{d}{dY^*} \left[\frac{dW^*}{dY^*} (I_2^*)^{1/2(\nu - 1)} \exp(-T^*) \right] \qquad (4.26)$$

$$S^2 Pe U^* \frac{\partial T^*}{\partial X^*} = \frac{\partial^2 T^*}{\partial Y^{*2}} + S^{-(\nu - 1)} Na(I_2^*)^{1/2(\nu + 1)} \exp(-T^*) \qquad (4.27)$$

The dimensionless form of I_2 is

$$I_2^* = \left(\frac{dU^*}{dY^*} \right)^2 + \left(\frac{dU^*}{dY^*} \right)^2 = \left(\frac{H}{V_x} \right)^2 I_2 \qquad (4.28)$$

The dimensionless downstream pressure gradient, π_P, is as defined in eqn (4.17), while the dimensionless transverse pressure gradient, π_T, is

$$\pi_T = \frac{G_z H}{\bar{P}} \qquad (4.29)$$

While π_P and π_T are defined in terms of the local channel depth, H, the Péclet and Nahme numbers are defined in terms of conditions at the channel inlet, $x = X^* = 0$. The boundary conditions for the dimensionless velocities are

$$U^* = W^* = 0 \qquad \text{at } Y^* = 0$$

$$U^* = 1, W^* = \tan \theta \qquad \text{at } Y^* = 1 \qquad (4.30)$$

and the following integral conditions must be satisfied:

$$\int_0^1 U^* \, dY^* = \pi_V \qquad \int_0^1 W^* \, dY^* = 0 \qquad (4.31)$$

A numerical method of solution may be outlined as follows. Equations (4.25) and (4.26) may be integrated subject to boundary conditions, eqn (4.30), to find the dimensionless velocity profiles at the channel inlet, where the initial temperature profile is known. Equation (4.27) expressed in finite-difference form may then be used to step a small distance downstream to find the temperature profile there. Repetition of the velocity analysis completes the calculations at the new channel section, and makes it possible to take a further downstream step in terms of temperatures, and so on until the end of the channel is reached.

Considering first the velocity analysis at a particular downstream position where the temperature profile is known, eqns (4.25) and (4.26) may be integrated to give

$$\pi_P(Y^* - Y_0^*) = \frac{dU^*}{dY^*} (I_2^*)^{1/2(v-1)} \exp(-T^*) \qquad (4.32)$$

$$\pi_T(Y^* - Y_1^*) = \frac{dW^*}{dY^*} (I_2^*)^{1/2(v-1)} \exp(-T^*) \qquad (4.33)$$

where $P_{xy} = 0$ at $Y^* = Y_0^*$, and $P_{zy} = 0$ at $Y^* = Y_1^*$. These results may be rearranged to give

$$\frac{dU^*}{dY^*} = \pi_P(Y^* - Y_0^*)F(Y^*) \qquad (4.34)$$

$$\frac{dW^*}{dY^*} = \pi_T(Y^* - Y_1^*)F(Y^*) \qquad (4.35)$$

where, using eqn (4.28), the function $F(Y^*)$ is given by

$$F(Y^*) = [(\pi_P Y^* - \pi_P Y_0^*)^2 + (\pi_T Y^* - \pi_T Y_1^*)^2]^{1/2(v-1)/v} \exp\left(\frac{T^*}{v}\right) \qquad (4.36)$$

Hence, using the boundary conditions given by eqns (4.30):

$$U^* = \pi_P \int_0^{Y^*} (\alpha - Y_0^*)F(\alpha) \, d\alpha \qquad (4.37)$$

$$1 = \pi_P \int_0^1 (\alpha - Y_0^*)F(\alpha) \, d\alpha = \pi_P(J_1 - Y_0^* J_0) \qquad (4.38)$$

where

$$J_m = \int_0^1 \alpha^m F(\alpha)\, d\alpha \tag{4.39}$$

Equation (4.37) may be integrated to find the dimensionless flowrate defined by eqns (4.31):

$$\pi_V = \pi_P \int_0^1 \int_0^{Y^*} (\alpha - Y_0^*)F(\alpha)\, d\alpha\, dY^*$$

$$= \pi_P \int_0^1 (1-\alpha)(\alpha - Y_0^*)F(\alpha)\, d\alpha$$

$$= \pi_P[(1 + Y_0^*)J_1 - J_2 - Y_0^* J_0] \tag{4.40}$$

Similar treatment of eqn (4.35) for transverse flow yields the following results, which are equivalent to eqns (4.38) and 4.40):

$$\tan\theta = \pi_T(J_1 - Y_1^* J_0) \tag{4.41}$$

$$0 = \pi_T[(1 + Y_1^*)J_1 - J_2 - Y_1^* J_0] \tag{4.42}$$

Equations (4.38), (4.40), (4.41) and (4.42), together with definitions (4.36) and (4.39), must be solved for the four unknowns π_P, Y_0^*, π_T and Y_1^*. As Y_0^*, which defines the position of the downstream flow stress neutral surface, can take any value between $-\infty$ and $+\infty$, it is more convenient to choose as an alternative variable the product $\pi_P Y_0^*$, which is much more limited in its range of possible values. Similarly, the product $\pi_T Y_1^*$ can be used in place of Y_1^*. Hence, the four equations may be rearranged for the new unknowns as

$$\pi_P = \beta(J_0 - J_1 - J_0\pi_V) \tag{4.43}$$

$$\pi_P Y_0^* = \beta(J_1 - J_2 - J_1\pi_V) \tag{4.44}$$

$$\pi_T = \beta(J_0 - J_1)\tan\theta \tag{4.45}$$

$$\pi_T Y_1^* = \beta(J_1 - J_2)\tan\theta \tag{4.46}$$

where

$$\beta = (J_0 J_2 - J_1^2)^{-1} \tag{4.47}$$

One method that has been used successfully to solve these simultaneous non-linear algebraic equations is the Newton–Raphson technique. All four are of the general form

$$f_n(x_1, x_2, x_3, x_4) = 0 \qquad n = 1, 2, 3, 4 \tag{4.48}$$

where $x_1 \equiv \pi_P$, $x_2 \equiv \pi_P Y_0^*$, $x_3 \equiv \pi_T$, $x_4 \equiv \pi_T Y_1^*$. For example

$$f_1 = \pi_P - \beta(J_0 - J_1 - J_0 \pi_V) = 0 \tag{4.49}$$

Given some initial estimates, x_m^0, for the unknowns, the corresponding function values, f_n^0, can be determined, and in general are not zero. Improved values of the x_m may be obtained with the aid of the Taylor series expansions of the f_n

$$f_n \simeq f_n^0 + \sum_{m=1}^{4} \delta x_m \frac{\partial f_n}{\partial x_m} = 0 \tag{4.50}$$

where the partial derivatives are evaluated at $x_m = x_m^0$, and the δx_m are the estimated corrections to the x_m:

$$x_m = x_m^0 + \delta x_m \tag{4.51}$$

Equations (4.50) can be rearranged in the form of four linear equations for the δx_m:

$$\left[\frac{\partial f_n}{\partial x_m} \right] [\delta x_m] = -[f_n^0] \tag{4.52}$$

The square matrix of derivatives on the left-hand side of this result contains the partial derivatives of each of the f_n with respect to each of the four unknowns, a total of 16 derivatives. These can be found numerically; for example

$$\frac{\partial f_1}{\partial x_2} = \frac{f_1(x_1^0, x_2^0 + \delta, x_3^0, x_4^0) - f_1(x_1^0, x_2^0, x_3^0, x_4^0)}{\delta} \tag{4.53}$$

where δ is some suitably small increment, say 10^{-4}.

Having found the δx_m from eqns (4.52), the improved x_m are obtained from eqns (4.51), and the whole iterative process repeated until there are no significant changes in the values of the unknowns between successive iterations. In practice, provided reasonably good initial estimates are available, the process converges satisfactorily in three or four iterations. For all but the first velocity analysis, good initial values are provided by the converged solutions from the analysis at the previous step along the channel. For the first velocity analysis, however, acceptable initial values are provided by the relevant isothermal Newtonian flow solutions (Fenner, 1979):

$$\begin{aligned} \pi_P &= 6 - 12\pi_V & \pi_P Y_0^* &= 2 - 6\pi_V \\ \pi_T &= 6 \tan \theta & \pi_T Y_1^* &= 2 \tan \theta \end{aligned} \tag{4.54}$$

Turning now to the temperature analysis, eqn (4.27) is first expressed in finite-difference form. Figure 4.5 shows part of a finite-difference grid covering the channel cross-section illustrated in Fig. 4.4. The grid points at which the melt temperature is to be computed are distinguished by the counters i and j, i being used as a counter in the X^* direction, and j in the Y^* direction. At the typical grid point i,j, approximations to the derivatives involved in the energy equation are

$$\frac{\partial T^*}{\partial X^*} = \frac{T^*_{i+1,j} - T^*_{i,j}}{\delta X^*} \tag{4.55}$$

$$C_{i,j} = \frac{\partial^2 T^*}{\partial T^{*2}} = \frac{T^*_{i,j+1} - 2T^*_{i,j} + T^*_{i,j-1}}{(\delta Y^*)^2} \tag{4.56}$$

where $C_{i,j}$ represents the conduction term at the particular grid point. If $D_{i,j}$ is the corresponding dissipation term in eqn (4.27)

$$D_{i,j} = S_i^{-\nu} Na(I_2^*)^{1/2(\nu+1)} \exp(-T^*_{i,j}) \tag{4.57}$$

then the equation may be expressed as

$$S_i^2 Pe U^*_{i,j} \frac{(T^*_{i+1,j} - T^*_{i,j})}{\delta X^*} = C_{i,j} + D_{i,j} \tag{4.58}$$

Now, because the $U^*_{i,j}$, $T^*_{i,j}$, $C^*_{i,j}$ and $D^*_{i,j}$ are all known, this equation can be used to find $T^*_{i+1,j}$ at the new downstream position. In practice, however, this explicit approach to solving the parabolic differential energy equation is only stable for very small downstream increments, δX^*. An implicit method of the Crank-Nicholson type, in which the conduction term is

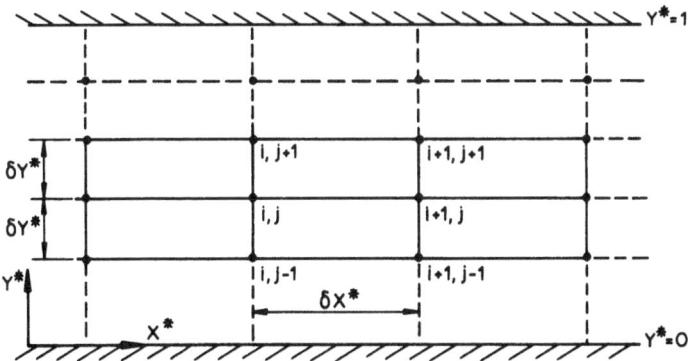

Fig. 4.5. Part of a finite-difference grid for computing temperature profiles.

averaged between the old and new downstream positions, is much more useful. Thus, in place of $C_{i,j}$ in eqn (4.58), the following term is used:

$$\frac{1}{2}\left[C_{i,j} + \frac{T^*_{i+1,j+1} - 2T^*_{i+1,j} + T^*_{i+1,j-1}}{(\delta Y^*)^2}\right] \tag{4.59}$$

The disadvantage of this implicit approach is that the temperatures $T^*_{i+1,j+1}$ and $T^*_{i+1,j-1}$ are introduced into the equation for $T^*_{i+1,j}$. However, the equations for temperatures at grid points within the flow at the new downstream position, together with dimensionless forms of the thermal boundary condition equations, eqns (4.7) and (4.22), namely

$$T^*_{i+1,j} = T^*_b(X^*_{i+1}) \qquad \text{at } Y^* = 1$$
$$T^*_{i+1,1} = T^*_b(X^*_{i+1}) \qquad \text{at } Y^* = 0 \tag{4.60}$$

$\hat{j}$ being the maximum value of j, form a tridiagonal linear set of equations. A well-established direct method of solution exists for such a set (see, for example, Fenner, 1974a, including a suitable computer subprogram).

Even using the implicit method of solving the energy equation, there is a limit to the size of the downstream increment for the procedure to be stable. This is of the order of

$$\delta X^* < 4(\delta Y^*)^2 Pe \tag{4.61}$$

For typical values of $\delta Y^* = 0.05$ (that is, with 21 grid points over the depth of the channel, which is ample) and $Pe = 10^3$, δX^* (that is, $\delta x/H_0$) should not exceed 10. This is rarely a significant limitation in practice.

Having computed the new temperature profile, the corresponding velocity profile, which will show only small changes from the previous one, can be obtained. Note that in evaluating the J_m defined by eqn (4.39), numerical integration of products involving the function $F(Y^*)$ given by eqn (4.36) must be performed. This is because the required temperature profile is only defined at the finite-difference grid points. Simpson's rule provides a suitable method of integration.

Figure 4.6 shows a flow chart for the overall developing flow analysis, while Fig. 4.7 shows a more detailed flow chart for the velocity analysis. In addition to the operations shown, a substantial amount of essentially straightforward pre-processing of input data to convert them into appropriate forms is required, together with some post-processing to translate the results into practically useful numbers. The main items of input data required are the extruder geometry, material viscous and thermal properties; also operating conditions which include flowrate, screw speed and thermal boundary and initial conditions. In addition to detailed temperature and velocity profiles, the information available from the analysis includes any convenient average temperatures (such as bulk mean temperatures), the pressure profile, screw drive power required and mean distributive mixing applied (see, for example, Fenner, 1979); also barrel heating or cooling required. A detailed example of the use of such an analysis in the design of extruders was given by Fenner (1975a).

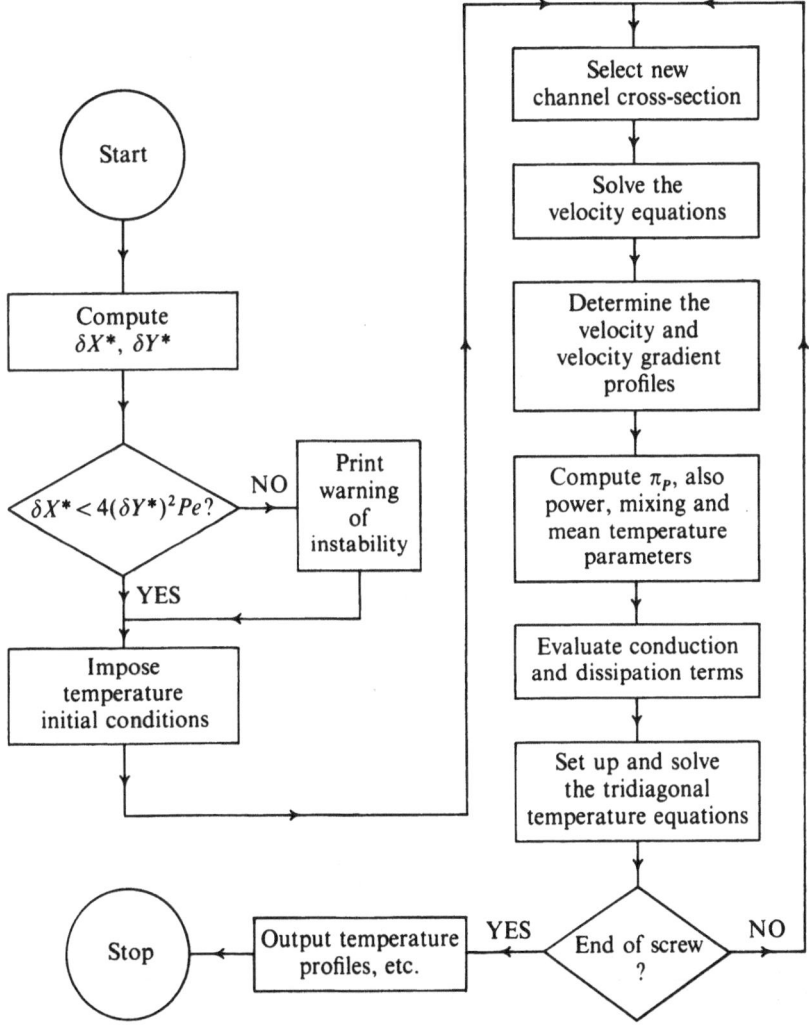

Fig. 4.6.   Flow chart for the analysis of developing flow.

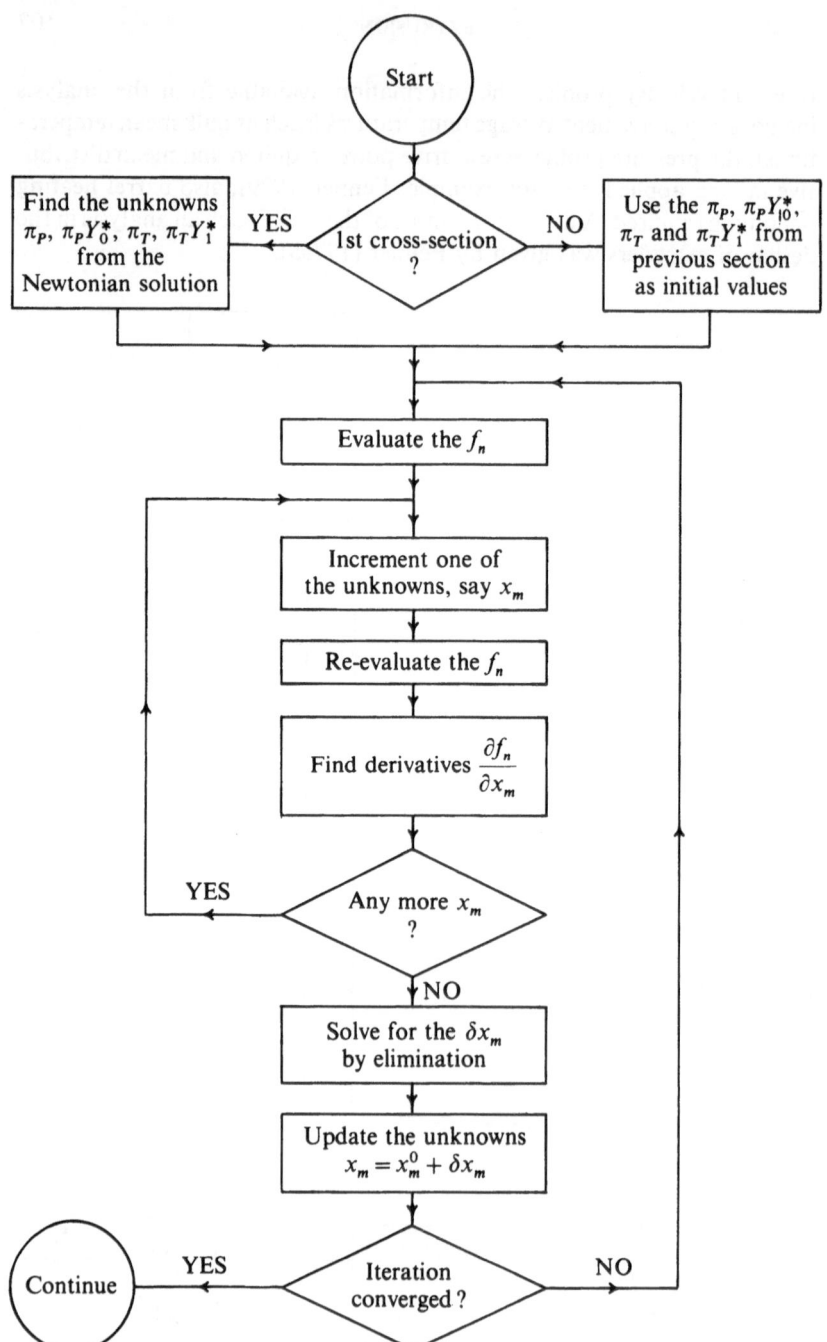

Fig. 4.7. Flow chart for the velocity analysis in developing flow.

General experience of using the present developing flow analysis to study either melt-fed machine performance or melt flow in a plasticating extruder after melting is complete may be summarised as follows. In general, it offers one of the best compromises for routine use in screw design work in terms of its ability to predict actual melt flow behaviour for an acceptable amount of numerical computation. It generally gives good predictions of melt temperatures (Martin, 1970), but may be less satisfactory for pressure profiles. The latter is because of the fact that many extruders operate under nearly drag flow conditions in their metering sections, with correspondingly small pressure gradients there. Under such conditions, the computed values of the gradients are particularly sensitive to some of the geometric assumptions adopted, such as the neglect of channel curvature. The ability to accurately predict pressure gradients in melt flow is not, however, of overriding importance. As the metering section is often operating under nearly drag flow conditions, much of the delivery pressure required at the end of the screw is provided by the feeding and melting processes.

4.2   SOLIDS FEEDING IN SINGLE-SCREW EXTRUDERS

In comparison with the large amount of research on melt flow in screw extruders, relatively little attention has been given to the conveying of solid granules or powder from the feed hopper of a plasticating extruder to the position at which a significant amount of melting has occurred. There are several reasons for this. Solids conveying is usually a relatively minor part of the overall extrusion process, and occurs only over the first few turns of the screw. It rarely controls directly the output from the machine, which is normally determined by the later melting and melt-pumping processes.

A set of continuum mechanics equations can be established for solids conveying in screw channels similar to those governing melt flow. Even though such a continuum approach ignores the particulate nature of polymer feedstocks, it is still difficult to obtain useful solutions. Most analyses therefore make the further simplification of treating material movement as plug flow in the channel, ignoring the effects of stresses and deformations within the bulk of the solid.

4.2.1 One-dimensional plug flow solutions
Many of the assumptions made in melt flow analyses are also relevant to solid plug flow. Important differences are, however, that the screw channel cannot be unrolled for analysis purposes without taking account of channel

curvature, and that conditions at the boundaries of the solid plug are assumed to be determined by a Coulomb friction mechanism.

Darnell and Mol (1956) were the first to obtain useful plug flow solutions. They presented their analytical results in terms of the feed angle at which the solid plug moves relative to planes perpendicular to the screw axis. Figure 4.8 shows a portion of solid plug in a screw channel, ϕ being the feed angle measured at the barrel surface and the arrow showing the direction of motion of the plug relative to the stationary barrel. Assuming $e \ll W$, the

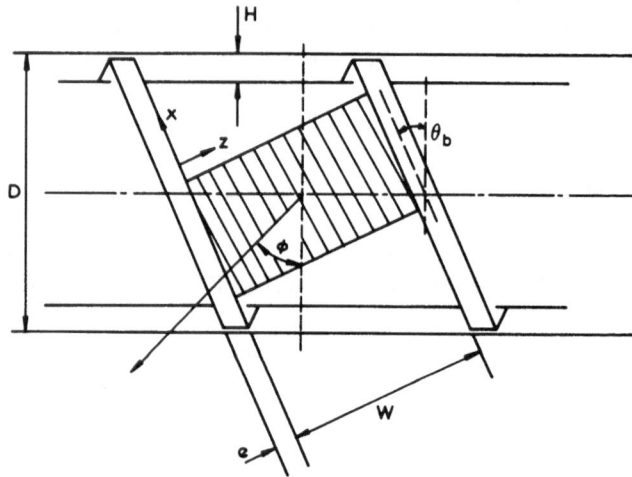

Fig. 4.8.   Direction of motion of the solid plug in the screw channel.

following relationship between volumetric flowrate per screw revolution and the feed angle may be derived from a kinematic argument:

$$\frac{\dot{V}}{N} = \frac{\pi^2 DH(D-H)\tan\phi\tan\theta_b}{(\tan\phi + \tan\theta_b)} \tag{4.62}$$

where θ_b is the flight helix angle at the barrel surface. Schneider (1969) applied a correction factor to this result to allow for the effect of finite flight width.

It is convenient to define a pressure in the solid plug, p, as the direct compressive stress acting in the downstream direction. The compressive stresses at the screw and barrel surfaces are in general not equal to p, but may be expressed as $k_1 p$ acting on the barrel surface, $k_2 p$ on the sides of the channel and $k_3 p$ on the screw root. After a considerable amount of manipulation of the equilibrium equations, Schneider (1969) arrived at the following form of equation for determining the feed angle:

$$\cos \phi = K \sin \phi + \frac{\omega_f}{\omega_b} \frac{k_2}{k_1} \frac{2HE}{A} (K \tan \bar{\theta} + E)$$

$$+ \frac{\omega_s}{\omega_b} \frac{k_3}{k_1} C \cos \theta_s (K \tan \theta_s + C)$$

$$+ \frac{HE \cos \bar{\theta}}{\omega_b k_1} (K \tan \bar{\theta} + E) \frac{1}{p} \frac{dp}{dx} \qquad (4.63)$$

where

$$K = \frac{E(\tan \bar{\theta} + \omega_f)}{(1 - \omega_f \tan \bar{\theta})} \qquad E = \frac{D - H}{D} \qquad C = \frac{D - 2H}{D}$$

and $A = W \sec \theta_b$ is the axial channel width. θ_s and $\bar{\theta}$ are the flight helix angles at the screw surface and mean channel depth, respectively, while ω_s, ω_f and ω_b are the coefficients of friction at the screw root, flight sides and barrel surfaces, respectively, and are assumed to be independent of pressure and rubbing speed. The downstream coordinate, x, is measured at the mean channel depth. The use of pressure ratios k_1, k_2 and k_3 was based on the observation that when a granular material is compressed in a tube, the radial stress generated is substantially less than the applied stress.

Equation (4.63) can be expressed as a differential equation for pressure in the form

$$\frac{1}{p} \frac{dp}{dx} = \lambda \qquad (4.64)$$

where λ is independent of p and x. Hence

$$p = p_0 \exp(\lambda x) \qquad (4.65)$$

where p_0 is the initial pressure at $x = 0$. The need to specify an initial pressure is a difficulty not readily overcome in simple plug flow analyses whose assumptions imply a zero value for p_0.

One of the main difficulties in applying any analysis of solids conveying in an extruder lies in assigning appropriate numerical values to the coefficients of friction. Schneider (1969), for example, showed that for a given combination of polymer and metal surface, the coefficient of friction was significantly dependent on both temperature and the time for which relative motion occurred. More recently, Huxtable et al. (1981) confirmed the dependence of friction on temperature, and also demonstrated the importance of both solid particle form and small quantities of additives.

There have been various refinements applied to the one-dimensional plug flow analysis, for example, by allowing the solid bulk density to vary with pressure, and by introducing viscous drag in thin melt films in place of Coulomb friction at the plug boundaries. Kacir and Tadmor (1972) have suggested that the region where surface melt films exist, but before the later melting mechanism is established, be called the delay zone of the extruder. Lovegrove (1972) and Lovegrove and Williams (1973, 1974) demonstrated the importance of gravity and centrifugal forces in initiating the plug flow mechanism. The amount of computation involved in any one-dimensional plug flow analysis is relatively trivial, and it is only when more sophisticated treatments are attempted that fully numerical techniques need to be introduced.

4.2.2 Two-dimensional plug flow solutions
In one-dimensional plug flow solutions, the pressure or direct compressive stress in the downstream direction is assumed to vary only in this direction. The relative motions of the screw, solid plug and channel are such, however, that frictional forces generate not only a pressure gradient in the downstream direction but also one in the transverse direction. Since in a typical extruder the length of screw channel over which Coulomb friction conditions exist is of the order of at most 10 times the channel width, the pressure difference across this width is likely to be significant compared to the pressure built up in the downstream direction.

If pressure is assumed to be a two-dimensional function of x and z, then Lovegrove (1972) showed that a pair of differential equations of the following form govern the plug flow:

$$\frac{\partial P_{zx}}{\partial x} + k_2 \frac{\partial p}{\partial z} = F_1 p + G_1(x, z) \qquad (4.66)$$

$$\frac{\partial P_{zx}}{\partial z} + \frac{\partial p}{\partial x} = F_2 p + G_2(x, z) \qquad (4.67)$$

The parameters F_1, F_2, G_1 and G_2 are determined by the screw geometry, coefficients of friction, pressure ratios and the feed angle. G_1 and G_2 only appear when gravity or centrifugal forces are included in the analysis. Neglecting such body forces, P_{zx} may be eliminated to give a hyperbolic differential equation:

$$k_2 \frac{\partial^2 p}{\partial z^2} - \frac{\partial^2 p}{\partial x^2} = F_1 \frac{\partial p}{\partial z} - F_2 \frac{\partial p}{\partial x} \qquad (4.68)$$

which yields solutions of the form

$$p = p_0 \exp(\lambda_1 x) \exp(\lambda_2 z) \qquad (4.69)$$

The pressure profiles in both the downstream and transverse directions tend to be of the exponential form. If body force terms are retained, solutions to the full governing equations, eqns (4.66) and (4.67), can generally only be obtained numerically. Lovegrove (1972) obtained such solutions by the method of characteristics. The main use for such solutions is for studying in detail both pressure initiation and the cyclic pressure variations, rather than for direct use in screw design.

More sophisticated analyses of solids feeding in extruders are possible if the plug flow assumption is discarded, allowing for both deformation of the plug under local variations in the state of stress and the possibility of internal slipping between layers of particles leading to loose granular flow. Lovegrove (1974) has made a preliminary study of the stresses and deformations in elastic solids being conveyed and has examined the validity of the plug flow assumption.

4.3 MELTING IN SINGLE-SCREW EXTRUDERS

The onset of melting in an extruder occurs when a thin film of melt is formed between the hot barrel surface and the compacted plug of material in the screw channel. The thickness of this film increases as the plug moves along the screw channel and similar films are in due course formed at the screw root and flight surfaces. Experimental evidence suggests that the film at the barrel surface increases in thickness to several times the size of the clearance between the screw flight and barrel before there is any observable change in the mechanism of melting. Analysis of flow in this delay zone can be accomplished with the aid of viscous drag boundary conditions in the analysis for solids conveying, together with a non-isothermal treatment of the solid plug.

4.3.1 Observed melting mechanisms

Whereas analyses of solids conveying and melt flow in extruders can be developed from fundamental mechanical principles with little or no reference to observed behaviour, attempts to treat the intervening melting process are usually based on observed mechanisms. Once past the delay zone, typical sections through the contents of the screw channel are as shown schematically in Fig. 4.9 (a) and (b), views from a stationary

114   R. T. FENNER

screw equivalent to Fig. 4.4. In Fig. 4.9(a), the screw surface temperature, T_s, is below the melting temperature, T_m, of the polymer and
no melt film is present at the screw. Such a situation usually exists for a
short distance before T_s exceeds T_m, and a film is formed at the screw as
shown in Fig. 4.9(b). Both sectional views are somewhat idealised, in that
the channel is actually curved, and the corners of the solid bed become
rounded. The term solid bed customarily used in connection with melting is
equivalent to the term solid plug in|conveying analyses.

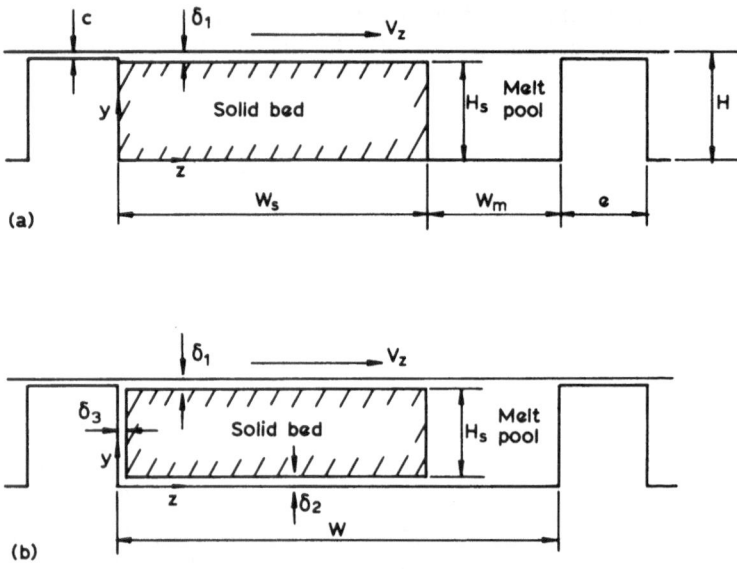

Fig. 4.9.   Melting mechanisms: (a) without lower film (cool screw); (b) with lower
melt film formed (hot screw).

There are up to four distinct regions in the screw channel cross-section—
the upper and lower melt films at the barrel and screw surfaces, respectively,
the solid bed and the melt pool. The lower melt film could be regarded as
two regions, the film along the screw root and the one on the trailing side of
the flight. Owing to the proximity of the heated barrel and the intense shear,
much of the melting occurs in the upper melt film. The motion of the barrel
relative to the solid bed sweeps the melt so formed into the melt pool region
between the bed and the leading side of the flight. It is to be expected that
the upper film thickness, δ_1, varies with both x and z, although in practice it
also remains small as newly melted material is transferred to the pool. The

flow in this pool is a combination of clockwise circulation and downstream motion owing to the dragging actions of both the barrel and bed moving relative to the screw, together with the effect of the downstream pressure gradient.

If melt films are formed at the screw surfaces, their thicknesses, δ_2 at the screw root and δ_3 at the flight, are independent of z and y, respectively. As the motion of the solid bed relative to the screw is in the downstream direction, the lower films grow in thickness in this direction. Usually, the main change exhibited by successive cross-sections along the screw is a reduction in the width of the solid bed, with a corresponding increase in the melt pool width.

Completion of melting, which is often not achieved until well into the metering section, can occur in various ways. If the process is stable and the bed remains continuous in the downstream direction, the corners of the bed become rounded and either its width or depth diminish to negligible proportions. Very often the melting process becomes unstable, in the sense that the bed suffers breaks along planes normal to the downstream direction. This phenomenon is known as solid bed break-up. The resulting pieces of compacted solid then decrease in size, often as much by thermal conduction as by shearing in thin melt films.

4.3.2 The simple Tadmor model for melting

Dimensions not previously defined in Fig. 4.9 are the melt pool width, W_m, solid bed width, W_s, and depth, H_s. The downstream velocity of the solid bed relative to the screw, V_{sx}, is its dominant velocity component, and is assumed to be independent of y and z. Many of the assumptions introduced earlier for melt flow analyses are also applicable in the derivation of melting models. The polymer is assumed to have constant specific heats in both the melt and solid phases, γ_m and γ_s, respectively, together with a latent heat of fusion, ξ, at the sharp melting point, T_m.

One of the simplest theoretical models for melting, based on the mechanism shown in Fig. 4.9(a), is often referred to as the Tadmor model (Tadmor, 1966). The downstream bed velocity, V_{sx}, is assumed to be independent of x, and its value is calculated as that necessary for the solid plug prior to melting to give the prescribed total mass flowrate along the channel. As a consequence of this considerable simplification, the motion of the bed is prescribed, and conditions for its equilibrium need not be examined. A further simplification is the omission of the lower melt film although, given the constant V_{sx} assumption, this merely implies that the rate of melting is determined by the upper melt film alone. In its simplest form, the Tadmor model assumes that the thickness of the upper melt film, δ_1, is independent of z and that the flow there can be treated as fully developed isothermal Newtonian drag flow. The analysis of heat transfer in the solid bed treats the bed as a semi-infinite solid with temperature variations in the y direction only. Due to the simplicity of the model, there is no need to analyse flow in the melt pool in detail, or to take account of pressure gradients in the screw channel.

Various refinements have been added to the Tadmor model, and these have been reviewed by Fenner (1977). One of the most significant was proposed by Donovan (1971a, b) who questioned the assumption of constant solid bed velocity, and introduced a 'solid bed acceleration parameter' to allow the velocity to increase in a prescribed manner. It was shown that V_{sx} does vary, but not in a way that can be readily prescribed.

4.3.3 Improved melting models
In Tadmor-type melting models, the solid bed velocity is a prescribed function of x, and in most cases is assumed to be constant. Although this simplification often allows satisfactory predictions to be made concerning overall melting performance, it does not provide a realistic description of solid bed motion, particularly in the later stages of melting. Clearly, the bed suffers substantial deformation in order that its width may decrease while most of the melting is taking place at its upper surface. This deformation need not be confined to the y, z plane: elongation of the bed in the x direction may contribute to its reduction in width. The bed velocity should be regarded as a slowly varying function of x. Consequently, equilibrium of the bed must be considered, and the lower melt films at the screw surface become important in terms of the shear stresses exerted on the bed in addition to the melting caused.

The first melting model to treat V_{sx} as an unknown function of x was described by Shapiro (1973) and has since been developed by Shapiro *et al.* (1976a, b) and Halmos *et al.* (1978). A somewhat simplified version of this model was used by Edmondson (1973) and Edmondson and Fenner (1975), the principal differences lying in the treatment of the upper melt film. Although it is the latter version of the model which is described in detail here, the two versions are compared and contrasted.

With variable downstream bed velocity, mass balances must be established for the various regions of the flow. For the upper melt film, the rate of change of downstream mass flow is equal to the difference between the rate of melting over the interface with the solid bed and the net transverse flowrate out of the melt film. If conditions in the film are assumed not to vary with z across the width of the bed, this can be expressed symbolically as

$$\frac{d}{dx}(\dot{m}_{1x}W_s) = \dot{M}_1 W_s - (\dot{m}_{1z} - \dot{m}_{fz}) \qquad (4.70)$$

where $\dot{m}_{1x}$ and $\dot{m}_{1z}$ are the mass flowrates in the upper film in the x and z directions, respectively, per unit width of film in the z and x directions, respectively, and $\dot{M}_1$ is the rate of melting per unit area from the top of the bed. Assuming drag flow and a linear velocity profile in the clearance between screw flight and barrel, the leakage mass flowrate over the flight in the transverse direction per unit downstream distance is

$$\dot{m}_{fz} = \tfrac{1}{2}\rho_m V_z c \qquad (4.71)$$

where ρ_m is the melt density.

In the cases of the lower melt films, which are each assumed to be of uniform thickness at a particular channel cross-section, all the melt is retained, and the mass balance conditions are

$$\frac{d}{dx}(\dot{m}_{2x}W_s) = \dot{M}_2 W_s \qquad \frac{d}{dx}(\dot{m}_{3x}H_s) = \dot{M}_3 H_s \qquad (4.72)$$

$\dot{m}_{2x}$ and $\dot{m}_{3x}$ are the downstream mass flowrates per unit width of film in the screw root and flight films, respectively, $\dot{M}_2$ and $\dot{M}_3$ being the corresponding melting rates per unit area. Neglecting melting at the interface between the bed and melt pool, the mass balance for the solid bed is

$$\frac{d}{dx}(\rho_s V_{sx} W_s H_s) = -\dot{M}_1 W_s - \dot{M}_2 W_s - \dot{M}_3 H_s \qquad (4.73)$$

where ρ_s is the solid density, and $H_s = H - \delta_1 - \delta_2$. The downstream mass flowrate in the melt pool, $\dot{M}_{mx}$, depends on its size, the velocities of the barrel and solid bed relative to the screw and the downstream pressure gradient, G_x. An overall mass balance is used to equate the total flowrate along the screw channel, $\dot{M}$, which is assumed to be a known quantity and, for steady flow, to be independent of both time and x, to the sum of the flowrates in the melt films, solid bed and melt pool:

$$\dot{M} = \dot{m}_{1x}W_s + \dot{m}_{2x}W_s + \dot{m}_{3x}H_s + \rho_s V_{sx} W_s H_s + \dot{M}_{mx} - \tfrac{1}{2}\rho_m W V_x c \qquad (4.74)$$

In addition to the mass balances, a balance between the forces due to the

downstream pressure gradient acting on the cross-section of the bed and
the shear forces acting on its surfaces gives

$$G_x H_s W_s = P_{x1} W_s - P_{x2} W_s - P_{x3} H_s \qquad (4.75)$$

where P_{x1}, P_{x2} and P_{x3} are the shear stresses in the x direction at the bed
surfaces in the upper and lower melt films. In order to determine the
parameters $\dot{m}_{1x}$, $\dot{m}_{1z}$, $\dot{m}_{2x}$, $\dot{m}_{3x}$, $\dot{M}_1$, $\dot{M}_2$, $\dot{M}_3$, P_{x1}, P_{x2}, P_{x3} and $\dot{M}_{mx}$ in
governing eqns (4.70) and (4.72) to (4.75), detailed analyses of flow and heat
transfer in the films, bed and melt pool regions must be performed. In the
case of the solid bed, no attempt is made to analyse internal deformations.
The temperature profile within the bed may be obtained with the aid of a
simplified form of energy equation, eqn (4.3), which is analogous to
eqn (4.15) used to describe developing melt flow:

$$\rho_s \gamma_s V_{sx} \frac{\partial T}{\partial x} = \alpha_s \frac{\partial^2 T}{\partial y^2} \qquad (4.76)$$

where α_s is the thermal conductivity of the solid bed. The boundary
conditions are $T = T_m$ at the upper and lower faces of the bed, together with
an initial condition of ambient temperature throughout at the beginning of
melting. The solid bed temperature profile can be obtained by a finite-
difference treatment of eqn (4.76) very similar to that described in Section
4.1.2 for thermally developing melt flow, but with the simplification of
having no dissipation term. As a result of neglecting bed temperature
variations in the z direction, it is not possible to analyse realistically the
melting into the small film at the screw flight. It is therefore convenient to
assume $\delta_3 = \delta_2$, $\dot{m}_{3x} = \dot{m}_{2x}$, $\dot{M}_3 = \dot{M}_2$ and $P_{x3} = P_{x2}$. The side film can be
treated as an extension of the one at the screw root, and eqns (4.72) become

$$\frac{d}{dx}[\dot{m}_{2x}(W_s + H_s)] = \dot{M}_2(W_s + H_s) \qquad (4.77)$$

A thorough analysis of flow and heat transfer in the melt pool presents
problems at least as great as those already discussed for flow in a melt-filled
channel. In addition to a net influx of melt and two moving boundaries
instead of one, the application of the lubrication approximation in the
transverse direction is much less justifiable. In order to avoid an excessive
expenditure of computational effort on the melt pool analysis, the rather
crude assumption of isothermal Newtonian flow is made. The volumetric
flowrates in the pool owing to the various drag and pressure flow effects can
be expressed in terms of shape factors F_D and F_P that are functions only

of the flow channel depth-to-width ratio (see, for example, Fenner, 1979):

$$F_D\left(\frac{H}{W}\right) = \frac{16}{\pi^3}\frac{W}{H} \sum_{i=1,3,5,\cdots}^{\infty} \frac{1}{i^3} \tanh\left(\frac{i\pi}{2}\frac{H}{W}\right) \quad (4.78)$$

$$F_P\left(\frac{H}{W}\right) = 1 - \frac{192}{\pi^5}\frac{H}{W} \sum_{i=1,3,5,\cdots}^{\infty} \frac{1}{i^5} \tanh\left(\frac{i\pi}{2}\frac{W}{H}\right) \quad (4.79)$$

For example, as a result of the motion of the barrel relative to the screw

$$\dot{V}_1 = \tfrac{1}{2} W_m H V_x F_D\left(\frac{H}{W_m}\right) \quad (4.80)$$

and as a result of the motion of the solid bed

$$\dot{V}_2 = \tfrac{1}{2} H W_m V_{sx} F_D\left(\frac{W_m}{H}\right) \quad (4.81)$$

Also, as a result of the downstream pressure gradient

$$\dot{V}_3 = -\frac{W_m H^3 G_x}{12\bar{\eta}} F_P\left(\frac{H}{W_m}\right) \quad (4.82)$$

where $\bar{\eta}$ is a suitable mean viscosity, evaluated at the mean shear rate in the pool,

$$\bar{e} = \left[\left(\frac{V_x}{H}\right)^2 + \left(\frac{V_{sx}}{W_m}\right)^2\right]^{1/2} \quad (4.83)$$

and|the bulk mean temperature of the melt entering from the upper film. Finally, the downstream mass flowrate in the melt pool may be found by superimposing the three components:

$$\dot{M}_{mx} = \rho_m(\dot{V}_1 + \dot{V}_2 + \dot{V}_3) \quad (4.84)$$

The analysis of melt flow in the intensely sheared thin films on three sides of the solid bed can be carried out with the aid of a largely analytical solution for non-isothermal, non-Newtonian drag flow developed by Martin (1967). The following forms of results are those derived by Fenner (1979) for a flow between parallel boundaries a distance δ apart, the lower boundary maintained at temperature T_m, and the upper at T_b, and the upper boundary moving at a relative velocity of V_r. A dimensionless shear stress, S, for the film is defined as the ratio between the actual constant shear stress there and the shear stress which would exist in the material concerned when sheared at the mean rate, V_r/δ, and the boundary temperature, T_b. S may be obtained from a pair of non-linear algebraic equations:

$$E\cosh^2 C = \cosh^2 F \tag{4.85}$$

$$B = S^{-1/2(v-1)/v}\cosh C(\tanh F - \tanh C) \tag{4.86}$$

where

$$B = \left(\frac{Na}{2v}\right)^{1/2} \qquad E = \exp\left[-\frac{\zeta(T_b - T_m)}{v}\right]$$

$$F = BS^{1/2(v+1)/v}\cosh C + C$$

and Na is the Nahme number for the flow

$$Na = \frac{\bar{\eta}V_r^2\zeta}{\alpha} \qquad \bar{\eta} = \left|K\left(\frac{V_r}{\delta}\right)^{v-1}\exp\left[-\zeta(T_b - T_0)\right]\right. \tag{4.87}$$

The parameter C is a second unknown to be found from eqns (4.85) and (4.86). A method of the Newton–Raphson type, similar to that described in Section 4.1.2 for solving the velocity equations in developing melt flow, is appropriate. Having computed S and C numerically, other results may be obtained. The dimensionless flow rate in the direction of the relative velocity is

$$\pi_V = \frac{\dot{V}}{V_r\delta} = \frac{AS^{-1/2(v-1)/v}}{B}\left[\frac{1}{ABS^{1/2(v+1)/v}}\ln\left(\frac{\cosh F}{\cosh C}\right) - \tanh C\right] \tag{4.88}$$

where $A = \cosh C$, and $\dot{V}$ is the volumetric flow rate per unit width. Also, the dimensionless temperature gradient at the lower boundary, which is needed for calculating melting rates there, is given by

$$\left(\frac{dT^*}{dY}\right)_{Y=0} = \zeta\delta\left(\frac{dT}{dy}\right)_{y=0} = -2vBS^{1/2(v+1)/v}|\sinh C \tag{4.89}$$

Consider first the flow in the lower melt film of local thickness $\delta \equiv \delta_2$, associated with the relative velocity $V_r \equiv V_{sx}$. The downstream mass flow rate per unit width is

$$\dot{m}_{2x} = \rho_m V_{sx}\delta_2(1 - \pi_{V2}) \tag{4.90}$$

where π_{V2} is the dimensionless volumetric flowrate obtained from eqn (4.88) for the conditions prevailing in the lower film. Note that $(1 - \pi_{V2})$ is used rather than π_{V2} because the analytical solution relates to a film in which the melting interface is fixed and the other boundary moves, whereas in the present case the reverse is true. In addition to mass flowrate, the temperature gradient at the melting interface and the shear stress in the lower film are required:

$$\left(\frac{dT}{dy}\right)_{y=0} = \frac{1}{\zeta\delta_2}\left(\frac{dT^*}{dY}\right)_{Y=0} \qquad P_{x2} = \bar{\eta}_2\left(\frac{V_{sx}}{\delta_2}\right)S \qquad (4.91)$$

where S and the dimensionless temperature gradient are obtained from the analytical solution, and y is in the outward normal direction to the solid bed surface, which is the inward radial direction of the screw. There is some difficulty in defining the lower film boundary temperature at the screw surface. Although in the metering section of an extruder screw, the screw temperature is approximately equal to the barrel temperature, at the feed end of the screw temperature is usually much lower than that of the heated barrel. Practical experience shows that an empirical function can be used to describe the variation of screw temperature with downstream position, which allows this temperature to rise exponentially to that of the barrel (Cox and Fenner, 1980; Edmondson, 1973).

In the upper melt film of local thickness $\delta \equiv \delta_1$, the relative velocity between the barrel and solid bed is V_r at an angle ψ to the downstream direction, as shown in Fig. 4.10. As freshly melted material is entrained in the direction of V_r, there is a corresponding increase in film thickness. Since the magnitude of V_{sx} rarely exceeds half that of V_x, V_r is not predominantly in either the downstream or transverse directions. While δ_1 should be treated as a function of both x and z, the resulting solutions would be very costly to compute. It is at this point that the Shapiro model deviates significantly from the present one in that it assumes that δ_1 depends on z only rather than x only. The local mass flowrate per unit width, melting interface temperature gradient and shear stress in the upper film may be obtained from the dimensionless results of the analytical solution for the conditions prevailing in this film as

$$\dot{m}_{1r} = \rho_m V_r \delta_1 \pi_{V1} \qquad \left(\frac{dT}{dy}\right)_{y=0} = \frac{1}{\zeta\delta_1}\left(\frac{dT^*}{dY}\right)_{Y=0} \qquad P_{r1} = \bar{\eta}_1\left(\frac{V_r}{\delta_1}\right)S \qquad (4.92)$$

where both $\dot{m}_{1r}$ and P_{r1} are in the direction parallel to V_r. Hence, the required mass flowrate and shear stress components may be obtained as

$$\dot{m}_{1x} = \dot{m}_{1r}\cos\psi + \rho_m\delta_1 V_{sx} \qquad (4.93)$$

$$\dot{m}_{1z} = \dot{m}_{1r}\sin\psi \qquad (4.94)$$

$$P_{x1} = P_{r1}\cos\psi \qquad (4.95)$$

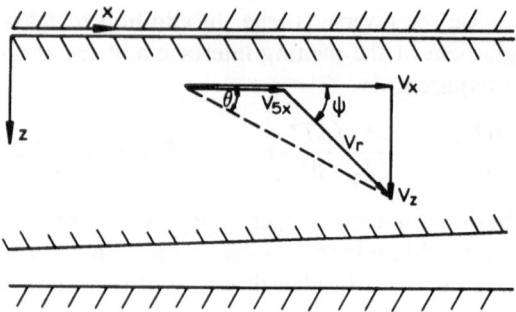

Fig. 4.10.   Relative velocity diagram for the solid bed.

Finally, the melting rates at the interfaces between the solid bed and melt films may be determined from the known temperature gradients. For the upper melt film

$$\dot{M}_1 \zeta_1 = \alpha_m \left(\frac{\partial T}{\partial y}\right)_m - \alpha_s \left(\frac{\partial T}{\partial y}\right)_s \tag{4.96}$$

where the temperature gradients are at the melting interface, for the melt and solid, respectively, the latter being obtained from the finite-difference solution for the bed temperature profile. Similarly, for the lower film

$$\dot{M}_2 \zeta_2 = \alpha_m \left(\frac{\partial T}{\partial y}\right)_m - \alpha_s \left(\frac{\partial T}{\partial y}\right)_s \tag{4.97}$$

where y is still the local outward normal to the surface of the solid bed. Heating up of the solid as it moves towards the melting interface, and heating of the freshly melted material to the mean temperatures of the films—both effectively forms of convection in the y direction—can be accounted for by defining the latent heats of fusion, ζ_1 and ζ_2, not as the natural value ζ but as

$$\zeta_1 = \zeta + \gamma_s (T_m - \bar{T}_s) + \gamma_m (\bar{T}_1 - T_m) \tag{4.98}$$

$$\zeta_2 = \zeta + \gamma_s (T_m - \bar{T}_s) + \gamma_m (\bar{T}_2 - T_m) \tag{4.99}$$

where $\bar{T}_s$ is the bulk mean temperature of the solid bed, and $\bar{T}_1$ and $\bar{T}_2$ are the bulk mean melt temperatures in the upper and lower films. For a polymer that does not exhibit a sharp melting point, eqns (4.98) and (4.99) can be interpreted as the overall changes in specific enthalpy between the mean temperature of the solid and those of the melt films.

The equations which govern downstream development of the physical variables δ_1, δ_2, V_{sx}, W_s and G_x in the present melting model are eqns (4.70), (4.73) and (4.77), together with mass and force balance equations, eqns (4.74) and (4.75). The three differential equations are of the form

$$\frac{dy_n}{dx} = f_n(z_1, z_2, z_3) \qquad n = 1, 2, 3 \tag{4.100}$$

where $z_1 \equiv \delta_1$, $z_2 \equiv \delta_2$, $z_3 \equiv V_{sx}$ and

$$f_1 = \dot{M}_1 W_s - (\dot{m}_{1z} - \dot{m}_{fz}) \tag{4.101}$$

$$f_2 = \dot{M}_2(W_s + H_s) \tag{4.102}$$

$$f_3 = -\dot{M}_1 W_s - \dot{M}_2(W_s + H_s) \tag{4.103}$$

$$y_n = g_n(z_1, z_2, z_3) \tag{4.104}$$

Expressing the y_n functions explicitly,

$$y_1 = \dot{m}_{1x} W_s \qquad y_2 = \dot{m}_{2x}(W_s + H_s) \qquad y_3 = \rho_s V_{sx} W_s H_s \tag{4.105}$$

The melting model equations are to be solved by finding the unknowns at a series of channel cross-sections in the downstream direction. Provided the distance, δx, between successive cross-sections is small, the values of the mass flowrates at the new cross-section, denoted by y'_n, can be obtained from those at the previous section by expressing eqns (4.100) in approximate finite-difference form:

$$y'_n = y_n + f_n(z_1, z_2, z_3)\delta x \qquad n = 1, 2, 3 \tag{4.106}$$

Since these new values must satisfy eqns (4.104) with the new values of δ_1, δ_2 and V_{sx} denoted by z'_n,

$$y'_n = g_n(z'_1, z'_2, z'_3) \qquad n = 1, 2, 3 \tag{4.107}$$

which represents a set of non-linear algebraic equations. These equations can be solved for δ_1, δ_2 and V_{sx} by a Newton–Raphson method of the type described in Section 4.1.2 for solving the velocity equations in developing melt flow. In order to evaluate the g_n during this iterative procedure, it is necessary to find both W_s and G_x, using eqns (4.74) and (4.75), respectively.

A computational procedure for solving the overall melting problem may be outlined as follows.

1. Given the current values of δ_1, δ_2, V_{sx}, W_s and G_x, and also the temperature profile in the solid bed at a particular channel cross-section, obtain the mass flowrates $\dot{m}_{1x}$, $\dot{m}_{1z}$ and $\dot{m}_{2x}$ from eqns (4.93), (4.94) and (4.90), also the melting rates $\dot{M}_1$ and $\dot{M}_2$ from

eqns (4.96) and (4.97), and the shear stresses P_{x1} and P_{x2} from eqns
(4.95) and (4.91).

2.   Calculate the solid bed temperature profile at the next channel
     cross-section, one small finite-difference step downstream, by the
     type of method outlined in Section 4.1.2 for developing melt flow.

3.   In order to determine the values of δ_1, δ_2, V_{sx}, W_s and G_x at the new
     cross-section, use the method outlined above.

4.   Compute the pressure at the new position using the mean value of
     its gradient over the downstream step.

5.   Repeat items 1–4 for successive downstream positions until melting
     is complete, the end of the screw is reached or the calculation
     procedure breaks down.

Figure 4.11 shows a flow chart for this procedure. Melting is terminated,
often well into the metering section, when either W_s decreases to zero, or δ_2
and to some extent δ_1 grow to make H_s zero. Sometimes the solid bed
velocity increases so rapidly towards the end of melting that the calculation
scheme becomes unstable, a behaviour which is usually associated with the
physical instability of solid bed break-up. The size of the downstream steps
should be kept small to reduce the risk of numerical instability, about
100–200 over the length of a typical screw being appropriate.

The initial conditions required to start the solution procedure may be
difficult to define. Owing to the cooling usually applied to the feed pocket,
the temperatures of both the screw and barrel surfaces are below the
melting point of the polymer, and a solids conveying analysis is appropriate. Normally after about three or four turns of the screw, the barrel
temperature has increased sufficiently for melting to start. Formation of the
melt film on an uncooled screw usually occurs one or two turns later. At the
first channel cross-section to be included in the melting analysis, initial
values are required for V_{sx} and W_s. Assuming that the solid bed is the only
flow region present, V_{sx} is found from the overall mass balance equation,
eqn (4.74), and W_s is equal to the channel width.

The main items of input data required for a melting analysis are the
extruder geometry, material properties and operating conditions which
include mass flow rate, screw speed and boundary temperatures at the
barrel and screw surfaces. Information available from the computations
includes the variations in the downstream direction of the solid bed width
and velocity, also upper and lower film thicknesses, pressure gradient and
pressure. The screw drive power and barrel heating or cooling required can
also be obtained, together with any appropriate temperature data.

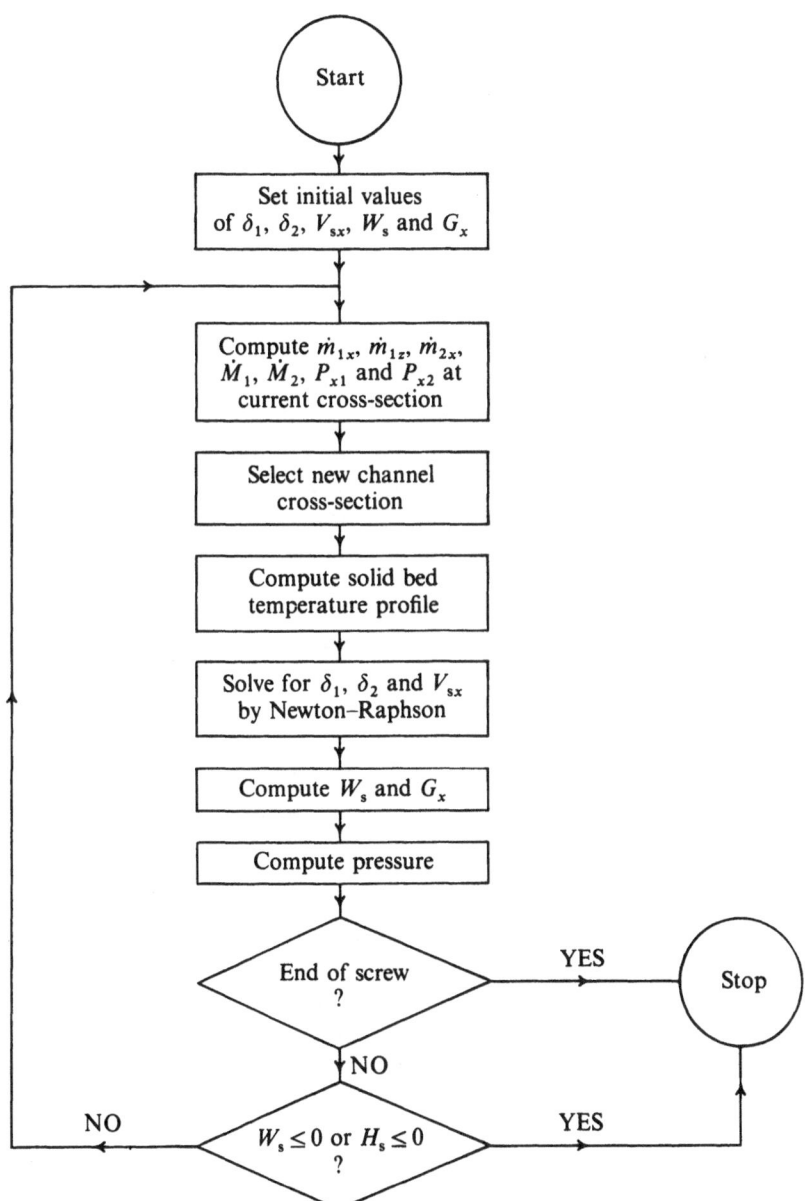

Fig. 4.11. Flow chart for the analysis of melting.

A considerable amount of experience has been gained in the application of the present melting model to practical problems of extruder design and operation. Edmondson (1973) based his work on an extensive series of experiments, some results of which were also published by Edmondson and Fenner (1975). Further comparisons with experiments have been published by Cox and Fenner (1980). The main conclusions may be summarised as follows. Taking solid bed width first, the agreement between experiment and the theoretical model are generally very good, particularly in the early stages of melting. In the later stages, however, it is generally less good. The observed width of the bed starts to fluctuate, and in most cases the bed breaks along planes normal to the downstream direction. Turning to solid bed velocity, the agreement between the theoretical model and experimental results where they are available are again very good. It is worth noting that the velocity ratio V_{sx}/V_x is generally found to be of the order of 0·2 during much of the melting process, implying a value for the angle ψ in Fig. 4.10 of about 20°. This means that downstream development of the upper melt film is more significant than its transverse development. After remaining nearly constant for some distance, both the observed and predicted solid bed velocities start to increase, the region of substantial acceleration usually corresponding with the onset of bed break-up. This break-up is observed to be a periodic instability, normally occurring at frequencies of the order of five to ten times lower than that of screw rotation, and tending to cause surging (Edmondson and Fenner, 1975; Fenner et al., 1979). The main deficiency of the present melting model is its inability to make reliable predictions of pressure, which is apparently a consequence of the rather crude model used to describe flow in the melt pool.

Only a limited number of (unpublished) comparisons have been made between melting performance predictions based on the Edmondson and Shapiro type models. Their abilities to compute the main parameters such as solid bed dimensions and velocity are similar, and neither predict pressure profiles well. The Shapiro model suffers from the disadvantage, however, of requiring much more computing time, and the currently available form of numerical implementation (Halmos et al., 1978) is very prone to failure of convergence in the downstream stepping procedure.

4.4 MELT FLOW IN EXTRUSION DIES

Extrusion die channels are often of comparatively simple shape, typically of circular, flat or annular cross-section. A further simplification which is frequently possible in the analysis of die flows is that of treating the flow as isothermal, either because the Nahme number is small, or because the flow channel is too short for a significant amount of thermal development to take place. Some practical examples justifying the isothermal assumption were provided by Fenner (1979), who also reviewed the analytical solutions available for unidirectional flows in simple channel geometries. Much of the published work on more complicated die flows is based on the work of Pearson (1962, 1963, 1964). A die flow analysis is usually carried out with one of two possible objectives in view. The first of these is simply to calculate the pressure drop across the die and the extrudate shape to be expected from a given die geometry run under specified conditions which include the mass flowrate. The alternative objective is to establish a flow channel geometry that gives the required extrudate shape, a problem for which a number of different solutions are possible. In practical die design, a further consideration is that the die body must be robust enough for deformations produced by the internal pressures to cause insignificant changes in the flow channel dimensions.

4.4.1 Flow in a channel of arbitrary cross-section

As an introduction to die flows, and also to finite-difference and finite-element methods for analysing them, consider isothermal flow along a channel of arbitrary but uniform cross-section, such as that shown in Fig. 4.12. The only non-zero velocity component is $v_x(y, z)$ which, assuming no slip at the channel walls, is zero at the boundaries. The elliptic equilibrium equation may be obtained from eqns (4.2) as

$$\frac{\partial}{\partial y}\left(\eta \frac{\partial v_x}{\partial y}\right) + \frac{\partial}{\partial z}\left(\eta \frac{\partial v_x}{\partial z}\right) = \frac{\partial p}{\partial x} = G_x \qquad (4.108)$$

the downstream pressure gradient, G_x, being independent of y and z. For a Newtonian fluid, this result takes the harmonic form

$$\frac{\partial^2 v_x}{\partial y^2} + \frac{\partial^2 v_x}{\partial z^2} = \nabla^2 v_x = \frac{G_x}{\eta} \qquad (4.109)$$

of Poisson's equation, ∇^2 being the harmonic operator.

A finite-difference approach to the solution of this problem may be outlined as follows. Figure 4.12 shows part of a grid in the two-dimensional solution domain. The lines of the grid are uniformly spaced in the y and z directions, the distances between them being δy and δz, respectively. The points to be used in the finite-difference analysis are located at the

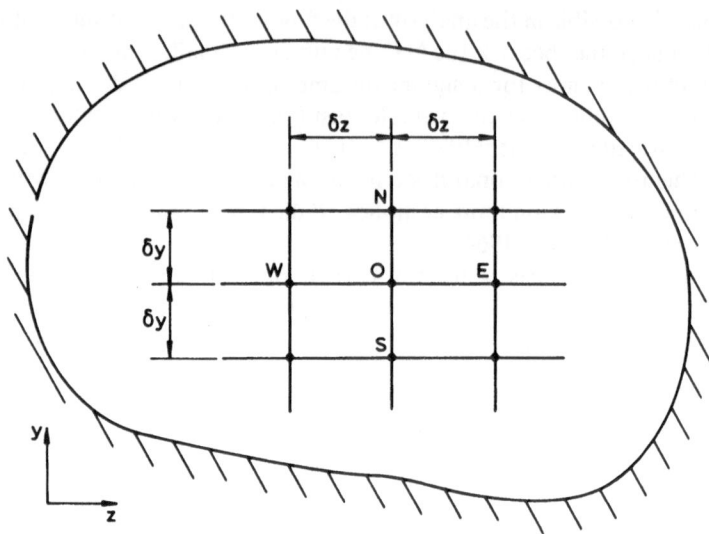

Fig. 4.12.   Channel of arbitrary cross-section showing part of a rectangular finite-difference grid.

intersections of the grid lines. The point labelled O may be regarded as a typical point within the solution domain, and the compass-point labels N, S, E and W are used for the four adjacent grid points. The central difference approximation of eqn (4.109) for the point O is

$$\frac{(v_x)_S - 2(v_x)_O + (v_x)_N}{(\delta y)^2} + \frac{(v_x)_W - 2(v_x)_O + (v_x)_E}{(\delta z)^2} = \frac{G_x}{\eta} \qquad (4.110)$$

An equivalent difference equation for treating the non-Newtonian equation, eqn (4.108), is

$$\frac{1}{(\delta y)^2}\left\{\frac{(\eta_N + \eta_O)}{2}[(v_x)_N - (v_x)_O] - \frac{(\eta_S + \eta_O)}{2}[(v_x)_O - (v_x)_S]\right\}$$

$$+ \frac{1}{(\delta z)^2}\left\{\frac{(\eta_E + \eta_O)}{2}[(v_x)_E - (v_x)_O] - \frac{(\eta_W + \eta_O)}{2}[(v_x)_O - (v_x)_W]\right\} = G_x$$

$$(4.111)$$

which requires viscosities at the grid points to be evaluated in terms of the local rates of shear by means of a constitutive equation such as eqn (4.4). While either eqn (4.110) or (4.111) may be used to define velocities at grid points within the solution domain, the condition $v_x = 0$ is imposed at points on the boundary. The resulting set of algebraic equations for the velocities is most conveniently solved by an iterative method of the Gauss–Seidel type, particularly in the non-Newtonian case, where the grid point viscosities must be updated periodically during the iteration process in the light of the current estimates of velocities. The volumetric flowrate can be found by numerically integrating the computed velocity profile as

$$\dot{V} = \int \int v_x \, dy \, dz \qquad (4.112)$$

over the channel cross-section. Martin (1969) provided a more thorough discussion of finite-difference methods for problems of this type in connection with his work on flow in extruder screw channels. Fenner (1974a) published a computer program for solving the Newtonian form of governing equation.

In most finite-difference techniques, the grid lines are required to be parallel to the coordinate axes. Therefore, using Cartesian coordinates, the technique is best suited to solving problems with rectangular solution domains, so that rows of points can always terminate with points on the domain boundary. Irregular boundary shapes are more difficult to accommodate, and points near the boundary require special treatment. Such problems are more conveniently solved by the alternative finite-element approach.

Finite-element methods involve dividing the region of interest into a number of subregions, known as finite elements. Associated with each element are a number of nodal points at which the unknowns, the velocities in the present case, are to be determined. Elements of varying complexity have been employed for solving flow problems of the types encountered in polymer processing (see, for example, Atkinson et al., 1970; Caswell and Tanner, 1978; Kiparissides and Vlachopoulos, 1976). The simplest type of two-dimensional element is, however, the three-noded triangle, which was used to solve channel flow problems of the present type by Palit and Fenner (1972a).

Figure 4.13 shows a typical triangular element numbered m in the y, z plane (local coordinates parallel to the channel coordinates in Fig. 4.12). The downstream velocity is assumed to vary linearly over the element:

$$v_x = C_1 + C_2 z + C_3 y \qquad (4.113)$$

where the three constants may be found in terms of the three nodal point velocities:

$$C_1 = (v_x)_i \qquad [C_2 C_3]^T = \frac{1}{2\Delta_m} [B][(v_x)_i (v_x)_j (v_x)_k]^T \qquad (4.114)$$

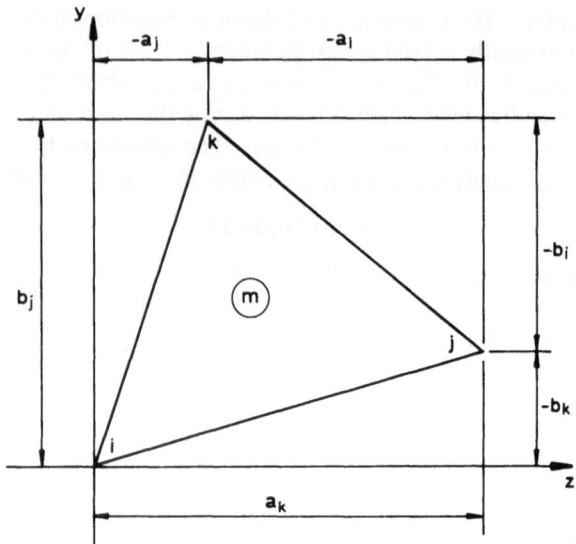

Fig. 4.13. A typical triangular element.

where

$$[B] = \begin{bmatrix} b_i & b_j & b_k \\ a_i & a_j & a_k \end{bmatrix}$$

and $\Delta_m = \frac{1}{2}(a_j b_k - a_k b_j)$ is the area of the element. Hence, the shear rates are constant over the element:

$$\frac{\partial v_x}{\partial z} = C_2 \qquad \frac{\partial v_x}{\partial y} = C_3 \qquad (4.115)$$

so that I_2, η and consequently the shear stresses, are also constant.

While finite-difference methods are usually derived directly from the governing differential equations, finite-element methods frequently employ an equivalent variational principle. A stationary value is sought for a functional, which is defined by an appropriate integral of the unknowns over the solution domain. In the present problem, the relevant functional is the rate of dissipation of total potential energy

$$\dot{E} = \int\int (2\eta I_2 - v_x G_x)\, dy\, dz \qquad (4.116)$$

which may be expressed as the sum of the integrals over the individual elements. The contribution of the typical element of Fig. 4.13 is

$$\dot{E}^{(m)} = \frac{1}{2}\eta_m \Delta_m (C_2^2 + C_3^2) - \frac{1}{3} G_x \Delta_m ((v_x)_i + (v_x)_j + (v_x)_k) \qquad (4.117)$$

Hence, differentiating with respect to, say, $(v_x)_i$

$$\frac{\partial \dot{E}^{(m)}}{\partial (v_x)_i} = \eta_m \Delta_m \left(C_2 \frac{\partial C_2}{\partial (v_x)_i} + C_3 \frac{\partial C_3}{\partial (v_x)_i} \right) - \tfrac{1}{3} G_x \Delta_m \qquad (4.118)$$

The condition to be satisfied for every point i is

$$\frac{\partial \dot{E}}{\partial (v_x)_i} = 0 = \sum_m \frac{\partial \dot{E}^{(m)}}{\partial (v_x)_i} \qquad (4.119)$$

where the summation need only be performed for elements which involve the point i. In other words

$$\sum_m \frac{\eta_m}{4\Delta_m} [B]^T [B] \begin{bmatrix} (v_x)_i \\ (v_x)_j \\ (v_x)_k \end{bmatrix} = \sum_m [k]_m \begin{bmatrix} (v_x)_i \\ (v_x)_j \\ (v_x)_k \end{bmatrix} = \sum_m \tfrac{1}{3} G_x \Delta_m \qquad (4.120)$$

or

$$[K][v_x] = [F] \qquad (4.121)$$

where $[k]_m$ is the element viscous stiffness matrix, $[K]$ the overall stiffness matrix, $[F]$ the overall vector of pressure forces, and $[v_x]$ the vector containing all the unknown nodal point velocities. More details of the stiffness assembly process, together with a computer program for the whole analysis, were given by Fenner (1975b).

After imposing the required boundary conditions, eqns (4.121) can be solved for the nodal point velocities. Although direct elimination methods of solution are the most commonly used in finite-element methods, in the present case the iterative Gauss–Seidel technique proves to be very satisfactory, and is particularly advantageous for solving non-Newtonian flow problems. As in the finite-difference approach, element viscosities and stiffnesses can be recomputed periodically during the solution process. A significant disadvantage of more sophisticated finite elements is that the resulting equations cannot be solved iteratively, and repeated use of a direct solution procedure for non-Newtonian problems is expensive in terms of computing time.

The principal advantage of finite-element methods over finite-difference methods is the geometric flexibility of meshes of elements. For example, a mesh of triangular elements of the type shown in Fig. 4.13 can be readily designed to suit any shape of solution domain, provided the boundary can be adequately approximated by a series of short straight lines which form sides of elements. Similarly, within the solution domain, since the variations of the unknowns are assumed to be linear over each element, it is desirable to have a concentration of relatively small elements in regions of the domain where the unknowns are likely to change rapidly.

4.4.2 Flow in a converging channel

Figure 4.14 shows a typical converging channel flow. This view could be either a typical cross-sectional plane parallel to the direction of flow, as in a sheet extrusion die, or a typical radial plane in an axisymmetric die such as a pipe die. Also, there could be relative motion between the upper and lower boundaries, as in a wire-coating die. If the angle of convergence of the channel is small, the lubrication approximation is applicable, and the flow can be treated as unidirectional. Otherwise, it must be treated as two-dimensional in the plane shown.

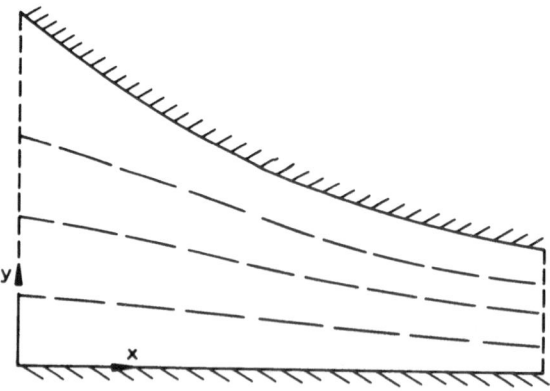

Fig. 4.14.   A converging flow channel.

The equilibrium equations in the plane Cartesian coordinates shown may be obtained from eqns (4.2) as

$$\frac{\partial p}{\partial x} = \frac{\partial P_{xx}^{E}}{\partial x} + \frac{\partial P_{xy}^{E}}{\partial y} \tag{4.122}$$

$$\frac{\partial p}{\partial y} = \frac{\partial P_{xy}^{E}}{\partial x} + \frac{\partial P_{yy}^{E}}{\partial y} \tag{4.123}$$

In order to automatically satisfy mass conservation in incompressible flow, it is convenient to introduce a stream function, Q, from which the velocity components may be derived as

$$v_x = -\frac{\partial Q}{\partial y} \qquad v_y = \frac{\partial Q}{\partial x} \tag{4.124}$$

Using eqns (4.4), the viscous stress components are given by

$$P^E_{xx} = 2\eta \frac{\partial v_x}{\partial x} = -2\eta \frac{\partial^2 Q}{\partial x \, \partial y} \qquad P^E_{yy} = 2\eta \frac{\partial v_y}{\partial y} = 2\eta \frac{\partial^2 Q}{\partial x \, \partial y}$$

$$P^E_{xy} = \eta \left(\frac{\partial v_x}{\partial y} + \frac{\partial v_y}{\partial x} \right) = \eta \left(-\frac{\partial^2 Q}{\partial y^2} + \frac{\partial^2 Q}{\partial x^2} \right) \tag{4.125}$$

Substitution of these expressions into eqns (4.122) and (4.123), and elimination of the pressure yields the elliptic fourth-order differential equation

$$\frac{\partial^2}{\partial x^2} \left(\eta \frac{\partial^2 Q}{\partial x^2} \right) + \frac{\partial^2}{\partial y^2} \left(\eta \frac{\partial^2 Q}{\partial y^2} \right) + 4 \frac{\partial^2}{\partial x \partial y} \left(\eta \frac{\partial^2 Q}{\partial x \, \partial y} \right)$$

$$- \frac{\partial^2}{\partial x^2} \left(\eta \frac{\partial^2 Q}{\partial y^2} \right) - \frac{\partial^2}{\partial y^2} \left(\eta \frac{\partial^2 Q}{\partial x^2} \right) = 0 \tag{4.126}$$

which for a Newtonian fluid reduces to the biharmonic form

$$\nabla^4 Q = \nabla^2 (\nabla^2 Q) = \left(\frac{\partial^2}{\partial x^2} + \frac{\partial^2}{\partial y^2} \right)^2 Q = 0 \tag{4.127}$$

While these governing equations can be directly approximated by finite-difference expressions for the derivatives, it is more usual to employ the equivalent pair of second-order equations formed by the introduction of a vorticity, V, which for Newtonian flow become

$$\nabla^2 Q = V \qquad \nabla^2 V = 0 \tag{4.128}$$

This pair of simultaneous harmonic differential equations can be solved in much the same way as that described in the last section for a single harmonic equation. From the results of the numerical analysis may be obtained the flow rate versus pressure drop characteristic for the channel, together with the distribution of stream function. This allows streamline patterns describing the flow, such as that sketched in Fig. 4.14, to be plotted.

Owing to the boundary shapes of many channel geometries, a finite-element method is often more convenient to use. For example, Palit and Fenner (1972b) described a method employing three-noded triangular elements, and Fenner (1974b) reported successful comparison of the results with experimental measurements for flow in a wire-coating die. Atkinson *et al.* (1970), Kiparissides and Vlachopoulos (1976) and Caswell and Tanner (1978) used higher-order elements to solve similar problems.

4.4.3 Narrow channel flows in dies

The flows in converging channels and channels of arbitrary cross-section discussed in the last two sections provide examples of flow domains which may be treated as two-dimensional. Another of this type occurs when the channel is narrow in the sense that the dimension normal to the plane of flow, namely, the channel depth, is small compared to the other two dimensions, and only varies slowly over the region of interest. Practical examples include flat and tubular film dies, wire- and cable-covering crossheads and pipe dies.

In such narrow channel flows, the lubrication approximation is applicable to velocities, and it is often reasonable to assume isothermal conditions, thus avoiding the need to consider temperature profiles. Consider the flow illustrated in Fig. 4.15, in which the local resultant direction of flow is r, at some arbitrary direction to the x and z axes in the local plane of the boundaries. The third coordinate, y, is normal to this plane. Let $\dot{V}_r$ be the volumetric flowrate in the r direction per unit width of channel. As the flow is locally one-dimensional between flat parallel boundaries, the pressure gradient in the r direction is

$$\frac{\partial p}{\partial r} = \frac{\pi_P \bar{\eta} \dot{V}_r}{H^3} \tag{4.129}$$

where π_P is a dimensionless pressure gradient, and $\bar{\eta}$ is the viscosity evaluated at the local mean shear rate, $\dot{V}_r/H^2$ (Fenner, 1979):

$$\pi_P = -2^{2v+1}\left(\frac{2v+1}{2v}\right)^v \qquad \bar{\eta} = K\left(\frac{\dot{V}_r}{H^2}\right)^{v-1} \tag{4.130}$$

Although the shape of the resultant velocity profile does not change, its direction may vary slowly with position. Let $\dot{V}_x$ and $\dot{V}_z$ be the component flowrates per unit width in the x and z directions. The mass conservation requirement can be expressed in terms of these components as

$$\frac{\partial \dot{V}_x}{\partial x} + \frac{\partial \dot{V}_z}{\partial z} = 0 \tag{4.131}$$

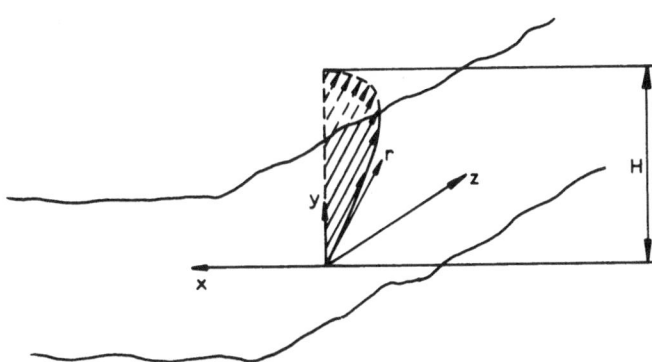

Fig. 4.15.   Geometry and coordinates for melt flow in a narrow channel.

an equation which is satisfied by the following form of stream function, $Q(x, z)$:

$$\dot{V}_x = \frac{\partial Q}{\partial z} \qquad \dot{V}_z = -\frac{\partial Q}{\partial x} \qquad (4.132)$$

From eqn (4.129), the pressure gradients in the coordinate directions are

$$\frac{\partial p}{\partial x} = \frac{\pi_P \bar{\eta} \dot{V}_x}{H^3} = \frac{\pi_P \bar{\eta}}{H^3} \frac{\partial Q}{\partial z} \qquad \frac{\partial p}{\partial z} = \frac{\pi_P \bar{\eta} \dot{V}_z}{H^3} = -\frac{\pi_P \bar{\eta}}{H^3} \frac{\partial Q}{\partial x} \qquad (4.133)$$

and therefore, since $p(x, z)$ is a continuous and differentiable function

$$\frac{\partial}{\partial x}\left(\frac{\bar{\eta}}{H^3} \frac{\partial Q}{\partial x}\right) + \frac{\partial}{\partial z}\left(\frac{\bar{\eta}}{H^3} \frac{\partial Q}{\partial z}\right) = 0 \qquad (4.134)$$

is the governing equation for the flow. Note that $\bar{\eta}$ is still given in terms of $\dot{V}_r$ by eqn (4.130), where

$$\dot{V}_r^2 = \dot{V}_x^2 + \dot{V}_z^2 = \left(\frac{\partial Q}{\partial x}\right)^2 + \left(\frac{\partial Q}{\partial z}\right)^2 \qquad (4.135)$$

Equation (4.134) is almost identical in quasi-harmonic form to eqn (4.108), the main differences being the presence of the H^3 terms, where H may vary in a prescribed way with x and z, and the zero on the right-hand side. Typical boundary conditions include prescribed values of Q, implying no flow across the boundary concerned.

Numerical finite-difference or finite-element methods of solution of the types outlined in Section 4.4.1 are readily applicable to shallow channel flows. For example, Pearson (1962, 1963) analysed flows in cross-heads and past spiders in pipe dies by a finite-difference method, and also discussed ways of inverting the analysis to determine the distribution of channel depth required to produce the desired form of extrudate. Fenner (1979) described a finite-element approach to the same types of problems, including some examples. He also compared some finite-element predictions and experimental measurements for flow in a pipe die, the agreement being excellent (Fenner, 1974b). Fenner and Nadiri (1979) described a modified form of the finite-element method suitable for analysing flow in cable-covering crossheads and dies, and Nadiri and Fenner (1980) reported satisfactory agreement with some practical trials of three-layer covering of high voltage electrical cable.

4.5   OVERALL PROCESS SIMULATION

If analyses of solids feeding, melting and melt flow in extruder screw channels and dies are to be useful in studying machine design and performance, they must be combined in order to predict overall behaviour. The transitions between successive stages of the process cause some difficulties, particularly the one between melting and melt flow. Even in the absence of breaks in the solid bed, existing melting models become increasingly unrealistic towards the end of melting, particularly in the way they treat flow in the melt pool, which eventually increases in size to fill the entire screw channel. At some point well before the end of melting, a more sophisticated melt pool analysis of the developing flow type needs to be introduced to effect the transition to fully molten flow.

Although it is not always made clear in analyses of extruder performance, the mass flowrate along the screw and through the die must be prescribed for a particular process simulation. It is this parameter which provides the link between solids conveying, melting, and melt flow: providing the machine is operating steadily, the mass flowrate is independent of both time and position along the screw and the die. In practice, the flowrate is often determined by a balance between the performance of the screw and the following characteristics of the breaker plate, screen pack, die and any other restriction at the delivery end of the extruder. These characteristics provide relationships between flowrates and delivery pressures which must be generated by the screw. Given a flow rate, an analysis of overall extruder performance can predict the pressure profile along the screw, and hence the delivery pressure. With the aid of an iterative procedure, the flowrate can be adjusted until the pressures balance.

The above mode of operation is often described as melt-controlled, although this term has also been used in the more restrictive sense of a balance being achieved between the melt flows in the die and metering section alone. As the melting process is capable of generating large pressure rises, its contribution to the overall pressure balance cannot be ignored. The contribution of the solids conveying process is generally less important in melt-controlled operation. The other main mode is that of feed-controlled operation, where the flowrate is determined by the capacity of either the hopper supplying the extruder or the feed section of the screw. An accurate analysis of solids conveying is clearly much more important in such a situation. In the case of melt-fed machines, feed-controlled operation appears to be much more common.

With the aid of the methods of analysis described in this chapter, it is possible to estimate the performance of a given extruder and die processing a particular polymer under known operating conditions. The next step towards improving design or performance of this machine is to investigate the effects of changing, for example, the screw dimensions or the operating conditions. In order to do this effectively, the appropriate design criteria must be prescribed. Although the relative importance of these vary according to the application, they are likely to include at least some of the following: (a) maximum output; (b) adequate mixing; (c) delivery temperature as uniform as possible within prescribed limits; (d) minimum machine size and cost; (e) minimum power consumption; (f) maximum machine life. Such requirements are of course very interdependent, and the last is concerned with mechanical design problems. Given the design constraints, however, it is possible in principle to determine an optimum design and set of operating conditions.

REFERENCES

Atkinson, B., Card, C. C. H. and Irons, B. M. (1970). *Trans. Inst. Chem. Engr.*, **48**, T276.
Caswell, B. and Tanner, R. I. (1978). *Polym. Eng. Sci.*, **18**, 416.
Choo, K. P., Hami, M. L. and Pittman, J. F. T. (1981). *Polym. Eng. Sci.*, **21**, 100.
Cox, A. P. D. and Fenner, R. T. (1978). *Plastics and Rubber: Processing*, **3**, 95.
Cox, A. P. D. and Fenner, R. T. (1980). *Polym. Eng. Sci.*, **20**, 562.
Darnell, W. H. and Mol, E. A. J. (1956). *SPE J.*, **12**, 20.
Donovan, R. C. (1971a). *Polym. Eng. Sci.*, **11**, 247.
Donovan, R. C. (1971b). *Polym. Eng. Sci.*, **11**, 484.
Edmondson, I. R. (1973). Ph.D. Thesis, University of London.
Edmondson, I. R. and Fenner, R. T. (1975). *Polymer*, **16**, 49.

Fenner, R. T. (1970). *Extruder Screw Design*, Iliffe, London.
Fenner, R. T. (1974a). *Computing for Engineers*, Macmillan, London.
Fenner, R. T. (1974b). *Plastics and Polymers*, **42**, 114.
Fenner, R. T. (1975a). *Polymer*, **16**, 298.
Fenner, R. T. (1975b). *Finite Element Methods for Engineers*, Macmillan, London.
Fenner, R. T. (1977). *Polymer*, **18**, 617.
Fenner, R. T. (1979). *Principles of Polymer Processing*, Macmillan, London.
Fenner, R. T. and Nadiri, F. (1979). *Polym. Eng. Sci.*, **19**, 203.
Fenner, R. T., Cox, A. P. D. and Isherwood, D. P. (1979). *Polymer*, **20**, 733.
Halmos, A. L., Pearson, J. R. A. and Trottnow, R. (1978). *Polymer*, **19**, 1199.
Hami, M. L. (1977), Ph.D. Thesis, University College of Swansea.
Huxtable, J., Cogswell, F. N. and Wriggles, J. D. (1981). *Plastics and Rubber Processing and Applications*, **1**, 87.
Kacir, L. and Tadmor, Z. (1972). *Polym. Eng. Sci.*, **12**, 387.
Kiparissides, C. and Vlachopoulos, J. (1976). *Polym. Eng. Sci.*, **16**, 712.
Lovegrove, J. G. A. (1972). Ph.D. Thesis, University of London.
Lovegrove, J. G. A. (1974). *J. Mech. Eng. Sci.*, **16**, 281.
Lovegrove, J. G. A. and Williams, J. G. (1973). *J. Mech. Eng. Sci.*, **15**, 114.
Lovegrove, J. G. A. and Williams, J. G. (1974). *Polym. Eng. Sci.*, **14**, 589.
Martin, B. (1967). *Int. J. Non-Linear Mech.*, **2**, 285.
Martin, B. (1969). Ph.D. Thesis, University of Cambridge.
Martin, B. (1970). Comparison of Theoretical Predictions for Single-Screw Extrusion with Experimental Results, Report to EFChE's working party on Non-Newtonian Liquid Processing.
Martin, B., Pearson, J. R. A. and Yates, B. (1969). Cambridge University, Department of Chemical Engineering Polymer Processing Research Centre, Report No. 5.
Nadiri, F. and Fenner, R. T. (1980). *Polym. Eng. Sci.*, **20**, 357.
Nebrensky, J., Pittman, J. F. T. and Smith, J. M. (1973). *Polym. Eng. Sci.*, **13**, 209.
Palit, K. (1972). Ph.D. Thesis, University of London.
Palit, K. and Fenner, R. T. (1972a). *AIChE J.*, **18**, 628.
Palit, K. and Fenner, R. T. (1972b). *AIChE J.*, **18**, 1163.
Pearson, J. R. A. (1962). *Trans. Plast. Inst., Lond.*, **30**, 230.
Pearson, J. R. A. (1963). *Trans. Plast. Inst., Lond.*, **31**, 125.
Pearson, J. R. A. (1964). *Trans. Plast. Inst., Lond.*, **32**, 239.
Schneider, K. (1969). Technical Report on Plastics Processing—Processes in the Feeding Zone of an Extruder, Institute of Plastics Processing (IKV), Aachen.
Shapiro, J. (1973). Ph.D. Thesis, University of Cambridge.
Shapiro, J., Halmos, A. L. and Pearson, J. R. A. (1976a). *Polymer*, **17**, 905.
Shapiro, J., Halmos, A. L. and Pearson, J. R. A. (1976b). *Polymer*, **17**, 912.
Tadmor, Z. (1966). *Polym. Eng. Sci.*, **6**, 185.
Yates, B. (1968). Ph.D. Thesis, University of Cambridge.
Zamodits, H. J. (1964). Ph.D. Thesis, University of Cambridge.

CHAPTER 5

Moulding

S. M. RICHARDSON

Department of Chemical Engineering and Chemical Technology,
Imperial College of Science and Technology, London, UK

5.1 INTRODUCTION

Computers have three principal roles in polymer moulding processes. The first is control of the moulding cycle (see Chapter 9 and also, for example, Parnaby *et al.*, 1978), the second is numerical machining of the mould network and the third is computer-aided design of the mould network, and hence simulation of the various stages of the moulding cycle. It is this third role which is of interest here and more specifically, the computer simulation of the injection-moulding cycle (as opposed to the compression, transfer and reaction injection-moulding cycles, which will be discussed in Section 5.5). Although it is to the mechanics of the moulding machine itself, as opposed to those of the process, that attention is normally paid, and although machine and process mechanics cannot be separated completely, it is the process mechanics alone that will be discussed here, in so far as this is possible.

Mould network design is traditionally an expensive multi-stage make-and-test (that is trial-and-error) affair: a typical mould network goes through several redesign stages before an acceptable network is obtained. Making each of these networks is very costly. The aim of computer simulation is, then, to eliminate these redesign stages, assuming (usually quite justifiably) that use of a computer is much cheaper than mould network manufacture. It would, of course, be unreasonable to expect computer simulation to replace all the make-and-test redesign stages: the accuracy with which the simulations can reasonably be performed precludes this. Instead, the aim of computer simulation is to eliminate most of the redesign stages and hence most of the mould network manufacturing costs.

5.2   PROCESS DESCRIPTION

The essentially unsteady process of injection moulding is one of the more complicated of the common polymer processing operations. Molten polymer is forced into mould cavities under pressure, allowed to solidify and then removed; the cycle is then repeated. Fuller descriptions of the process are given, for example, by Fenner (1979), Middleman (1977), Pearson (1966), Rubin (1972), and Tadmor and Gogos (1979). The principal stages of the cycle are (see Fig. 5.1):

(i)   preplasticisation (and hence melting) of the polymer to be injected, usually both mechanically (commonly by a screw extruder) and thermally (commonly by electric heaters);

(ii)   injection of the molten polymer (commonly by a reciprocating screw with a valve to prevent backflow) at a relatively high flowrate from the reservoir, through the nozzle, through the sprue (the main flow channel in the mould network), through the runners (which lead to the openings in the mould cavities called gates, because they usually present a considerable flow restriction) and into the mould cavities;

(iii) packing of the mainly molten polymer in the mould network at high pressure to ensure complete filling;

(iv) holding, i.e. cooling (and hence freezing and crystallising, in the case of thermoplastics) or heating (and hence cross-linking, i.e. curing and scorching, in the case of thermosets and rubbers), under pressure in the mould network to minimise the effects of shrinkage;

(v)   ejection and hence removal of the mainly solidified products from the mould cavities and waste from the rest of the mould network.

Because the preplasticisation stage is commonly carried out in a screw extruder (albeit of smaller length to diameter ratio than is common in continuous extrusion operations, and operated in an unsteady manner and with a molten polymer reservoir built into its end), simulation of this stage will not be discussed here (instead, see Chapter 4). Moreover, the ejection stage involves mechanical operations alone and so, although computer methods are applicable and useful (see, for example, Bangert et al., 1980), simulation of this stage will not be discussed here. Computer simulation

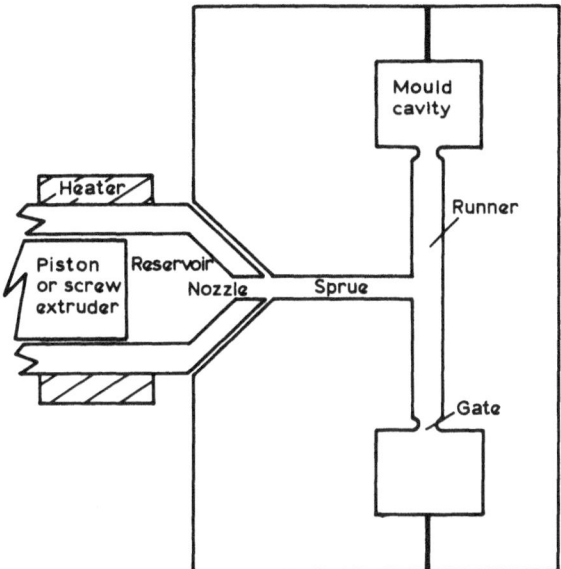

Fig. 5.1.   Flow passages for a two-part mould.

will, therefore, be taken here to comprise simulation of the injection, packing and holding stages.

5.3 MATHEMATICAL MODELLING

The injection, packing and holding stages of the moulding cycle can usually be thought of as comprising:

(i)   an initial period in which molten polymer is injected at a specified and an approximately constant flowrate $\dot{V}_{spec}$ until the pressure in the reservoir reaches a specified maximum P_{max};

(ii)   a subsequent period of injection at an approximately constant reservoir pressure P_{max} until a specified time τ_{inj};

(iii) a period of packing at an approximately constant reservoir pressure P_{pack} (usually $< P_{max}$) to ensure that the mould network is full until a specified time τ_{pack};

(iv)   a final period of holding (cooling or heating) at an approximately constant reservoir pressure P_{hold} (usually $< P_{pack}$) until a specified time τ_{hold} when the polymer in the mould network has solidified sufficiently for it to maintain its shape on ejection.

Although it is by no means necessary to do so, it is generally assumed that $\dot{V}_{spec}$, P_{max}, P_{pack} and P_{hold} are all constant, and that there are instantaneous changes between the four periods comprising the injection, packing and holding stages (see Fig. 5.2).

Because flow occurs in the mould network primarily during the injection stage of the cycle (the small amount of flow that occurs during the packing and (perhaps) the holding stages so as to ensure that the mould network is always full can be neglected for most purposes), it is convenient to consider the injection stage separately from the packing and holding stages. Furthermore, although solidification (caused by freezing or cross-linking)

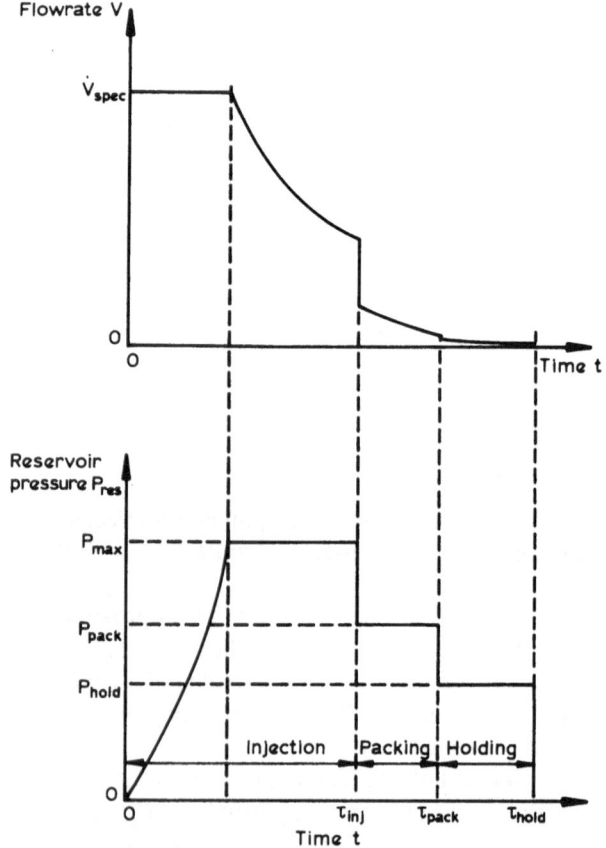

Fig. 5.2. Idealised mode of operation of the moulding machine.

occurs during the injection and packing stages, consideration of solidification will be postponed until consideration of the holding stage.

5.3.1 Injection

The injection stage of the moulding cycle is of an inherently unsteady nature, both because the moulding machine is operated in an unsteady manner and because the moulding process involves the motion of a melt front whose position is unknown *a priori*. Moreover, the geometry and topology of a mould network are usually rather complicated and the rheology of the molten polymer which is being injected is of an intricate (non-Newtonian, elastico-viscous) nature. In order that simulation be feasible, certain simplifying assumptions and hence approximations have to be made: these can conveniently be split into those of a geometrical and topological, a rheological, a kinematical, and a dynamical nature. Each will now be considered in turn.

5.3.1.1 Geometrical and topological approximations

Excluding for the present the mould cavities themselves, the other parts of the mould network, in particular the sprues and runners, are usually of a rather simple geometrical nature. Indeed, it is most common for the sprue and runners, for example, to be relatively long narrow ducts with a circular (or sometimes, for ease of manufacture, a semi-circular) cross-section which varies only slowly, if at all, with axial position. This suggests that a geometrical simplification be made and that the geometry of the sprue and runners might be idealised and supposed to consist of (perhaps a sequence of) circular pipes of length L and at most slowly varying ($|dD/dz| \ll 1$) diameter D ($\ll L$) (see Fig. 5.3(a)). It is, of course, much less common for the mould cavities themselves to have a simple geometrical nature: indeed, one of the chief advantages of the injection-moulding process is that polymeric objects with complicated shapes can be produced relatively easily. This does not, however, totally preclude simplification of the geometry of the mould cavities. Most injection-moulded products are thin in at least one direction across the mould cavities (so that the dimension along the mould cavities, in the direction of flow or injection, is much larger than at least one dimension across the direction of flow). This has a variety of implications, not only for the geometry of the mould network, but also for the polymer melt rheology and for the kinematics and dynamics of injection, as will be seen later. The relevant implication here is that simplifying assumptions can often be made about the geometry of the mould cavities which are analogous to those made about the geometry of

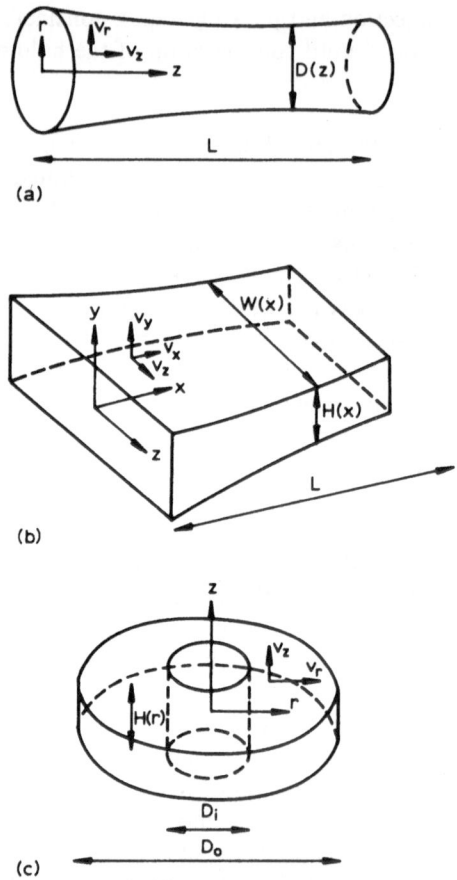

Fig. 5.3. Geometrically simple units. a, Long pipe; b, long channel; c, thin disc.

the sprue and runners, except that a rather larger class of geometrically simple ducts can usefully be used. These include, for example, rectangular channels of length L and at most slowly varying ($|dW/dx|$, $|dH/dx| \ll 1$) width W and height H ($\ll L$) (see Fig. 5.3(b)) and discs of inner diameter D_i, outer diameter D_o and at most slowly varying ($|dH/dr| \ll 1$) height H ($\ll D_o - D_i$) (see Fig. 5.3(c)).

If such geometrical simplifications can be made, then the mould network can be replaced by a sequence of units (pipes, channels, discs, etc.) of relatively simple geometry connected at junctions (where the effects of tees and bends in the mould network and of gates can be suitably incorporated).

A topological difficulty does arise, however, in the case of multi-gated mould cavities: because the positions of the weld lines (or, strictly, weld surfaces) are, except in the case of rather symmetrical mould cavities, unknown *a priori*, splitting each of the mould cavities into several units of a simple geometrical nature, one to each gate, can only be done on a trial-and-error basis. This point will be discussed further in Section 5.4.

If such geometrical simplifications cannot, however, be made (in particular, if the mould cavities do not have a simple geometrical nature) then more sophisticated methods must be used. These will be discussed in Section 5.4.

5.3.1.2 Rheological approximations

Because, as was noted earlier, the sprue, runners and, in many cases, the mould cavities are long narrow ducts, there is essentially simple shear flow within these parts of the mould network. Hence normal stress, extensional and elastic effects are largely negligible and the relevant rheological melt property is a (shear) viscosity η. Adequate agreement with experimental results is usually assured by assuming that η has a power-law dependence on shear rate $\dot{\gamma}$, an exponential dependence on temperature T and no dependence on pressure p (though the large pressure differences encountered in injection moulding mean that the assumption of independence of η from p may be rather poor):

$$\eta = K\dot{\gamma}^{\nu-1} \exp\left(-\zeta(T - T_0)\right) \tag{5.1}$$

It is clear that normal stress, extensional and elastic effects are important at gates, at tees and bends in the mould network and in mould cavities which are not long and narrow. Such effects are usually ignored, though they can be dealt with, albeit crudely, by associating an effective length with each (see Richardson *et al.*, 1980). The first serious attempt at the simulation of injection of elastico-viscous fluids has been made by Isayev and Hieber (1980), but it is still at a very early-stage.

5.3.1.3 Kinematical approximations

There are two kinematical aspects of the flow that occurs during injection to which attention must be paid. The first concerns the flow within the mould network at some distance from the melt front. If the mould network is such that it can adequately be represented as a sequence of geometrically simple units, then it is reasonable to assume that the principal component of the velocity **v** is along the units and that the principal component of the velocity gradient ∇**v** is the derivative across the units of the principal velocity

component. Thus, for long pipes (see Fig. 5.3(a)), long channels (see
Fig. 5.3(b)) and thin discs (see Fig. 5.3(c)), the principal velocity components are v_z, v_x and v_r, respectively, and the principal velocity gradient
components are $\partial v_z/\partial r$, $\partial v_x/\partial y$ and $\partial v_r/\partial z$, respectively. It is then for many
purposes reasonable to assume that the shear rate $\dot{\gamma}$ is given by the
magnitude of the principal velocity gradient component.

The second kinematical aspect concerns the flow near the melt front.
Here, two points need to be considered. The first is that, if the mould
network is such that it can adequately be represented as a sequence of
geometrically simple units, for some period of time each of these units is
only partially full (when they are empty, they can, of course, be ignored).
Hence for a shorter period of time, the filled portion of these units does not
correspond to a long narrow duct. It is usual for this complication, which
arises for only a rather small fraction of the total injection time, to be
ignored. The second point is the way in which the melt front advances,
particularly in the mould cavities. It is usual to assume both that no jetting
occurs and that, if the mould network is such that it can adequately be
represented as a sequence of geometrically simple units, each of these units
fills in a well-defined way (see Fig. 5.4) (so that, for example, the so-called
fountain effect is ignored). This has obvious implications for weld-line
geometry in the case of multi-gated mould cavities. Experimental evidence
(see White and Dee, 1974) suggests that the latter is a reasonable
assumption; the former assumption should be reasonable if the mould
network is not too badly designed.

5.3.1.4 Dynamical approximations

The assumptions and hence approximations of a dynamical nature can
conveniently be split into those that bear on the fluid mechanics and those
that bear on the heat transfer that occur during injection. Of course, the
fluid mechanics and heat transfer cannot be completely isolated: the
processes are intimately coupled.

Because the viscosity η of a molten polymer is rather large, injection is
generally a low Reynolds number process and the momentum conservation
equations become stress equilibrium equations:

$$\nabla p = \nabla . \mathbf{T}^{\mathrm{E}} \tag{5.2}$$

If the mould network is such that it can adequately be represented as a
sequence of geometrically simple units, so that the mould network
comprises a sequence of long narrow ducts, then the lubrication approximation may be made, so that the flow is locally fully developed from the

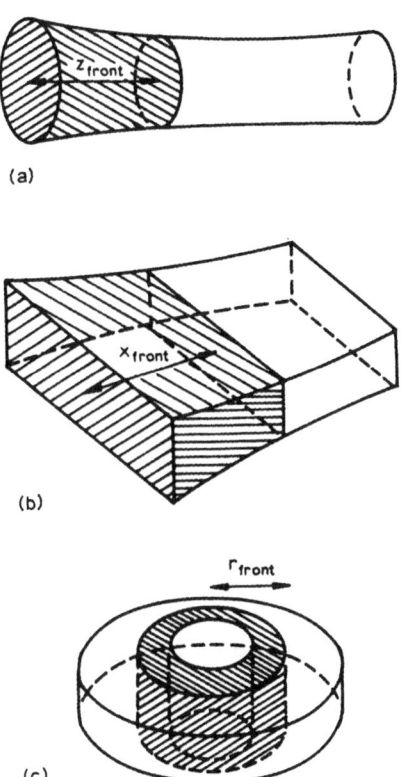

Fig. 5.4. Partially full geometrically simple units. a, Long pipe; b, long channel; c, thin disc.

point of view of the velocity field **v**. Making the kinematical approximations noted earlier, this means that the principal velocity component is v_1 and the principal velocity gradient component is $\partial v_1/\partial x_2$ (see Fig. 5.5), assuming for present purposes (and henceforth, unless otherwise implied) that the x_3 dimension is much longer than the x_2 dimension, and hence that

$$\frac{\partial p}{\partial x_1} = \frac{\partial t_{12}^{E}}{\partial x_2} \qquad (5.3)$$

$$\frac{\partial p}{\partial x_2} = 0 \qquad (5.4)$$

$$\frac{\partial p}{\partial x_3} = 0 \qquad (5.5)$$

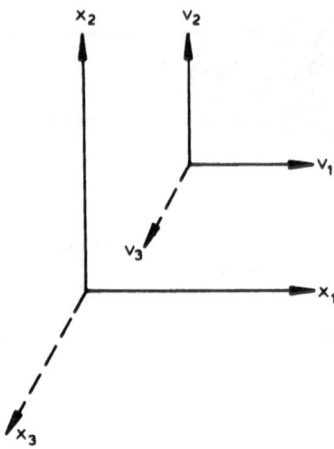

Fig. 5.5.   Alignment of coordinates. (x_1 is aligned along, and x_2 and x_3 across, the direction of flow; x_1 is aligned in the direction of the longest dimension and x_2 is aligned in the direction of the shortest dimension in a geometrically simple unit.)

The density ρ of a molten polymer is generally assumed to be constant, so that the mass conservation equation is

$$\nabla . \mathbf{v} = 0 \qquad (5.6)$$

which implies that the flowrate $\dot{V}$ is spatially constant throughout the mould network. The large pressure and temperature differences encountered in injection moulding mean, however, that ρ is not constant. Nevertheless, an assumption of constant ρ is usually adequate as far as mass conservation is concerned; it is also usually adequate for bulk viscosity effects to be neglected so that, making the rheological and kinematical approximations noted earlier,

$$t_{12}^{\mathrm{E}} = \eta \, \frac{\partial v_1}{\partial x_2} \qquad (5.7)$$

where

$$\eta = K \left| \frac{\partial v_1}{\partial x_2} \right|^{\nu - 1} \exp\left(-\zeta(T - T_0)\right) \qquad (5.8)$$

On the other hand, an assumption of constant ρ is not necessarily adequate as far as energy conservation is concerned, as will be seen later.

It is generally assumed that molten polymers do not slip at the walls of the mould network during injection, so that at walls

$$v_1 = v_2 = v_3 = 0 \qquad (5.9)$$

Because the thermal conductivity α of a molten polymer is rather small, injection is generally a high Péclet number process; it is, moreover, usually a high Graetz number process. As a result:

(i)   the flow is not usually fully developed from the point of view of the temperature field T;

(ii)   unsteadiness of the temperature field T cannot immediately be ignored;

(iii) if the mould network is such that it can adequately be represented as a sequence of geometrically simple units, then the principal temperature gradient component is $\partial T/\partial x_2$ (see Fig. 5.5);

(iv)   axial conduction along streamlines (in the x_1 direction, approximately) is negligible compared with transverse conduction (in the x_2 direction, approximately).

Hence the energy conservation equation:

$$\rho\gamma\left(\frac{\partial T}{\partial t} + \mathbf{v}\cdot\nabla T\right) = \alpha\nabla^2 T + \mathbf{T}^{\mathrm{E}}\!:\!\nabla\mathbf{v} \qquad (5.10)$$

becomes

$$\rho\gamma\left(\frac{\partial T}{\partial t} + v_1\frac{\partial T}{\partial x_1} + v_2\frac{\partial T}{\partial x_2}\right) = \alpha\frac{\partial^2 T}{\partial x_2^2} + t_{12}^{\mathrm{E}}\frac{\partial v_1}{\partial x_2} \qquad (5.11)$$

Note that:

(i)   Although $|v_1| \gg |v_2|$, $|\partial T/\partial x_1| \ll |\partial T/\partial x_2|$ so that, to an order of magnitude, axial and transverse convection are equally important.

(ii)   Heat generation by viscous dissipation is usually important in injection moulding: the term $t_{12}^{\mathrm{E}}(\partial v_1/\partial x_2)$ is not negligible and, as a result, the momentum and energy conservation equations are coupled, not only through the temperature-dependent viscosity, but also through the dissipation term.

(iii) Variations in density ρ are usually too large for γ to be taken to be the specific heat evaluated at constant volume, c_v; on the other hand, pressure variations are usually assumed to be small enough for γ to be taken to be the specific heat evaluated at constant pressure, c_p. (Strictly, however, if γ is taken to be c_p then a term

$$\left(\frac{\partial \ln(1/\rho)}{\partial \ln T}\right)_p \left(\frac{\partial p}{\partial t} + v_1 \frac{\partial p}{\partial x_1} + v_2 \frac{\partial p}{\partial x_2}\right)$$

should, to within the other approximations made here, be added to the right-hand side of eqn (5.11). Here, this term will be neglected, though calculations have been made (see Toor, 1957, 1958) where this term is incorporated.)

Although there are not always good reasons for doing so, it is often assumed that the injection operating conditions change only slowly with time so that the local temperature is approximately that given by a steady flow: the flow is assumed to be quasi-steady, and eqn (5.11) becomes

$$\rho\gamma\left(v_1 \frac{\partial T}{\partial x_1} + v_2 \frac{\partial T}{\partial x_2}\right) = \alpha \frac{\partial^2 T}{\partial x_2^2} + t_{12}^{\mathrm{E}} \frac{\partial v_1}{\partial x_2} \qquad (5.12)$$

In order to complete the description of the temperature field T, the temperature T_{res} of the molten polymer in the reservoir (which may be a function of time) must be specified, as must the thermal conditions at the walls of the mould network. Because the thermal conductivity $\tilde{\alpha}$ of the (metal) mould walls is usually much higher than α, a first approximation that is commonly made is to specify the wall temperature T_{wall} (which may be a function of position and time). If the (transient) heat conduction through the walls cannot be ignored then, assuming that the principal temperature gradient component in the walls is $\partial T/\partial x_2$, the energy conservation equation is

$$\tilde{\rho}\tilde{\gamma} \frac{\partial T}{\partial t} = \tilde{\alpha} \frac{\partial^2 T}{\partial x_2^2} \qquad (5.13)$$

This requires specification of the initial wall temperature T_{init} and some wall temperature T_{spec} (which may be a function of position and time) to complete the description of the wall temperature field T. An alternative, and much simpler approach is to lump all the effects of heat transfer within the mould walls into a heat transfer coefficient h so that at walls

$$-\alpha \frac{\partial T}{\partial x_2} = h(T - T_{\mathrm{spec}}) \qquad (5.14)$$

5.3.2 Packing and holding

The common feature of the packing and holding stages of the moulding cycle is that there is a small, indeed often negligible, flow during them so that heat transfer is essentially a transient conduction process (modified in the holding stage, perhaps, by an energy change associated with freezing or cross-linking). The common feature of the two possible holding modes, freezing and cross-linking, is that partial freezing or cross-linking occurs during the injection and packing stages of the moulding cycle. This is because the mould walls must be maintained at a temperature T_{wall} below the effective melting point T_{melt} (in the case of freezing) or above some temperature T_{react} (in the case of cross-linking): the thermal capacity of the walls is usually too large to change T_{wall} rapidly enough between the injection, packing and holding stages.

5.3.2.1 Packing

Because the flow that occurs during packing is very small (compared with the flow that occurs during injection), the momentum conservation equations do not necessarily need to be solved for this stage of the moulding cycle. Furthermore, the convection and dissipation terms in the energy conservation equation are negligible. Hence, if the mould network is such that it can adequately be represented as a sequence of geometrically simple units, the energy conservation equation is

$$\rho\gamma \frac{\partial T}{\partial t} = \alpha \frac{\partial^2 T}{\partial x_2^2} \tag{5.15}$$

In order to complete the description of the temperature field T, the melt temperature $T_{start\ 1}$ at the end of the injection stage (and hence the start of the packing stage) and also a thermal boundary condition on the wall must be specified.

The small flow that occurs because of thermal expansion and compressibility effects can now either be ignored or be determined in one of two ways:

(i)   If the packing pressure P_{pack} is high enough for volume changes to be instantaneously made up by extra molten polymer from the reservoir, then all that is needed, to a first approximation, is that the rate of change (with time) of the mean polymer temperature be determined: from this rate of change of the mean density can be calculated, and hence the flowrate $\dot{V}$ of extra molten polymer required to ensure that the mould network is full can be determined.

(ii)   Alternatively, if P_{pack} is not high enough, the mean temperature
       and hence the mean polymer density can be determined, from
       which the mean pressure on the polymer can be calculated: the
       difference between this mean pressure and the packing pressure
       P_{pack} enables the flowrate $\dot{V}$ of extra molten polymer to be
       determined by solution of the momentum conservation equations.

5.3.2.2 Cooling (freezing)

In order to simulate the freezing of thermoplastics, it is necessary to be able
to predict the thickness δ of the frozen layer (see Fig. 5.6). (Of course, δ is
not in fact well-defined: for present purposes, it is assumed that the polymer
has an effective melting point T_{melt} and that the interface between molten
and frozen polymer may be identified with the T_{melt} isotherm.) The
crystallisation process that occurs upon freezing, and which can be
described by, say, the Avrami equation (see, for example, Tadmor and
Gogos, 1979), is usually ignored. Assuming that the mould network is such
that it can adequately be represented as a sequence of geometrically simple
units, the energy conservation equations are

$$\rho\gamma \frac{\partial T}{\partial t} = \alpha \frac{\partial^2 T}{\partial x_2^2} \tag{5.16}$$

for the molten polymer (for the freezing that occurs during injection,
eqn (5.11)—or perhaps eqn (5.12)—holds) and

$$\hat{\rho}\hat{\gamma} \frac{\partial T}{\partial t} = \hat{\alpha} \frac{\partial^2 T}{\partial x_2^2} \tag{5.17}$$

for the frozen polymer. In order to complete the description of the
temperature field T, the melt temperature $T_{start\ 2}$ at the start of freezing and
also a thermal boundary condition on the wall must be specified;
furthermore, the boundary conditions

$$T|_{\text{molten polymer}} = T|_{\text{frozen polymer}} = T_{melt} \tag{5.18}$$

$$\alpha \frac{\partial T}{\partial x_2}\bigg|_{\text{molten polymer}} = \hat{\alpha} \frac{\partial T}{\partial x_2}\bigg|_{\text{frozen polymer}} + \hat{\rho}\hat{\lambda} \frac{\partial \delta}{\partial t} \tag{5.19}$$

and, for the freezing that occurs during injection,

$$v_1 = v_2 = v_3 = 0 \tag{5.20}$$

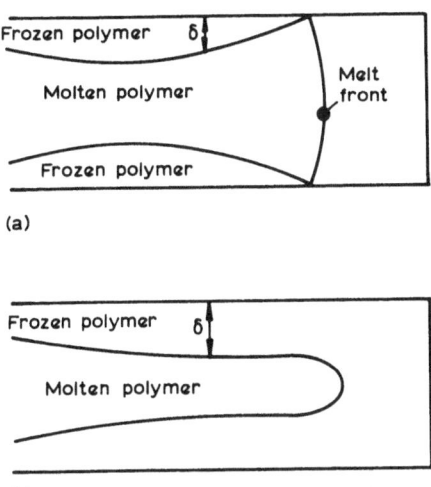

Fig. 5.6. Freezing in a duct. a, During injection; b, after injection.

(instead of eqn (5.9)) must be satisfied at the interface. Note that the latent heat term $\rho\hat{\lambda}(\partial\delta/\partial t)$ in eqn (5.19) can be neglected—provided the Stefan number is not too large—if the flow is assumed to be quasi-steady.

5.3.2.3 Heating (cross-linking)

In order to simulate the cross-linking of thermosets and rubbers, it is necessary to be able to predict the degrees (or fractions) of cure **C** and scorch **S** of material points in the polymer. Assuming that the mould network is such that it can adequately be represented as a sequence of geometrically simple units, then the energy conservation equation is

$$\rho\gamma\frac{\partial T}{\partial t} = \alpha\frac{\partial^2 T}{\partial x_2^2} + \Delta H_c\frac{\partial \mathbf{C}}{\partial t} + \Delta H_s\frac{\partial \mathbf{S}}{\partial t} \qquad (5.21)$$

where ΔH_c and ΔH_s denote the (commonly neglected) energies liberated on curing and scorching, respectively (for the cross-linking that occurs during injection, eqn (5.11)—or perhaps eqn (5.12)—suitably modified, holds). In order to complete the description of the temperature field T, the melt temperature $T_{\text{start 3}}$ at the start of cross-linking and also a thermal boundary condition on the wall must be specified.

Curing and scorching are usually assumed to involve n_cth and n_sth order, respectively, irreversible reaction kinetics and an Arrhenius temperature dependence (see, for example, Lee and Macosko, 1980; Mussatti and Macosko, 1973)

$$\frac{\partial C}{\partial t} = k_c (1 - C)^{n_c} \exp(-E_c/RT) \qquad (5.22)$$

$$\frac{\partial S}{\partial t} = k_s (1 - S)^{n_s} \exp(-E_s/RT) \qquad (5.23)$$

(Equations (5.22) and (5.23) must be modified for the cross-linking that occurs during injection, with the local derivatives $\partial/\partial t$ replaced by material derivatives $(\partial/\partial t) + v_1(\partial/\partial x_1) + v_2(\partial/\partial x_2)$. Note that so far an exclusively Eulerian approach has been adopted in formulating mathematical models of the moulding process: for cure and scorch calculations, however, a Lagrangian approach is more appropriate, in particular for the cross-linking that occurs during injection.) Note that build-up of solid material in the mould network when C and/or $S \rightarrow 1$ is usually ignored.

5.4 COMPUTER SIMULATION

The information required for computer simulation of the injection moulding cycle comprises:

(i)   geometrical and topological information defining the mould network;
(ii) physical (rheological and thermal) and—if a thermoset or rubber—chemical (kinetic) properties of the polymer;
(iii) operating parameters defining the manner in which the moulding machine is run.

The information that a computer simulation must produce comprises:

(i)   estimates of the minimum injection, packing and holding times, τ_{inj}, τ_{pack} and τ_{hold}, respectively (see Fig. 5.2);
(ii) estimates of the flowrate $\dot{V}$ and pressure drop ΔP within the mould network and hence of the required reservoir pressure P_{res} ($\simeq \Delta P$ because injection pressures are very high) and of the forces F that develop within the mould network;
(iii) estimates of the polymer temperature T and hence of the flow-average (or bulk) temperature difference $\overline{\Delta T}$ within the mould network and of the frozen layer thickness δ or of the degree of cure C and scorch S.

Much of the information required for simulation is fixed *a priori* and hence is not at the disposal of the designer. Indeed, the geometry of the mould cavities and the physical properties of the polymer are normally pre-specified. Moreover, the maximum operating pressures, forces and flow-rates, and maximum and minimum operating temperatures within the moulding machine are fixed. Thus it is usually the geometry and topology of the mould network (excluding the mould cavities) and, to some extent, the thermal conditions on the walls of the mould network that are at the disposal of the designer.

In practice, several simulations have to be performed before a satisfactory design, giving mouldings of an acceptable quality, is evolved. At present, while the simulation itself is usually handled completely automatically by the computer, the redesign stages between simulations involve, in all except rather trivial cases, personal intervention by the designer (see Fig. 5.7).

It is convenient in the discussion of simulation of the moulding cycle to

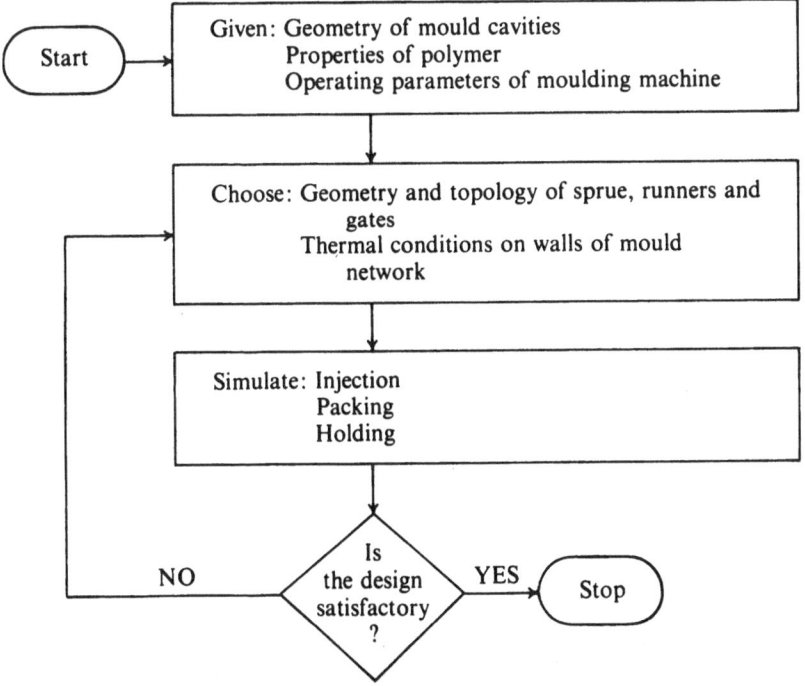

Fig. 5.7. Simulation flowchart.

consider the three stages—injection, packing and holding—separately, although some computer packages perform simulations of all stages. Because such simulations are essentially sequential, however, no loss of generality results from consideration of each stage in turn.

5.4.1 Injection

Schemes for simulation of the injection stage of the moulding cycle can usefully be classified into:

(i)   those of a simpler (usually analytical) nature, where full use is made of the approximations—particularly the geometrical approximations—of Section 5.3: the computer is used essentially for book-keeping purposes only;

(ii) those of a more complicated (usually numerical) nature, where only partial use is made of the approximations of Section 5.3: the computer is used to solve—usually partial—differential equations, as well as for book-keeping purposes.

Each type of scheme is now discussed in turn.

5.4.1.1 Simpler schemes

The common feature of the simulation schemes of a simpler nature is that the greatest possible simplifications are made to the geometry of the mould network, so that the network is assumed to be such that it can adequately be represented as a sequence of geometrically simple units (see Fig. 5.3) filling in a well-defined way (see Fig. 5.4). The principal difference between the schemes lies in the assumptions about the temperature field T; Byam et al. (1979) (see also Byam and Colbert, 1980) assume that T is a function of axial position x_1 (and time t) alone (see Fig. 5.5) whereas Richardson et al. (1980) make no a priori assumptions about T.

If, as is assumed by Byam et al. (1979), T is a function of x_1 alone, so that, in particular, $\partial T/\partial x_2 = 0$, then variations of viscosity η across the flow (in the x_2 direction) are solely due to variations of shear rate $|\partial v_1/\partial x_2|$. As a result, the momentum conservation equations may be integrated independently of the energy conservation equation: the equations are (partially) uncoupled. For the sake of example (and because it has an obvious direct relevance to the geometry of many sprues and runners), the archetypal geometrically simple unit will here be taken to be a long ($L \gg D$), partially full ($0 \leq z_{\text{front}} \leq L$), circular pipe of constant diameter ($\mathrm{d}D/\mathrm{d}z = 0$) (see Figs 5.3(a) and 5.4(a)). Time t is discretised into intervals Δt, which need not be constant. Suppose that $z_{front}(t)$ is known. Then the mass conservation equation

$$\dot{V} = \frac{\pi D^2}{4} \frac{dz_{front}}{dt} \qquad (5.24)$$

may be integrated approximately to yield

$$z_{front}(t + \Delta t) = z_{front}(t) + \Delta t \frac{4\dot{V}}{\pi D^2} \qquad (5.25)$$

(eqn (5.25) is exact if $\dot{V}$ is independent of t, of course). Assuming that $\partial T/\partial r = 0$, the (axial) momentum conservation equation may be integrated approximately to yield

$$\frac{dp}{dz} = -\frac{128\dot{V}\bar{\eta}}{\pi D^4} \qquad (5.26)$$

where $\bar{\eta}$ is the viscosity η evaluated at the mean shear rate $4\dot{V}/\pi D^3$ and flow-average temperature $\bar{T}$. Heat transfer to the walls is modelled using a heat transfer coefficient h: as a result, the (quasi-steady) energy conservation equation becomes

$$\frac{s\bar{T}}{dz} = -\frac{1}{\rho\gamma} \frac{dp}{dz} + \frac{\pi Dh}{\rho\gamma\dot{V}} (T_{spec} - \bar{T}) \qquad (5.27)$$

The filled length z_{front} of the pipe is now discretised into sections Δz, which need not be constant: eqns (5.26) and (5.27) may be integrated approximately to yield

$$p(z + \Delta z, t + \Delta t) = p(z, t + \Delta t) - \Delta z \frac{128\dot{V}\bar{\eta}}{\pi D^4} \qquad (5.28)$$

$$\bar{T}(z + \Delta z, t + \Delta t) = \bar{T}(z, t + \Delta t) + \Delta z \frac{128\dot{V}\bar{\eta}}{\rho\gamma\pi D^4}$$

$$+ \Delta z \frac{\pi Dh}{\rho\gamma\dot{V}} (T_{spec} - \bar{T}(z, t + \Delta t)) \qquad (5.29)$$

respectively, where $\bar{\eta}$ is evaluated at $\bar{T}(z, t + \Delta t)$. This means that pressures and flow-average temperatures can be determined at all discretised positions in the mould network for all discretised times until the mould network is full. The computer is used not so much to solve the equations (which are explicit and trivial) as to store all the information about z_{front}, p and $\bar{T}$, i.e. the computer has a book-keeping role.

A somewhat more elaborate simulation scheme has been devised by Richardson *et al.* (1980), although the book-keeping role of the computer is essentially maintained. Here no *a priori* assumptions are made about the temperature field T (though the quasi-steady approximation is made and it is assumed that the temperature of the walls of the mould network T_{wall} is specified). In order to avoid having to solve the highly non-linear coupled set of partial differential conservation equations numerically (analytical solution is impossible), approximate—usually analytical—asymptotic solutions are used based on the values taken by the three dimensionless groups that characterise injection, which are:

(i)   The Graetz number

$$Gz \equiv \frac{4\rho\gamma\dot{V}}{\pi\alpha z_{front}} \qquad (5.30)$$

(which may be interpreted as the ratio of heat convected along the flow to heat conducted across it).

(ii) The Nahme (or Griffith) number

$$Na \equiv \frac{16\eta^*\dot{V}^2\zeta}{\pi^2 D^4\alpha} \qquad (5.31)$$

(which may be interpreted as the ratio of heat generated by viscous dissipation to heat required to significantly alter the viscosity).

(iii) The Brinkman number

$$Br \equiv \frac{16\eta^*\dot{V}^2}{\pi^2 D^4\alpha|\bar{T}_{entry} - T_{wall}|} \qquad (5.32)$$

(which may be interpreted as the ratio of heat generated by viscous dissipation to heat flux due to the difference between the flow-average entry temperature $\bar{T}_{entry}$ to a unit and the wall temperature T_{wall} of a unit). Here

$$\eta^* = K\left(\frac{4\dot{V}}{\pi D^3}\right)^{\nu-1}\exp\left(-\zeta(T^* - T_0)\right) \qquad (5.33)$$

and

$$T^* = \begin{cases} \bar{T}_{entry} & \text{if } Gz \gg 1 \\ \frac{1}{2}(\bar{T}_{entry} + T_{wall}) & \text{(say) if } Gz = 0(1) \\ T_{wall} & \text{if } Gz \ll 1 \end{cases} \qquad (5.34)$$

It is also convenient to use:

(iv)   the (anonymous) number

$$Pn \equiv \zeta |\bar{T}_{entry} - T_{wall}| \qquad (5.35)$$

(so that $Pn \equiv Na/Br$).

Then, depending on the magnitudes of Gz, Na, Br and Pn (see Fig. 5.8), one of the following asymptotic solutions is selected automatically by the computer (see also Pearson, 1978).

Solution 1 is the case of thermally fully developed flow with essentially constant viscosity: $\dot{V}$ and ΔP are given by a generalisation of the Hagen–Poiseuille law (see, for example, Middleman, 1977) while an elementary solution of the energy conservation equation gives $\overline{\Delta T}$.

Solution 2 is the case of thermally fully developed flow with significant viscosity variations due to viscous heating: $\dot{V}$, ΔP and $\overline{\Delta T}$ can be obtained analytically (see Martin, 1967) for flow in a pipe; for flow in ducts of other cross-sections, solution of a two-point boundary value problem (say, by a combined Runge-Kutta/*regula falsi* (false-position secant) method) is necessary.

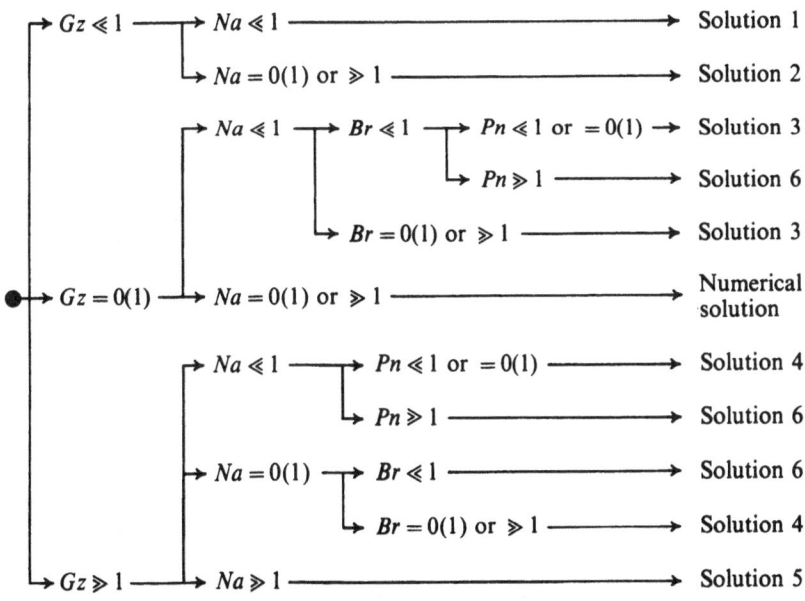

Fig. 5.8.   Selection of asymptotic solutions.

Solution 3 is the case of thermally developing flow with essentially constant viscosity: $\dot{V}$ and ΔP are given as in Solution 1; $\overline{\Delta T}$ is given by solution of the generalised Graetz–Nusselt problem (see Toor, 1957).

Solution 4 is the case of thermally undeveloped flow with essentially constant viscosity: $\dot{V}$ and ΔP are given as in Solution 1; $\overline{\Delta T}$ is given by an adiabatic analysis or, more accurately, by a Lévêque type solution (see Richardson, 1979).

Solution 5 is the case most often encountered in injection moulding of thermally undeveloped flow with significant viscosity variations due to viscous heating: $\dot{V}$, ΔP and $\overline{\Delta T}$ can be obtained analytically (see Pearson, 1977) when $Na \gg Gz$ and also more generally (see Ockendon, 1979).

Solution 6 is the case of thermally undeveloped flow with significant viscosity variations due to the difference between the entry and wall temperatures: $\dot{V}$, ΔP and $\overline{\Delta T}$ can be obtained analytically (see Ockendon and Ockendon, 1977), albeit in a rather tenuous way.

If $Gz = 0(1)$ and $Na = 0(1)$ or $\gg 1$ and numerical solution is required (see Fig. 5.8) the methods to be discussed later should strictly be used; if, however, rather approximate estimates of $\dot{V}$, ΔP and $\overline{\Delta T}$ are acceptable then it is possible either to match $Gz \ll 1$ and $Gz \gg 1$ estimates or else to use Solution 5 estimates alone.

Although the principal use of the computer in these simpler simulation schemes is that of book-keeping, other uses arise if:

(i) the mould network branches at any point into several (N, say) runners to feed (perhaps) N one-gated mould cavities or one N-gated mould cavity: it is then necessary to guess the flowrates $\dot{V}_i$ in each of the N runners and subsequent mould cavity or cavities such that

$$\sum_{i=1}^{N} \dot{V}_i = \dot{V} \tag{5.36}$$

and to check that the pressure drops ΔP_i (from the reservoir to each melt front) are equal to within some acceptable tolerance: if they are not, the $\dot{V}_i$ must be reguessed, and so on;

(ii) the reservoir pressure P_{res}, rather than the flowrate $\dot{V}$, is specified (see Fig. 5.2): it is then necessary to guess $\dot{V}$ and to check that the pressure drop (from the reservoir to the melt front (or fronts)) equals P_{res} to within some acceptable tolerance: if it does not, $\dot{V}$ must be reguessed, and so on.

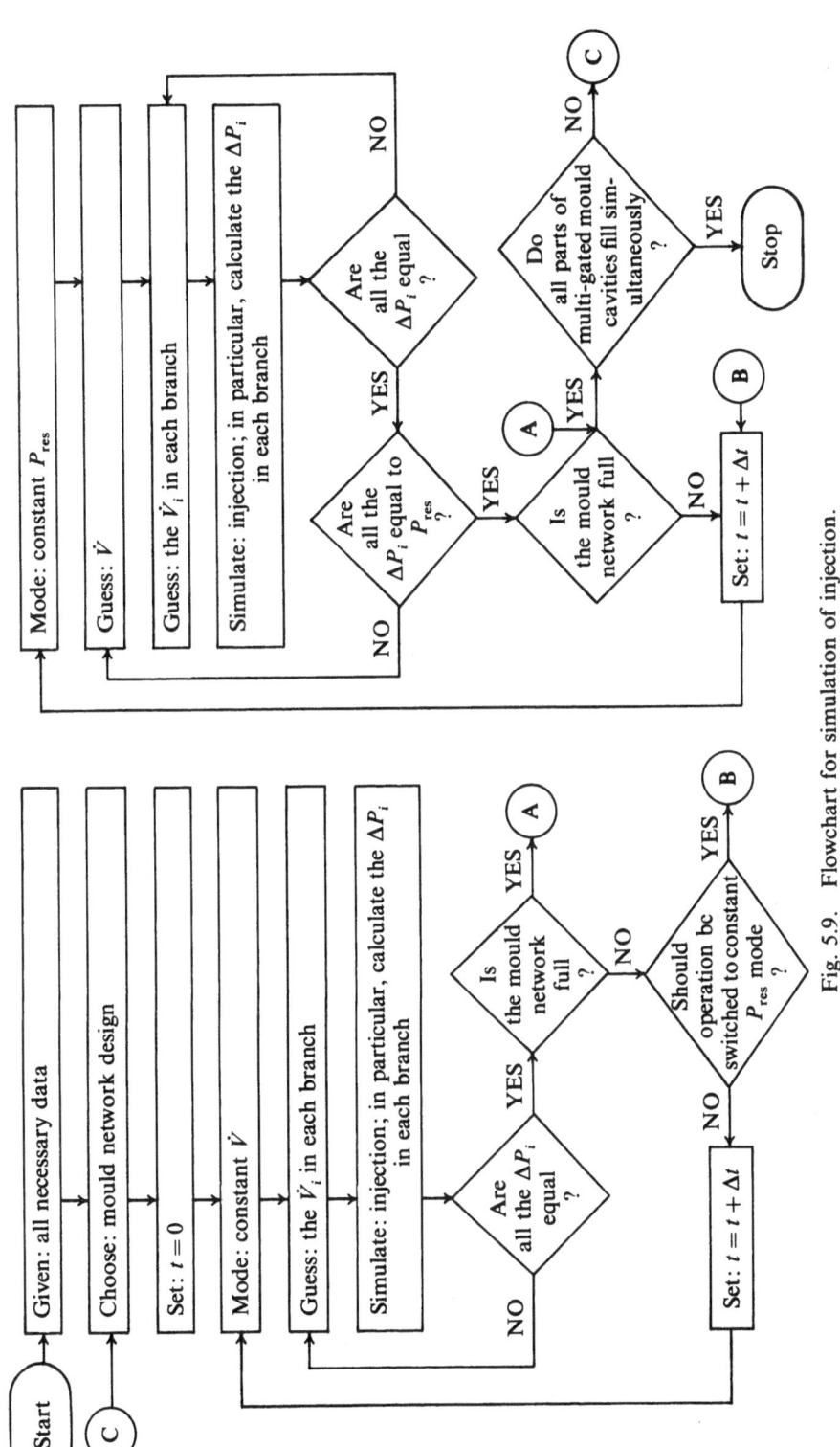

Fig. 5.9. Flowchart for simulation of injection.

Both of these situations can be handled entirely automatically by the computer (see Fig. 5.9), the *regula falsi* method being an appropriate way of reguessing both the $\dot{V}_i$ and $\dot{V}$. Another potential use of the computer, but one which, as yet, cannot be handled automatically by the computer (and requires intervention by the designer instead), arises if:

(iii)   the mould cavities are multi- (N-, say) gated: it is then necessary to choose the N geometrically simple units representing each part of each mould cavity not only such that the pressure drops ΔP_i are equal but also so that all N units fill simultaneously, to within some acceptable tolerance (so that the guessed positions of the weld lines are correct): if they do not, the N geometrically simple units must be rechosen, and so on.

It should, of course, be noted that there is no guarantee of convergence of any of these iterative reguessing/rechoosing schemes. Thus, for example, if $Gz \ll 1$ and $Na = 0(1)$ or $\gg 1$ (which is, admittedly, only rarely encountered in injection moulding) the $\dot{V}$–ΔP relation is not unique (see Fig. 5.10): a given value of $\dot{V}$ corresponds to a single value of ΔP whereas a given value of ΔP corresponds to two values of $\dot{V}$ (or, perhaps, to none): iteration involving specified values of the reservoir pressure P_{res} and hence of ΔP

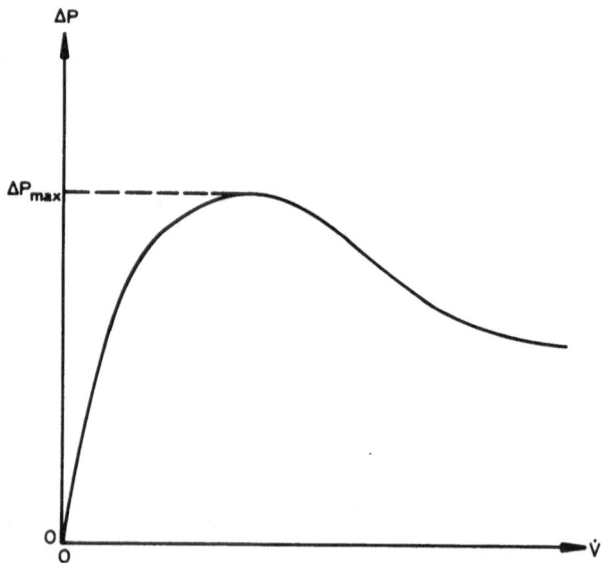

Fig. 5.10.   Flowrate–pressure drop curve for $Gz \ll 1$ and $Na = 0(1)$ or $\gg 1$.

near the maximum allowable pressure drop ΔP_{max} would be most unlikely to converge. Fortunately, convergence failures appear to be relatively infrequent in practice.

5.4.1.2 More complicated schemes

The common feature of all the simulation schemes of a more complicated nature is that rather few simplifications are made to the geometry of the mould network, although it is still usual to assume that the various parts of the mould network are thin in at least one transverse direction. The result is that the computer has to be used to solve—usually partial—differential equations numerically.

There is a relatively small number of more complicated schemes for simulation of injection in the sprue and runners, mainly because of the geometrical simplicity of these parts of the mould network. Williams and Lord (1975) have devised a scheme for simulation of injection in a long $(L \gg D)$, circular pipe of constant diameter $(dD/dz = 0)$, though a pipe of slowly varying diameter can be handled (see Figs 5.3(a) and 5.4(a)). This involves solution of the partially integrated conservation equations:

$$\dot{V} = \int_0^{D/2} 2\pi r v_z \, dr \tag{5.37}$$

$$v_z = -\frac{\partial p}{\partial z} \int_r^{D/2} \frac{r \, dr}{\eta} \tag{5.38}$$

$$\rho \gamma v_z \frac{\partial T}{\partial z} = \alpha \frac{\partial}{\partial r} \left(r \frac{\partial T}{\partial r} \right) + \frac{r}{2} \frac{\partial p}{\partial z} \frac{\partial v_z}{\partial r} \tag{5.39}$$

where η is a specified function of $|\partial v_z/\partial r|$ and T, subject to the boundary conditions:

$$\left. \begin{array}{ll} T = T_{\text{entry}} & \text{at } z = 0 \\[2mm] \dfrac{\partial T}{\partial r} = 0 & \text{at } r = 0 \\[2mm] -\alpha \dfrac{\partial T}{\partial r} = h(T - T_{\text{spec}}) & \text{at } r = \dfrac{D}{2} \end{array} \right\} \tag{5.40}$$

Note that transverse convection is neglected and that the quasi-steady approximation is made; note also that, although the pipe is taken by Williams and Lord (1975) always to be full (so that $z_{\text{front}} = L$), there is no reason why filling should not be simulated by extending eqn (5.37) thus:

$$\dot{V} = \int_0^{D/2} 2\pi r v_z \, \mathrm{d}r = \frac{\pi D^2}{4} \frac{\mathrm{d}z_{\mathrm{front}}}{\mathrm{d}t} \tag{5.41}$$

To solve these equations, the radial direction is discretised (see Fig. 5.11) into N_r intervals Δr_i (which need not be constant: gradients of v_z and T are smaller near $r = 0$ and larger near $r = D/2$, so the Δr_i can advantageously be made larger near $r = 0$ and smaller near $D/2$). Then, if η is presumed to be

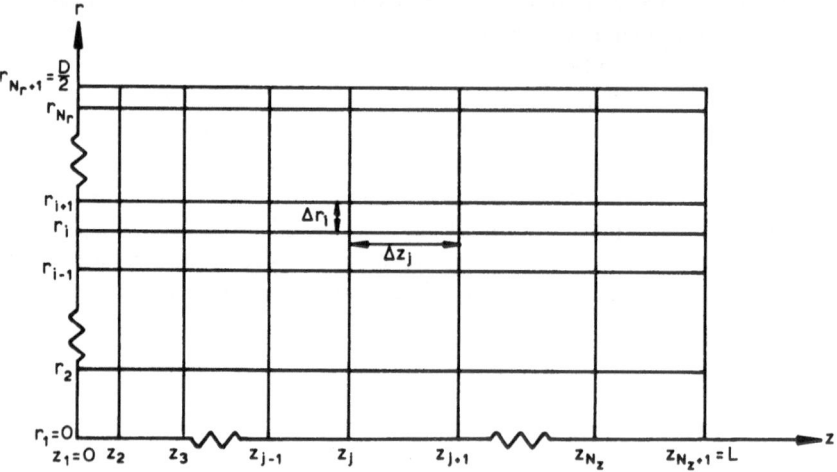

Fig. 5.11.   Mesh for flow in a pipe.

known as a function of r for given z, eqn (5.38) can be integrated numerically by, for example, trapezoidal quadrature to give $v_z/(-\partial p/\partial z)$. Numerical integration of eqn (5.37) with known $\dot{V}$ then gives $(-\partial p/\partial z)$ and hence v_z. Equation (5.39) is now cast into finite-difference form with (conservative) central differencing of the conduction term and central differencing of the dissipation term to yield

$$\frac{\mathrm{d}T_i}{\mathrm{d}z} = \frac{\alpha}{\rho \gamma v_{z_i} r_i} \left\{ \frac{(r_{i+1}+r_i)\left(\dfrac{T_{i+1}-T_i}{r_{i+1}-r_i}\right) - (r_i+r_{i-1})\left(\dfrac{T_i-T_{i-1}}{r_i-r_{i-1}}\right)}{(r_{i+1}-r_{i-1})} \right\}$$
$$- \frac{r_i}{2\rho \gamma v_{z_i}} \left(-\frac{\partial p}{\partial z}\right)\left(\frac{v_{z_{i+1}}-v_{z_{i-1}}}{r_{i+1}-r_{i-1}}\right) \tag{5.42}$$

for $i = 2 \ldots N_r$. Finite-difference forms of eqns (5.40) can then be used to yield equations for T_1 and T_{N_r+1}. Finally, the axial direction is discretised (see Fig. 5.11) into N_z intervals Δz_j (which need not be constant). The first-order ordinary differential equations, eqns (5.42), can then be integrated numerically using, for example, an explicit (forward) method so that dT_i/dz is approximated by $[(T_{i,j+1} - T_{i,j})/(z_{j+1} - z_j)]$. Stability problems (arising near $r = D/2$, where gradients are largest) mean, however, that unacceptably small values of Δz_j have to be used. An implicit (backward) method, with dT_i/dz approximated by $[(T_{i,j} - T_{i,j-1})/(z_j - z_{j-1})]$, is often claimed (falsely) to be unconditionally stable (though it is usually more stable than an explicit method), but then a set of simultaneous linear algebraic equations must be solved at each axial position z_j. A balance must therefore, be struck: Williams and Lord (1975) use an explicit method near $r = 0$ (say, for $2 \le i \le N_{r*}$) and an implicit method near $r = D/2$ (say, for $N_{r*} < i \le N_r$; in practice, N_{r*} is rather close to N_r). The only remaining problem is to determine η: Williams and Lord (1975) use a rather complicated predictor–corrector iterative scheme, which is incorporated into their scheme for marching from z_1 ($= 0$) to $z_{N_z + 1}$ ($= L$ or, perhaps, z_{front}). Wang et al. (1978) have devised a similar scheme to that of Williams and Lord (1975): the principal differences are, from the point of view of the model, that the temperature of the walls of the mould network T_{wall} is specified and, from the point of view of the numerical simulation, that a fully implicit finite-difference method is used with lagging of the velocity and pressure terms in the energy conservation equation.

There is a relatively large number of more complicated schemes for simulation of injection in the mould cavities: these can usefully be classified into those where the cavities are taken to be discs and those where they are taken to be two-dimensional channels. The latter are the geometrically most complicated mould cavities, and will be discussed later.

Kamal and Kenig (1972a, 1972b) have devised a scheme, based on one proposed by Pearson (1966), for simulation of injection in a disc of constant height H ($dH/dr = 0$) (see Figs 5.3(c) and 5.4(c)), representing a centre-fed one-gated mould cavity. This involves solution of the conservation equations

$$\frac{\partial}{\partial r}(rv_r) = 0 \qquad (5.43)$$

$$\frac{\partial p}{\partial r} = \frac{\partial}{\partial z}\left(\eta \frac{\partial v_r}{\partial z}\right) \qquad (5.44)$$

$$\rho\gamma\left(\frac{\partial T}{\partial t} + v_r \frac{\partial T}{\partial r}\right) = \alpha \frac{\partial^2 T}{\partial z^2} + \eta\left(\frac{\partial v_r}{\partial z}\right)^2 \qquad (5.45)$$

where

$$\eta = K \left| \frac{\partial v_r}{\partial z} \right|^{\nu - 1} \exp\left(-E/RT\right) \qquad (5.46)$$

subject to the boundary conditions

$$\left.\begin{array}{lll} T = T_{\text{entry}}, & & \text{at } r = \dfrac{D_i}{2} \\[2ex] \dfrac{\partial v_r}{\partial z} = 0, & \dfrac{\partial T}{\partial z} = 0 & \text{at } z = 0 \\[2ex] v_r = 0, & -\alpha \dfrac{\partial T}{\partial z} = h(T - T_{\text{spec}}) & \text{at } z = \dfrac{H}{2} \end{array}\right\} \qquad (5.47)$$

(symmetry means that attention can be confined to $0 \le z \le H/2$). Note that transverse convection is neglected and that the quasi-steady approximation is not made; note also that the (possibly time-dependent) pressure drop ΔP, rather than the flowrate $\dot{V}$, is specified, where

$$\Delta P = -\int_{D_i/2}^{r_{\text{front}}} \frac{\partial p}{\partial r} \, dr \qquad (5.48)$$

$$\dot{V} = 2 \int_0^{H/2} 2\pi r v_r \, dz = \pi H \frac{dr_{\text{front}}^2}{dt} \qquad (5.49)$$

Kamal and Kenig (1972a) also specify that

$$\frac{\partial T}{\partial r} = 0 \qquad \text{at } r = r_{\text{front}} \qquad (5.50)$$

implying that there are negligible heat losses to the atmosphere in that part of the mould cavity which is yet to be filled: although this is not an unreasonable condition physically, it is entirely superfluous mathematically and is inconsistent with the neglect of radial conduction in eqn (5.45). Note that the initial condition on eqn (5.45) is effectively provided by the entry ($r = D_i/2$) boundary condition, since the mould cavity is initially empty. These equations are now cast into finite-difference form by discretisation (see Fig. 5.12) of the radial and axial directions into N_r and N_z intervals Δr and Δz, respectively, and also by discretisation of time into intervals Δt (although it is not altogether necessary to do so, all these intervals will here be taken to be constant). The main computational problem is presented by the energy conservation equation, eqn (5.45): Kamal and Kenig (1972a) use a two-level (alternating direction explicit)

Saul'yev scheme

$$\left(\frac{T_{i,j,n+1} - T_{i,j,n}}{\Delta t}\right) = \frac{\alpha}{\rho\gamma}\left(\frac{T_{i,j-1,n+1} - T_{i,j,n+1} - T_{i,j,n} + T_{i,j+1,n}}{\Delta z^2}\right)$$

$$+ \text{(other terms)} \tag{5.51}$$

$$\left(\frac{T_{i,j,n+2} - T_{i,j,n+1}}{\Delta t}\right) = \frac{\alpha}{\rho\gamma}\left(\frac{T_{i,j-1,n+1} - T_{i,j,n+1} - T_{i,j,n+2} + T_{i,j+1,n+2}}{\Delta z^2}\right)$$

$$+ \text{(other terms)} \tag{5.52}$$

for $j = 2 \ldots N_z$. Assuming that $T_{i,j,n} = T(r_i, z_j, t_n)$ is known for all i and j, eqn (5.51) is used to determine $T_{i,j,n+1}$ for all i and j by sweeping from $j = 2$ to $j = N_z$; eqn (5.52) is then used to determine $T_{i,j,n+2}$ for all i and j by sweeping in the reverse direction from $j = N_z$ to $j = 2$. The so-called other terms referred to in eqns (5.51) and (5.52) are finite-difference forms of the convection and dissipation terms. Kamal and Kenig (1972a) do not specify how these forms are to be obtained. In fact, the presence of these terms causes numerical instabilities, so that a Courant number ($v_r \Delta t/\Delta r$) restriction becomes necessary (see also Roache, 1972). The algorithm for the whole scheme devised by Kamal and Kenig (1972a) is given below:

Step 1. Given: the $T_{i,j,n}$, $v_{r_{i,j,n}}$, $p_{i,n}$ and $r_{\text{front},n}$.

Step 2. Guess: the $p_{i,n+1}$ (assume a linear pressure drop from $D_i/2$ to $r_{\text{front},n}$).

Step 3. Calculate: the $v_{r_{i,j,n+1}}$ (from the momentum conservation equations).

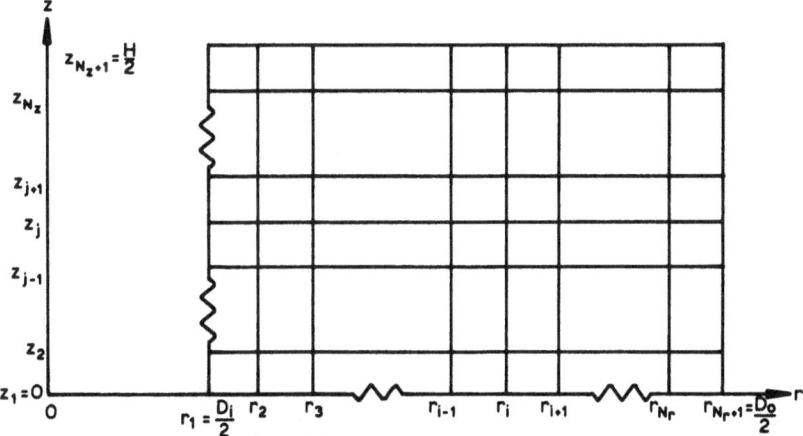

Fig. 5.12. Mesh for flow in a disc.

Step 4.   Calculate: the $v_{r_{i>1,j,n+1}}$ (from the mass conservation equation and the no slip boundary condition).

Step 5.   Calculate: the $p_{i,n+1}$ (from the momentum conservation equations).

Step 6.   Check: if $p_{N_r+1,n+1}$ is equal to atmospheric pressure (to within some acceptable tolerance) proceed; otherwise, correct the $p_{i,n+1}$ and return to Step 3.

Step 7.   Calculate: $\dot{V}_{n+1}$ and $r_{\text{front},n+1}$.

Step 8.   Calculate: the $T_{i,j,n+1}$ (from the energy conservation equation).

Step 9.   Set: $n = n + 1$ and (unless the mould cavity is full) return to Step 2.

Berger and Gogos (1973) and Wu *et al.* (1974) have devised similar schemes to that of Kamal and Kenig (1972a). The principal differences are, from the point of view of the model, that hoop stress effects are incorporated in the dissipation term in the energy conservation equation, that an inertial term $\rho v_r^2/r$ can be incorporated in the momentum conservation equation and that (because, for $r = 0(H)$, $|v_r/r|$ and $|\partial v_r/\partial r| = 0(|\partial v_r/\partial z|)$) the shear rate $\dot{\gamma}$ is taken to be $\sqrt{(\partial v_r/\partial z)^2 + 2(v_r/r)^2}$ in the dissipation term in the energy conservation equation. From the point of view of the numerical simulation, the principal differences are that a Runge-Kutta method is used to solve an integrated form of the momentum conservation equation

$$-\frac{\partial p}{\partial r} = \dot{V} \bigg/ 4\pi r \int_0^{H/2} \frac{z^2 \, dz}{\eta} \tag{5.53}$$

and that the implicit Crank–Nicolson scheme is used (by Berger and Gogos (1973); the similar O'Brien–Hyman–Kaplan scheme is used by Wu *et al.* (1974)) to solve the energy conservation equation:

$$\left(\frac{T_{i,j,n+1} - T_{i,j,n}}{\Delta t}\right) = \frac{1}{2}\left\{\frac{\alpha}{\rho\gamma}\left(\frac{T_{i,j+1,n} - 2T_{i,j,n} + T_{i,j-1,n}}{\Delta z^2}\right.\right.$$
$$+ \frac{T_{i,j+1,n+1} - 2T_{i,j,n+1} + T_{i,j-1,n+1}}{\Delta z^2}\bigg)$$
$$- v_{r_{i,j,n}}\left(\frac{T_{i,j,n} - T_{i-1,j,n}}{\Delta r} + \frac{T_{i,j,n+1} - T_{i-1,j,n+1}}{\Delta r}\right)\bigg\}$$
$$+ \frac{\eta_{i,j,n}}{\rho\gamma}\left[\left(\frac{v_{r_{i,j+1,n}} - v_{r_{i,j-1,n}}}{2\Delta z}\right)^2 + \left(\frac{v_{r_{i,j,n}}}{r_i}\right)^2\right] \tag{5.54}$$

for $j = 2 \ldots N_z$. Note that central differencing and lagging are used for the dissipation term and that upwind differencing and partial lagging are used for the convection term. Upwind differencing of the convection term leads to an artificial radial thermal diffusivity κ_{art} effect: κ_{art} is generally significantly larger than the real (physical) thermal diffusivity $\kappa = \alpha/\rho\gamma$. Because, however,

$$\left| \frac{1}{r} \frac{\partial}{\partial r} \left(r \frac{\partial T}{\partial r} \right) \right| \Big/ \left| \frac{\partial^2 T}{\partial z^2} \right| \ll 1 \qquad (5.55)$$

for high Péclet number flows, the effect of this artificial diffusivity is usually not too severe in practice. Note also that, while a conservative form is used for the conduction (diffusion) term, a non-conservative form is used for the convection term in eqn (5.54). This means that the energy conservation implicit in eqn (5.45) is simulated accurately as far as conduction is concerned, but not as far as convection is concerned. While the effects of this are not normally too severe in practice, a conservative form can be obtained for the convection term quite easily by (upwind) differencing of $(1/r)(\partial/\partial r)(rv_r T)$ (as opposed to the analytically identical $v_r(\partial T/\partial r)$). The set of simultaneous linear algebraic equations resulting from the implicit scheme can be written as a matrix equation

$$\mathbf{Ax} = \mathbf{y} \qquad (5.56)$$

where the matrix $\mathbf{A}$ and vector $\mathbf{y}$ are both known and the vector $\mathbf{x}$ (incorporating the unknown $T_{i,j,n+1}$) is to be determined. Because $\mathbf{A}$ is tridiagonal, the Thomas algorithm (see, for example, Ames (1969) or Roache (1972)) can be used to yield $\mathbf{x}$ in a two-sweep process. Wang et al. (1975a,b, 1976, 1978) have devised schemes which are similar to those of Kamal and Kenig (1972a), Berger and Gogos (1973) and Wu et al. (1974). All that needs to be noted here is that Wang et al. (1975a) use the following finite-difference form for the thermal wall boundary condition, eqn (5.47):

$$-\alpha \left(\frac{3T_{i,N_z+1,n+1} - 4T_{i,N_z,n+1} + T_{i,N_z-1,n+1}}{2\Delta z} \right) = h(T_{i,N_z+1,n+1} - T_{spec})$$
$$(5.57)$$

This is, however, inconsistent with the use of central differencing for the conduction term at $j = N_z$, in the sense that the heat transferred to the wall is calculated incorrectly (non-conservatively) (see also Roache, 1972). For consistency, what is required is

$$-\alpha \left(\frac{T_{i,N_z+1,n+1} - T_{i,N_z,n+1}}{\Delta z} \right) = h(T_{i,N_z+1,n+1} - T_{spec}) \qquad (5.58)$$

The most complicated schemes are those that simulate injection in a two-dimensional channel, representing an edge-gated thin mould cavity (see Fig. 5.13). Wang *et al.* (1975b, 1976, 1977, 1978, 1979) have devised a variety of related schemes for solving the conservation equations

$$\frac{\partial p}{\partial x} = \frac{\partial}{\partial y}\left(\eta\,\frac{\partial v_x}{\partial y}\right) \tag{5.59}$$

$$\frac{\partial p}{\partial z} = \frac{\partial}{\partial y}\left(\eta\,\frac{\partial v_z}{\partial y}\right) \tag{5.60}$$

$$\rho\gamma\left(\frac{\partial T}{\partial t} + v_x\frac{\partial T}{\partial x} + v_z\frac{\partial T}{\partial z}\right) = \alpha\frac{\partial^2 T}{\partial y^2} + \eta\left[\left(\frac{\partial v_x}{\partial y}\right)^2 + \left(\frac{\partial v_z}{\partial y}\right)^2\right] \tag{5.61}$$

$$\frac{\partial \bar{v}_x}{\partial x} + \frac{\partial \bar{v}_z}{\partial z} = 0 \tag{5.62}$$

where

$$\left.\begin{array}{l} \bar{v}_x = \dfrac{1}{H/2}\displaystyle\int_0^{H/2} v_x\,\mathrm{d}y \\[2ex] \bar{v}_z = \dfrac{1}{H/2}\displaystyle\int_0^{H/2} v_z\,\mathrm{d}y \end{array}\right\} \tag{5.63}$$

and

$$\eta = K\left(\sqrt{\left(\frac{\partial v_x}{\partial y}\right)^2 + \left(\frac{\partial v_z}{\partial y}\right)^2}\right)^{\nu-1}\exp\left(-\zeta(T - T_0)\right) \tag{5.64}$$

subject to the boundary conditions

$$\left.\begin{array}{llll} T = T_{\text{entry}}, & p = P_{\text{entry}}, & & \text{at } \sqrt{x^2 + z^2} = R_{\text{gate}} \\[1ex] v_x = 0, & \dfrac{\partial T}{\partial x} = 0 & & \text{at } x = 0 \\[2ex] \dfrac{\partial v_x}{\partial y} = 0, & \dfrac{\partial v_z}{\partial y} = 0, & \dfrac{\partial T}{\partial y} = 0 & \text{at } y = 0 \\[2ex] v_x = 0, & v_z = 0, & T = T_{\text{wall}} & \text{at } y = \dfrac{H}{2} \\[2ex] \dfrac{\partial v_x}{\partial z} = 0, & v_z = 0, & \dfrac{\partial T}{\partial z} = 0 & \text{at } z = 0 \\[2ex] v_z = 0, & \dfrac{\partial T}{\partial z} = 0 & & \text{at } z = \dfrac{W}{2} \end{array}\right\} \tag{5.65}$$

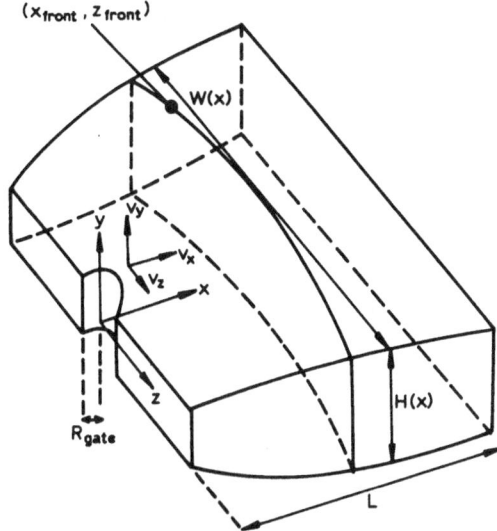

Fig. 5.13. Partially full two-dimensional channel.

(symmetry means that attention can be confined to $0 \leq y \leq H/2$ and $0 \leq z \leq W/2$). For the sake of example, the channel height H and width W are here taken to be constant ($\mathrm{d}H/\mathrm{d}x = \mathrm{d}W/\mathrm{d}x = 0$). Note that the Hele Shaw approximation implicit in these equations means that the no slip boundary condition can be imposed only at $y = H/2$ and not at $x = 0$ or $z = W/2$. Note also that transverse convection is neglected and that the quasi-steady approximation is not made. The (average) location $(x_{\text{front}}, z_{\text{front}})$ of the melt front is given by

$$\left.\begin{array}{l} \dfrac{\mathrm{d}x_{\text{front}}}{\mathrm{d}t} = \bar{v}_{x_{\text{front}}} \\[3mm] \dfrac{\mathrm{d}z_{\text{front}}}{\mathrm{d}t} = \bar{v}_{z_{\text{front}}} \end{array}\right\} \qquad (5.66)$$

The momentum conservation equations, eqns (5.59) and (5.60), can be integrated to yield

$$\left.\begin{array}{l} v_x = -\dfrac{\partial p}{\partial x} \displaystyle\int_y^{H/2} \dfrac{y\,\mathrm{d}y}{\eta} \\[4mm] v_z = -\dfrac{\partial p}{\partial z} \displaystyle\int_y^{H/2} \dfrac{y\,\mathrm{d}y}{\eta} \end{array}\right\} \qquad (5.67)$$

whence

$$\bar{v}_x = -\frac{\partial p}{\partial x}\frac{\sigma}{H/2} \left.\begin{array}{c} \\ \\ \\ \\ \end{array}\right\} \quad (5.68)$$

$$\bar{v}_z = -\frac{\partial p}{\partial z}\frac{\sigma}{H/2}$$

where

$$\sigma = \int_0^{H/2} \frac{y^2\,dy}{\eta} \quad (5.69)$$

Substitution of eqns (5.68) into the mass conservation equation, eqn (5.62), yields

$$\frac{\partial}{\partial x}\left(\sigma\frac{\partial p}{\partial x}\right) + \frac{\partial}{\partial z}\left(\sigma\frac{\partial p}{\partial z}\right) = 0 \quad (5.70)$$

which has to be solved subject to the boundary conditions

$$\begin{aligned} p &= P_{\text{entry}} && \text{at } \sqrt{x^2+z^2} = R_{\text{gate}} \\ p &= 0 && \text{at } x = x_{\text{front}},\ z = z_{\text{front}} \\ \frac{\partial p}{\partial n} &= 0 && \text{at solid walls} \end{aligned} \left.\begin{array}{c} \\ \\ \\ \\ \\ \end{array}\right\} \quad (5.71)$$

where n denotes the normal to the walls.

Note that if $\dot{V}$ is specified rather than P_{entry}, an iterative solution scheme is required. The earliest method of Wang *et al.* (1975b, 1976) involves discretisation of the x, y and z directions and time t into intervals Δx, Δy, Δz and Δt, respectively (these intervals are taken here to be constant, though it is not essential to do so). An under-relaxation method is used to solve the following finite-difference form of eqn (5.70)

$$(\sigma_{i+\frac{1}{2},k} + \sigma_{i,k+\frac{1}{2}} + \sigma_{i-\frac{1}{2},k} + \sigma_{i,k-\frac{1}{2}})P_{i,k}$$

$$= \sigma_{i+\frac{1}{2},k}P_{i+1,k} + \sigma_{i,k+\frac{1}{2}}P_{i,k+1} + \sigma_{i-\frac{1}{2},k}P_{i-1,k} + \sigma_{i,k-\frac{1}{2}}P_{i,k-1} \quad (5.72)$$

where

$$P_{i,k} = p(x_i, z_k) \quad \text{and} \quad \sigma_{i+\frac{1}{2},k} = \tfrac{1}{2}(\sigma_{i,k} + \sigma_{i+1,k}), \text{ etc.}$$

A predictor–corrector plus interpolation scheme is used to integrate finite-difference forms of eqns (5.66). A more recent method of Wang *et al.*

(1977, 1979) (see also Hieber and Shen, 1980) uses a hybrid finite-difference/ finite-element scheme. Finite-difference forms are used for y and t derivatives; finite-element forms for x and z derivatives. The advantage of using a finite-element scheme is usually claimed to be that injection in mould cavities of much greater geometrical complexity can easily be simulated than is generally possible using a finite-difference scheme. The finite-element part of the scheme involves use of triangular elements (see Fig. 5.14) with quadratic interpolation functions Q_l for pressure (because eqn (5.70) is second-order) and linear interpolation functions L_m for temperature (because eqn (5.61) is first-order (in the x and z directions)):

$$p(x, z) = \sum_{l=1}^{6} Q_l(x, z)\mathbf{P}_l \qquad (5.73)$$

$$T(x, z) = \sum_{m=1}^{3} L_m(x, z)\mathbf{T}_m \qquad (5.74)$$

where the l refer to the six nodes at the vertices and mid-points of the sides, and the m refer to the three nodes at the vertices, of a triangular element; the $\mathbf{P}_l$ are nodal pressures and the $\mathbf{T}_m$ are nodal temperatures. A Galerkin procedure is used, resulting in a set of simultaneous non-linear algebraic

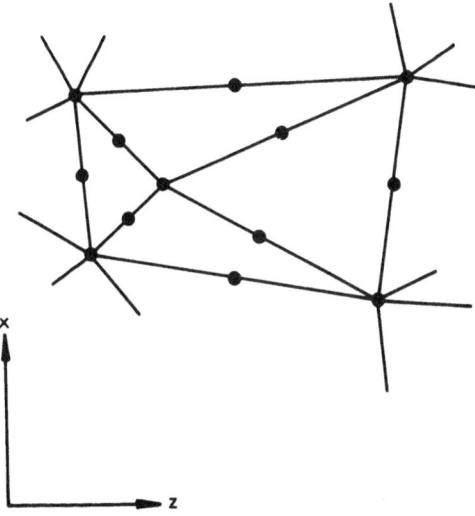

Fig. 5.14.   Finite-element configuration ($\bullet$ represents a node).

equations for pressure which are solved by an under-relaxation method. The finite-difference part of the scheme uses an implicit form with (conservative) central differencing of the conduction term and lagging of the convection and dissipation terms (which are, of course, obtained using a finite-element form). The resulting set of simultaneous linear algebraic equations for temperature can be written as a matrix equation in which the matrix is tridiagonal so that, for example, the Thomas algorithm can be used. The melt front is advanced using a predictor–corrector method. The most recent method of Wang *et al.* (1978) (see also Hieber and Shen, 1980) replaces pressure p as a dependent variable by an average streamfunction ψ (temperature T is retained as the other main dependent variable)

$$\left.\begin{aligned} \bar{v}_x &= \frac{\partial \psi}{\partial z} \\[2mm] \bar{v}_z &= -\frac{\partial \psi}{\partial x} \end{aligned}\right\} \tag{5.75}$$

whence

$$\frac{\partial}{\partial x}\left(\frac{1}{\sigma}\frac{\partial \psi}{\partial x}\right) + \frac{\partial}{\partial z}\left(\frac{1}{\sigma}\frac{\partial \psi}{\partial z}\right) = 0 \tag{5.76}$$

Implementation of the ψ–T method (using a hybrid finite-difference/finite-element scheme) is basically the same as that of the p–T method.

5.4.2 Packing and holding
By comparison with simulation of the injection stage of the moulding cycle, simulation of the packing and holding stages is relatively straightforward.

5.4.2.1 Packing
Kamal and Kenig (1972a, b) have devised a scheme for simulation of packing in a disc of constant height H ($dH/dr = 0$) (see Figs 5.3(c) and 5.4(c)) by solution of the (transient conduction) energy conservation equation

$$\rho\gamma \frac{\partial T}{\partial t} = \alpha \frac{\partial^2 T}{\partial z^2} \tag{5.77}$$

subject to the boundary conditions

$$\left.\begin{aligned} \frac{\partial T}{\partial z} &= 0 && \text{at } z = 0 \\[2mm] -\alpha \frac{\partial T}{\partial z} &= h(T - T_{\text{spec}}) && \text{at } z = \frac{H}{2} \end{aligned}\right\} \tag{5.78}$$

(symmetry means that attention can be confined to $0 \leq z \leq H/2$). Note that the initial condition at the start of the packing stage is provided by the temperature field at the end of the injection stage. By analogy with the injection stage, these equations are solved for temperature by a two-level Saul'yev (finite-difference) scheme (giving eqns (5.51) and (5.52) without the so-called other terms). The mean polymer temperature is then calculated; this enables the rather small flowrate $\dot{V}$ of extra molten polymer from the reservoir to be determined.

5.4.2.2 Cooling (freezing)

The main problem associated with simulation of freezing is determination of the frozen layer thickness δ (see Fig. 5.6). Kamal and Kenig (1972a, b) have devised a scheme for simulation of freezing in a disc of constant height H ($dH/dr = 0$) (see Figs 5.3(c) and 5.4(c)) by solution of the energy conservation equations in the molten polymer (whether moving, during injection, or (at least effectively) stationary, during packing and holding) and in the frozen polymer, subject to the boundary conditions

$$
\left.\begin{aligned}
&\left.\frac{\partial T}{\partial z}\right|_{\text{molten polymer}} = 0 && \text{at } z = 0 \\[2mm]
&\left.\begin{aligned}
& T|_{\text{molten polymer}} = T|_{\text{frozen polymer}} = T_{\text{melt}} \\[2mm]
& \left.\alpha \frac{\partial T}{\partial z}\right|_{\text{molten polymer}} = \left.\hat{\alpha}\frac{\partial T}{\partial z}\right|_{\text{frozen polymer}} + \hat{\rho}\hat{\lambda}\frac{\partial \delta}{\partial t}
\end{aligned}\right\} && \text{at } z = \frac{H}{2} - \delta \\[4mm]
&\left.-\hat{\alpha}\frac{\partial T}{\partial z}\right|_{\text{frozen polymer}} = h(T|_{\text{frozen polymer}} - T_{\text{spec}}) && \text{at } z = \frac{H}{2}
\end{aligned}\right\} \quad (5.79)
$$

From a computational point of view, the presence of the frozen layer does not introduce great problems: a two-level Saul'yev (finite-difference) scheme is used to solve the energy conservation equations both in the molten and in the frozen polymer. The only remaining point to be considered concerns motion of the interface, and hence updating of δ: a finite-difference analogue of the equation involving $\partial\delta/\partial t$ can be devised quite readily.

A rather different method for simulation of freezing has been devised by Wang et al. (1978): the frozen polymer is modelled as a fluid with the same density, specific heat and thermal conductivity as the molten polymer, but with infinite viscosity (so that it does not move):

$$
\eta = \begin{cases} \eta(\dot{\gamma}, T) & \text{if } T \geq T_{\text{melt}} \\ \infty & \text{if } T < T_{\text{melt}} \end{cases} \quad (5.80)
$$

(Note that latent heat effects are ignored.) Simulation of freezing during injection is then quite straightforward (it is, of course, trivial during packing and holding) and requires no modifications other than incorporation of eqn (5.80) into the computational scheme.

5.4.2.3 Heating (cross-linking)

The main problem associated with simulation of cross-linking is tracking material points in the polymer. Byam *et al.* (1979) (see also Byam and Colbert, 1980) have devised a scheme for simulation of cross-linking, but provide no details of it: presumably, an explicit finite-difference marching scheme is used to solve eqns (5.22) and (5.23):

$$\mathbf{C}(z + \Delta z, t + \Delta t) = \mathbf{C}(z, t) + \Delta t k_c (1 - \mathbf{C}(z, t))^{n_c} \exp\left(-E_c/RT(z, t)\right)$$
(5.81)

$$\mathbf{S}(z + \Delta z, t + \Delta t) = \mathbf{S}(z, t) + \Delta t k_s (1 - \mathbf{S}(z, t))^{n_s} \exp\left(-E_s/RT(z, t)\right)$$
(5.82)

where

$$\Delta z = \frac{4\dot{V}}{\pi D^2} \Delta t$$
(5.83)

for cross-linking in a pipe of constant diameter D ($\mathrm{d}D/\mathrm{d}z = 0$) (see Figs 5.3(a) and 5.4(a)). Initial conditions on $\mathbf{C}$ and $\mathbf{S}$ are provided by the degrees of cure and scorch, respectively, which are assumed to have occurred in the reservoir (based on a mean residence time therein) prior to the start of the moulding cycle.

5.5   CONCLUSION

There are obvious extensions of the schemes discussed here for the simulation of injection moulding to schemes for the simulation of compression and transfer moulding and also, it might be thought, RIM (reaction injection moulding). In the cases of compression and transfer moulding (which are described in, for example, Tadmor and Gogos, 1979), such extensions undoubtedly exist. Because compression and transfer moulding are in a sense simpler versions of injection moulding, much of the modelling discussed in Section 5.3 and of the simulation discussed in Section 5.4 is immediately applicable (if not capable of still further simplification). In the case of RIM (which is described in, for example,

Tadmor and Gogos, 1979), however, matters are not nearly so straight-forward. While a start has been made on RIM simulation schemes (see Domine and Gogos, 1980), RIM is sufficiently different from injection moulding for significantly different simulation schemes to be necessary. It is not the presence of the reactions themselves that causes the differences; the reactions can be modelled (see, for example, Lee and Macosko, 1980) and simulated in a similar way to the cross-linking reactions that occur in injection moulding of thermosets and rubbers. It is rather the fact that RIM involves high Reynolds number (perhaps turbulent) flows of low viscosity unreacted or partially reacted fluids, whereas injection moulding involves low Reynolds number (certainly laminar) flows of high viscosity molten polymers, that means that the processes are different. This inevitably means that simulation schemes for injection moulding are not readily extendable for RIM.

REFERENCES

Ames, W. F. (1969). *Numerical Methods for Partial Differential Equations*, Nelson, London.
Bangert, H., Döring, E., Lichius, U. and Kemper, W. (1980). 10th IKV Kunststofftechnisches Kolloquium held at Aachen, Paper 7.
Berger, J. L. and Gogos, C. G. (1973). *Polym. Eng. Sci.*, 13(2), 102.
Byam, J. D. and Colbert, G. P. (1980). *Plastics and Rubber: Processing*, 5(3/4), 95.
Byam, J. D., Colbert, G. P. and Ziegel, K. D. (1979). Contribution 365 presented at the Society of Rheology meeting held at Boston.
Domine, J. D. and Gogos, C. G. (1980). *Polym. Eng. Sci.*, 20(13), 847.
Fenner, R. T. (1979). *Principles of Polymer Processing*, Macmillan, London.
Hieber, C. A. and Shen, S. F. (1980). *J. Non-Newtonian Fluid Mechanics*, 7(1), 1.
Isayev, A. I. and Hieber, C. A. (1980). *Rheol. Acta*, 19, 168.
Kamal, M. R. and Kenig, S. (1972a). *Polym. Eng. Sci.*, 12(4), 294.
Kamal, M. R. and Kenig, S. (1972b). *Polym. Eng. Sci.*, 12(4), 302.
Lee, L. J. and Macosko, C. W. (1980). *Int. J. Heat and Mass Transfer*, 23, 1479.
Martin, B. (1967). *Int. J. Non-Linear Mech.*, 2, 285.
Middleman, S. (1977). *Fundamentals of Polymer Processing*, McGraw-Hill, New York.
Mussatti, F. G. and Macosko, C. W. (1973). *Polym. Eng. Sci.*, 13(3), 236.
Ockendon, H. (1979). *J. Fluid Mech.*, 93(4), 737.
Ockendon, H. and Ockendon, J. R. (1977). *J. Fluid Mech.*, 83(1), 177.
Parnaby, J., Battye, P. G., Hassan, G. A. and Hadwell, C. P. (1978). *Plastics and Rubber: Processing*, 3(3), 89.
Pearson, J. R. A. (1966). *Mechanical Principles of Polymer Melt Processing*, Pergamon, Oxford.
Pearson, J. R. A. (1977). *J. Fluid Mech.*, 83(1), 191.

178   S. M. RICHARDSON

Pearson, J. R. A. (1978). *Polym. Eng. Sci.*, **18**(3), 222.
Richardson, S. M. (1979). *Int. J. Heat and Mass Transfer*, **22**, 1417.
Richardson, S. M., Pearson, H. J. and Pearson, J. R. A. (1980). *Plastics and Rubber: Processing*, **5**(2), 55.
Roache, P. J. (1972). *Computational Fluid Dynamics*, Hermosa, Albuquerque.
Rubin, I. I. (1972). *Injection Molding: Theory and Practice*, Wiley, New York.
Tadmor, Z. and Gogos, C. G. (1979). *Principles of Polymer Processing*, Wiley, New York.
Toor, H. L. (1957). *Trans. Soc. Rheol.*, **1**, 177.
Toor, H. L. (1958). *AIChEJ*, **4**, 319.
Wang, K. K., Shen, S. F., Stevenson, J. F., Hieber, C. A. and Chung, S. (1975a). Computer-aided injection molding system, Progress report 1, Cornell University.
Wang, K. K., Shen, S. F., Stevenson, J. F., Hieber, C. A., Chung, S. and Galskoy, A. (1975b). Computer-aided injection molding system, Progress report 2, Cornell University.
Wang, K. K., Shen, S. F., Stevenson, J. F. and Hieber, C. A. (1976). Computer-aided injection molding system, Progress report 3, Cornell University.
Wang, K. K., Shen, S. F., Stevenson, J. F. and Hieber, C. A. (1977). Computer-aided injection molding system, Progress report 4, Cornell University.
Wang, K. K., Shen, S. F., Cohen, C., Hieber, C. A. and Jahanmir, S. (1978). Computer-aided injection molding system, Progress report 5, Cornell University.
Wang, K. K., Shen, S. F., Cohen, C., Hieber, C. A., Isayev, A. I., Jahanmir, S. and Tayler, A. (1979). Computer-aided injection molding system, Progress report 6, Cornell University.
White, J. L. and Dee, H. B. (1974). *Polym. Eng. Sci.*, **14**(3), 212.
Williams, G. and Lord, H. A. (1975). *Polym. Eng. Sci.*, **15**(8), 553.
Wu, P.-C., Huang, C. F. and Gogos, C. G. (1974). *Polym. Eng. Sci.*, **14**(3), 223.

CHAPTER 6

Fibre Spinning

M. M. DENN

Department of Chemical Engineering,
University of California, Berkeley, USA

6.1 THE PROCESS

Fibre spinning is a process in which an extruded liquid polymer filament is continuously drawn and simultaneously solidified to form a continuous synthetic fibre. There are three fundamental processes for the manufacture of synthetic fibres: melt spinning, wet spinning, and dry spinning. Melt spinning is restricted to polymers that are thermally stable above the melting temperature, and solidification is effected by cooling to below the glass transition temperature during drawing. In wet and dry spinning the liquid filament is extruded as a solution, and the solvent is removed and solidification effected in a coagulation bath or by evaporation to a hot gas, respectively. Polymer fibre drawing processes are discussed in detail in the monograph by Ziabicki (1976), and recent literature on the mechanics is reviewed by Denn (1980a) and White (1982).

The discussion in this chapter is limited to melt spinning, which is conceptually the simplest of the three processes. The melt spinning process is shown schematically in Fig. 6.1. Molten polymer is extruded through a *spinneret* hole into quench air which is below the solidification temperature. There may be some extrudate swell. The solidified filament is wound around a pair of *Godet* rolls which operate at a linear speed that is greater than the extrusion velocity, resulting in a net area reduction relative to the spinneret; the area reduction ratio is known as the *draw ratio, Dr*. Most of the draw is presumed to take place in the melt region prior to solidification. A liquid finish is usually applied to the solidified filament prior to the

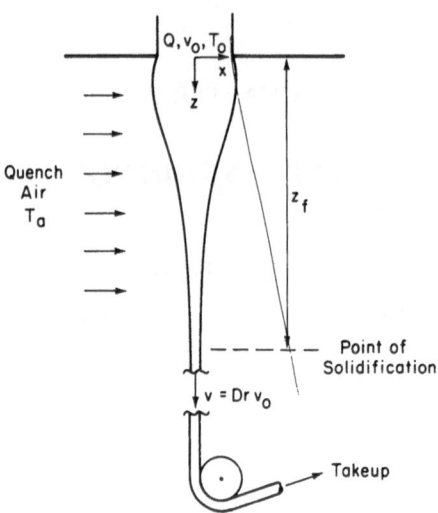

Fig. 6.1.   Schematic of the melt spinning process.

Godets, and the filament usually undergoes further solid-state processing for property development. Diameter and structural uniformity are the important factors for product quality. Structure determines final product properties such as tenacity, modulus, and dye uptake. The physical properties generally correlate with filament optical birefringence, which correlates in turn with the stress at solidification for polymers with low crystallinity like polyethylene terephthalate (PET).

Laboratory spinning facilities will typically spin a single filament, as shown in the schematic, but in pilot plant and commercial spinning a single spinneret will usually contain many holes. The polymer is metered through a melt pump to a spinning pack, which consists of a filter bed and a small flow readjustment region, followed by the spinneret plate. The individual holes, which may number several hundred, will typically be fabricated with tapered entries. There may be hole-to-hole variations in flow because of flow variations in the spinning pack and because of hole erosion. The extruded filaments are taken up together as a *yarn*. The quench air is blown across the yarn, causing filaments in different locations in the yarn bundle to undergo different quench temperatures and velocity fields, resulting in possible property variations across the yarn.

Typical commercial processing variables for the manufacture of PET fibres are shown in Table 6.1. Single-filament laboratory experiments rarely achieve take-up velocities approaching 1000 m/min or draw ratios

TABLE 6.1

TYPICAL PROCESSING VARIABLES FOR THE MANUFACTURE
OF PET FIBRES

Processing variable	Typical value
Extrusion temperature	295 °C
Solidification temperature	70 °C
Quench air temperature	30 °C
Spinneret hole diameter	0·25 mm
Take-up velocity	1 000–3 500 m/min
Air cross-flow velocity	0·2 m/s
Draw ratio	150

approaching 150. Furthermore, laboratory experiments are often carried out for research purposes under isothermal conditions, in which the liquid filament is drawn in an oven that is maintained at the extrusion temperature, followed by rapid solidification at a pre-determined point by passage into a cold water bath.† Air drag and inertia became important in determining deformation mechanics only at about 750 m/min take-up speed. Substantial crystallisation occurs for PET resins at take-up speeds in excess of about 5000 m/min, resulting in fibres with different properties. This stress-induced crystallisation probably results in solidification at temperatures above 70 °C, which is the approximate glass transition temperature.

Radial property variations within each filament in the yarn, which might manifest themselves as a radial birefringence gradient, are observed under some high speed spinning conditions and may be important in determining overall properties. These radial variations are probably caused by radial temperature gradients.

Several types of transients are of concern. The most important from a processing point of view is breakage of an individual filament, which can shut down a spinning position. Breakage is not well-understood, but is certainly related to excessive stress, perhaps caused by a local variation in temperature or a material property, or by the presence of an impurity. Amplification of external disturbances in the feed and quench air are likely to occur in some frequency ranges, causing continuous variations that are

† The term 'isothermal' has sometimes been misused in the literature, and applied to experiments in which the filament is only partially drawn in a chamber at extrusion temperature, with the remainder of the draw carried out in ambient air. Isothermal spinning theories cannot, and should not, be applied to such experiments.

observed as apparently random fluctuations in filament diameter. Finally, a self-sustained periodic oscillation known as *draw resonance* is observed in some laboratory experiments at low draw ratios, usually when the solidification point is fixed by a water bath. Draw resonance is not generally believed to occur under commercial spinning conditions, but it does occur in the dynamically similar process of extrusion coating. The random fluctuations resulting from amplification of external disturbances are distinct from the self-sustained oscillations of draw resonance, although the two have been confused in the literature by some authors.

There are hundreds of spinning positions in a modern textile fibre plant, so on-line feedback control of each spinning position is impractical. Thus, there is considerable incentive to use steady-state and dynamic simulation as a means of identifying sensitive process variables and operating conditions. The development of such simulation programs is the major thrust of this chapter. It is worth noting that two in-house corporate steady-state simulation programs for PET spinning have been publicly described, by Toyobo (Toyobo Co., Ltd, 1980; Matsuo *et al.*, 1977; Yasuda *et al.*, 1978, 1979) and Celanese (George, 1982), with claims of improved process performance. Both programs take the PET rheology to be Newtonian.

6.2   UNIFORM UNIAXIAL EXTENSION

The kinematics of continuous drawing are close to those of uniform, uniaxial extension, and useful insight into the expected behaviour in spinning is obtained by studying the uniform extension; see reviews by Denn (1977) and Petrie and Dealy (1980).

The flow is shown schematically in Fig. 6.2. The kinematics are

$$v_z = \Gamma_E z \qquad v_x = -1/2\Gamma_E x \qquad v_y = -1/2\Gamma_E y \qquad (6.1)$$

where the stretch rate, Γ_E, may depend on t but is independent of spatial position. The isotropic pressure is independent of position, leading to the result that the tension, T_E, at any time is equal to the primary normal stress difference, $t_{zz}^E - t_{xx}^E = t_{zz}^E - t_{yy}^E$. For a Newtonian fluid this leads to the result first obtained by Trouton,

$$T_E = 3\eta\Gamma_E \qquad (6.2)$$

where η is the shear viscosity. For an upper-convected Maxwell model with

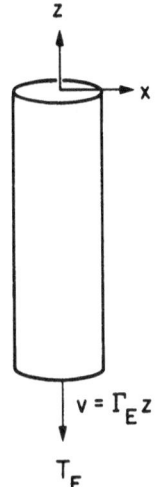

Fig. 6.2. Uniform, uniaxial extension.

relaxation time λ, starting from an unstressed state at $t = 0$, the corresponding result is

$$T_E = \eta \left\{ \frac{2}{1 - 2\lambda\Gamma_E} [1 - \exp(-[1 - 2\lambda\Gamma_E]t/\lambda)] \right.$$

$$\left. + \frac{1}{1 + \lambda\Gamma_E} [1 - \exp(-[1 + \lambda\Gamma_E]t/\lambda)] \right\} \Gamma_E \qquad (6.3)$$

This introduces a time scale of order λ, suggesting possible numerical stiffness when process time scales are large relative to λ. Furthermore, unbounded stresses will be approached asymptotically when $\lambda\Gamma_E > 1/2$. Finally, except for $\lambda\Gamma_E \ll 1$, the first of the two terms in the summation will dominate; this corresponds to $|t_{zz}^E| \gg |t_{xx}^E| = |t_{yy}^E|$.

If the fluid is characterised by a distribution of relaxation times and viscosities: $\{\lambda_i\}$, $\{\eta_i\}$, $i = 1, 2, \ldots N$, each element of which is Maxwellian, then the response is made up of a sum of terms, each of the form of the right-hand side of eqn. (6.3). Numerical stiffness is now ensured, since the breadth of the spectrum will comprise at least one decade in λ. The growth rate to an ultimate steady state or to an unbounded stress, and the magnitude of any steady state, is determined by the *maximum* relaxation time; this is in contrast to the more familiar behaviour in steady shear, where the mean relaxation time is the dominant quantity. Two relaxation

times are generally adequate to represent the qualitative uniaxial extensional response. Non-linear constitutive theories like those of Acierno and
coworkers (1976), Phan-Thien and Tanner (1977) and Phan-Thien (1978)
show a similar response up to the order of the dominant relaxation time,
but they contain terms that prevent the stress from becoming unbounded.
This can be important in the simulation of spinning.

6.3   SPINLINE MODEL

6.3.1 Basic equations

The mechanics of fibre spinning are always described by 'thin filament'
equations; the principal assumption in these equations is a small rate of
change of filament diameter with distance from the spinneret, $|dR/dz| \rightarrow 0$,
and they are clearly inapplicable at the spinneret, where curvature may be
substantial. The derivation is delicate, and the full range of assumptions
has never been carefully catalogued. The equations can be obtained by
averaging the field equations over a filament cross-section, or by expansion
in a small parameter. The equations obtained are those that follow from
macroscopic balances in which it is assumed that axial velocity and
temperature are uniform over a cross-section, and the flow is locally a
uniform, uniaxial extension, as follows:

   Mass:

$$\frac{\partial \rho A}{\partial t} + \frac{\partial}{\partial z}(\rho A v) = 0 \qquad (6.4)$$

Momentum:

$$\rho A \frac{\partial v}{\partial t} + \rho A v \frac{\partial v}{\partial z} = \frac{\partial}{\partial z}(A T_E) - (\pi A)^{1/2} \rho_a v^2 C_f + \rho A g + \frac{\partial}{\partial z}\left(\frac{A^{1/2}\sigma}{\pi^{1/2}}\right) \quad (6.5)$$

Energy:

$$\rho \gamma \frac{\partial T}{\partial t} + \rho \gamma v \frac{\partial T}{\partial z} = -\left(\frac{4\pi}{A}\right)^{1/2} h(T - T_a) - \left(\frac{4\pi}{A}\right)^{1/2} es(T^4 - T_a^4) + T_E \frac{\partial v}{\partial z}$$

$$(6.6)$$

$T_E = t_{zz}^E - t_{xx}^E = t_{zz}^E - t_{yy}^E$ is the cross-sectional averaged axial continuum
stress, and is dependent on fluid rheology. T_a is the ambient temperature.

   These equations contain two macroscopic transport coefficients, the
drag coefficient, C_f, and the convective heat transfer coefficient, h. There is

a question of consistency in neglecting shear stresses in the filament (explicit or implicit in all derivations) and including the air drag. The surface tension term ($A^{1/2}\sigma$) in the momentum equation is always neglected in polymer melt spinning calculations, as is the viscous dissipation term ($T_E(\partial v/\partial z)$)) in the energy equation. The radiation term ($es(T^4 - T_a^4)$)) is also normally neglected in the energy equation in polymer fibre spinning simulations, although it is important in glass spinning. A radiation contribution can be easily included in the convective heat transfer coefficient.

6.3.2 Transport coefficients

The phenomenological transport coefficients are difficult to measure under actual spinning conditions, and indirect experiments presumed to be equivalent have often been used. Correlation with the air Reynolds number, $\rho_a vd/\eta_a$, is nearly always assumed. Drag coefficient data correlate with

$$C_f = \beta(\rho_a vd/\eta_a)^{-0.61} \tag{6.7}$$

The midrange of reported data corresponds to a value of $\beta = 0.6$, but the best value is probably closer to 0.37 and is perhaps as low as 0.27 (Matsui, 1976; Gould and Smith, 1980).

Natural convection is not expected to be important at commercial spinning speeds, in which case the Nussult number should be a function of the Reynolds number only. Correlations for heat transfer into quiescent air are of the form

$$\frac{hd}{\alpha_a} = \delta(\rho_a vd/\eta_a)^a \tag{6.8}$$

where a ranges between 0.2 and 0.5. α_a is the thermal conductivity of air. The best published correlation seems to be that of Kase and Matsuo (1967), with $a = 1/3$ and $\delta = 0.42$, but these data were taken for air blown past a stationary cylinder. (Application of the Reynolds or Colburn analogies for heat and momentum transport, using the fact that the Prandtl number is approximately unity for air, causes eqn (6.8) to follow from eqn (6.7) with $a = 0.39$, $\delta = \beta/2$. The exponent is remarkably close, but the coefficient seems low.) Kase and Matsuo have included the effect of quench air flow normal to the fibre axis, and the final form is

$$\frac{hd}{\alpha_a} = 0.42(\rho_a vd/\eta_a)^{1/3}\left[1 + 8\left(\frac{v_x}{v}\right)^2\right]^{1/6} \tag{6.9}$$

6.3.3 Rheology

Most spinline simulations have taken the fluid rheology to be Newtonian or power-law; to within the thin filament approximation this leads to

$$T_E = 3\eta \frac{\partial v}{\partial z} \qquad (6.10)$$

For a power-law fluid, the viscosity–deformation rate equation in extensional flow is

$$\eta = K\left(\sqrt{3}\frac{\partial v}{\partial z}\right)^{\nu - 1} \qquad (6.11)$$

It should be noted that eqn (6.10) is not exact even for a Newtonian fluid, where the stress–deformation rate equation is linear, because of the neglect of terms associated with the changing filament area in the cross-sectional averaging. The consistency index, K, is temperature-dependent; some authors have used the form

$$K = \text{constant} \times \left[\exp\left(\frac{A}{T}\right) + \frac{B}{T - T_f}\right] \qquad (6.12)$$

The second term accounts for hardening near the solidification temperature and causes dv/dz to go to zero; it is otherwise unimportant.

Viscoelastic simulations have mostly been carried out for the Maxwell fluid and the Phan-Thien/Tanner generalisation; the latter has the following form in the thin filament approximation:

$$t_{zz}^E = \sum_{i=1}^{N} t_{zz(i)}^E \qquad t_{xx}^E = \sum_{i=1}^{N} t_{xx(i)}^E \qquad (6.13)$$

$$\Phi_i t_{zz(i)}^E + \lambda_i \left[\frac{\partial t_{zz(i)}^E}{\partial t} + v\frac{\partial t_{zz(i)}^E}{\partial z} - 2(1 - \chi)\frac{\partial v}{\partial z} t_{zz(i)}^E\right] = 2G_i\lambda_i\frac{\partial v}{\partial z} \qquad (6.14a)$$

$$\Phi_i t_{xx(i)}^E + \lambda_i \left[\frac{\partial t_{xx(i)}^E}{\partial t} + v\frac{\partial t_{xx(i)}^E}{\partial z} + (1 - \chi)\frac{\partial v}{\partial z} t_{xx(i)}^E\right] = -G_i\lambda_i\frac{\partial v}{\partial z} \qquad (6.14b)$$

$$\Phi_i = \exp\left[\varepsilon(t_{zz(i)}^E + 2t_{xx(i)}^E)/G_i\right] \qquad (6.15)$$

$\{\lambda_i\}$ and $\{G_i\}$ will be temperature-dependent. The only simulations including temperature change with this type of constitutive equation have been for PET, for which the temperature dependence of the relaxation time is the same as that of the viscosity, and the moduli are taken as constant. χ is a constant determined by viscous shear thinning, and ε is a constant associated with stress saturation in extension. There should be terms of the form $\lambda_i \mathbf{t}_{(i)}^E D \ln T/Dt$ in eqn (6.14); these were included by Fisher and Denn (1977), who found them to be negligible, and neglected in the simulations of Gagon and Denn (1981), discussed subsequently. Gupta and Metzner (1982) believe such terms to be important.

Some low speed, isothermal calculations have been carried out for special cases of the Kaye–BKZ model by Malkus (1981a, b). These were initially undertaken for the purpose of exploring the applicability of finite-element methods to memory integral constitutive equations, with the spinning equations representing a useful one dimensional example. This study is still underway at the time of writing, and the results are too preliminary for inclusion. The general memory integral formulation leads to problems in the specification of the flow upstream of the spinneret, as discussed subsequently.

6.3.4 Internal variables
A complete spinline model might contain several internal variables that are not accounted for in the conservation and constitutive equations outlined above. The most obvious is the degree of crystallinity, and the resulting effect of crystalline structure on the mechanics. Crystallisation can be included with an additional scalar differential equation if the kinetics are known. The enthalpy of crystallisation will then appear in the temperature form of the energy equation, and the expressions for stress will need to account for both crystalline and amorphous phases. Nakamura and coworkers (1972, 1974) have formulated the energy equation to include crystallisation kinetics, and they have obtained some solutions for slow speed spinning of a Newtonian fluid; however, they have not incorporated the presence of the crystalline phase into the stress. Crystallisation has not generally been included in simulations; the reason is probably that the focus has been on PET, and the degree of crystallinity is too small at normal PET spinning speeds, while the kinetics are unknown (but extremely fast) at high speeds.

Some constitutive equations, most notably that of Acierno and co-workers (1976) (see also Mewis and Denn, 1982), include a dynamically evolving scalar structural parameter that is related to the entanglement density and determines the instantaneous values of the rheological parameters. The only change in the formulation outlined above is the addition of one scalar partial differential equation that is coupled to the stress equation. Mewis and De Cleyn (1982) have analysed isothermal laboratory spinning data of aqueous solutions using such an equation, but no simulations with this formulation have been reported.

6.3.5 Boundary conditions

The downstream boundary conditions for the thin filament equations are the more easily formulated. The denier (g/9000 m) per filament will normally be specified, fixing the take-up area. For a fixed throughput rate, this is equivalent to fixing the take-up velocity; the velocity can be fixed mechanically, while the area cannot, so the take-up velocity is normally specified. It is generally assumed that the solidified filament is inextensible, in which case the take-up velocity is specified at the point of solidification, $z = z_f$. Solidification is always assumed to be instantaneous when the filament reaches the solidification temperature, T_f. Thus, at $z = z_f$,

$$v = \text{take-up velocity} \tag{6.16a}$$
$$T = \text{solidification temperature} \tag{6.16b}$$

It should be noted that z_f will not normally be specified *a priori*, and is in fact determined by eqns (6.16). During a transient, z_f will normally be a function of time, since T_f will not change. In some experimental situations, and in the mechanically similar process of extrusion coating, the length is fixed, either because solidification is imposed by a rapid, localized chill, or because the filament is taken up on the roll as a liquid; if z_f is specified, then the temperature boundary condition is not used.

Fluid emerging from the spinneret is in the midst of a transition from the shear flow in the spinneret to the essentially elongational flow far downstream. There may be extrudate swell. A small number of finite-element solutions of the complete momentum equation, discussed below, have been carried out to study the approach to the region of validity of the thin filament equations, and to define appropriate initial conditions. Consistency with the thin filament equations requires that the origin of the axial coordinate be taken at a point where conditions are radially uniform; the finite-element calculations indicate that this point is one-to-two diameters downstream of the spinneret. Nevertheless, the initial temperature is always taken to be the extrusion temperature. This discrepancy should not be important in a typical long spinline, but it could be significant in laboratory studies of very short spinlines. The flow rate will generally be given, but the initial area is unknown because of the extrudate swell and uncertainty as to the precise location of the coordinate origin. It is customary to specify the velocity and area at $z = 0$; some authors use the spinneret area, while others use correlations to estimate the amount of extrudate swell. Thus, at $z = 0$,

$$A = \text{given (usually spinneret area)} \qquad (6.17a)$$
$$v = \text{given (usually mean spinneret velocity)} \qquad (6.17b)$$
$$T = \text{extrusion temperature} \qquad (6.17c)$$

The momentum equation can sometimes be expressed in terms of the force on the filament at $z = 0$. The force is a constant of integration; if it is specified, then one of the other boundary conditions must be relaxed. For a fixed flow rate, the take-up velocity and the force cannot both be specified for physical reasons, and this observation forms the basis for iterative schemes; specifying the force converts the boundary value problem to an initial value problem, and the force is then adjusted until the take-up velocity condition is satisfied.

The boundary conditions in eqns (6.16) and (6.17) are sufficient for an inelastic liquid, eqn (6.10) or (6.11). Additional information is required for a viscoelastic liquid. For a rate-type fluid with a single relaxation time (eqns (6.13) and (6.14) with $N = 1$) the additional condition that is required is the ratio t_{xx}^E/t_{zz}^E at $z = 0$. This ratio must be -0.50 over the entire filament for an inelastic fluid; it can be shown to approach zero rapidly away from the spinneret for a Maxwell fluid, and the ratio is also zero in fully developed spinneret flow. Thus, the initial value of t_{xx}^E on the spinline has usually been taken as zero:

$$\text{at } z = 0, \qquad t_{xx}^E = 0 \qquad (6.18)$$

For an isothermal Maxwellian spinline the initial stress ratio will depend on the value of T_E/G at $z = 0$ (which, of course, depends on the solution); the finite-element calculations show that t_{xx}^E/t_{zz}^E at the point where the stresses become uniform radially goes monotonically from -0.50 to 0 as T_E/G increases. The solution to the thin filament equations is insensitive to the initial stress ratio as long as it is in the range from -0.50 to 0 (Denn et al., 1975), so the boundary condition, eqn (6.18), may be used with confidence.

For $N > 1$, it is necessary to specify the initial values of both the ratios $t_{xx(i)}^E/t_{zz(i)}^E$, which may be taken equal to zero, and the ratios $t_{zz(i)}^E/t_{zz(N)}^E$. The sensitivity to the latter has not been explored. The axial stress ratio in fully developed spinneret flow is $\lambda_i^2 G_i/\lambda_N^2 G_N$. $\lambda_i G_i$ is constant for a *wedge* spectrum, and the stress ratio is simply λ_i/λ_N; for a *box* spectrum, G_i is constant, and the stress ratio is λ_i^2/λ_N^2. The stresses corresponding to the shorter relaxation times should decay more rapidly in the rearrangement region, so these ratios probably represent an upper bound, with a lower bound of zero. In all cases, the stresses corresponding to the longest relaxation times dominate. Thus, we have at $z = 0$,

$$t_{zz(i)}^E / t_{zz(N)}^E = \text{given } \left(\text{probably between } 0 \text{ and } \frac{\lambda_i^2 G_i}{\lambda_N^2 G_N} \right) \qquad (6.19)$$

Boundary conditions at $z = 0$ for the thin filament equations present a somewhat different computational and theoretical problem for memory integral constitutive equations that are not equivalent to a rate equation. The constitutive equation requires integration over the entire history of deformation, but the kinematical assumptions leading to the thin filament equations are not valid for that portion of the history prior to the spinneret exit. Thus, an artificial velocity function must be specified (Malkus, 1981a, b). The consequences of the use of incorrect kinematics upstream of the spinneret have not been studied. (This problem does not arise for integral equations that are equivalent to rate equations. The linearity in strain history inherent in such equations enables a factorisation that makes specification of the stress boundary conditions described above equivalent to specification of the deformation history.)

If the spinline description contains an internal variable, such as entanglement density or degree of crystallinity, then a value for this variable must also be specified at $z = 0$. There is no reported experience regarding degree of crystallinity. Mewis and De Cleyn (1982), having considered results from low speed experiments using various spinneret geometries, suggest calculating the structural parameter by assuming converging flow in the spinneret and then averaging values across the exit plane.

6.4 ASYMPTOTIC SOLUTIONS

Asymptotic solutions to the thin filament equations are available, and they are useful both as a consistency check on numerical solutions and as a means of determining possible parameter ranges of numerical sensitivity; indeed, there are parameter values for which a solution to the boundary value problem posed here does not exist.

6.4.1 Isothermal, low speed, steady spinning
When the filament temperature is constant and z_f is fixed, many solutions exist. For conditions in which inertia, gravity, air drag, and surface tension

can be neglected, the solution for a Newtonian fluid is

$$v(z)/v_0 = \exp(z \ln Dr/z_f) \tag{6.20}$$

$$F = 3\eta A_0 v_0 \ln Dr/z_f \tag{6.21}$$

Here, the subscript '0' refers to the value specified at $z = 0$. The *draw ratio*, Dr, is defined

$$Dr \equiv v(z_f)/v_0 \tag{6.22}$$

This is equivalent to the area reduction ratio at steady state as long as the liquid density is constant. The definition in terms of velocities is preferable, since initial and final velocity may be presumed to be constant even under transient conditions, when the take-up area is known to be varying. Corresponding solutions exist for a power-law fluid, and solutions are available for the Newtonian fluid with any one of inertia, gravity, or surface tension included.

Perturbation solutions exist for the Maxwell fluid in the absence of gravity, inertia, air drag, and surface tension for small and large values of the ratio F/A_0G (Denn *et al.*, 1975; Petrie, 1979). The former, a singular perturbation, provides a correction to the Newtonian solution. The latter, to first order in A_0G/F, is

$$\frac{v(z)}{v_0} = 1 + \frac{z}{\lambda v_0} + \frac{3}{2}\left(\frac{GA_0}{F}\right)\left\{\ln\left(1 + \frac{z}{\lambda v_0}\right) + 1/2\left[1 - \frac{1}{(1 + z/\lambda v_0)^2}\right]\right\} + \cdots \tag{6.23}$$

$$\frac{F}{GA_0} \cong \frac{3\lambda v_0}{4}\left[\frac{2\ln Dr + 1 - Dr^{-2}}{z_f - \lambda v_0(Dr - 1)}\right] \tag{6.24}$$

Note that the force becomes infinite at a finite value of the draw ratio, equal to $1 + z_f/\lambda v_0$, and that the velocity profile becomes linear in this limit. Solutions do not exist for larger draw ratios. This non-existence of a solution does not occur when inertia is included, nor does it occur for the Phan-Thien/Tanner fluid even in the absence of inertia.

6.4.2 Temperature profile

The steady-state energy equation, eqn (6.6), neglecting dissipation and radiation, with eqn (6.8) for spinning into quiescent air, simplifies to the following form:

$$\frac{dT}{dz} = \left[\delta\left(\frac{\alpha_a}{\rho\gamma}\right)\left(\frac{2\rho_a}{\eta_a}\right)^a\left(\frac{Q}{\pi}\right)^{(a/2)-1}\right]v^{a/2}(T - T_a) \tag{6.25}$$

The dependence on the kinematics is extremely weak; for $a = 1/3$, as in the Kase–Matsuo correlation, the dependence is only $v^{1/6}$, which will vary over the spinline length by a factor of no more than two or three. We may therefore solve eqn (6.25) with $v^{a/2}$ as a constant to obtain an estimate of the length to solidification:

$$1 \geq z_{\mathrm{f}} \frac{\left[\delta\left(\dfrac{\alpha_{\mathrm{a}}}{\rho\gamma}\right)\left(\dfrac{2\rho_{\mathrm{a}}}{\eta_{\mathrm{a}}}\right)^{a}\left(\dfrac{Q}{\pi}\right)^{(a/2)-1} v_{0}^{a/2}\right]}{\ln\left(\dfrac{T_{\mathrm{f}} - T_{\mathrm{a}}}{T_{0} - T_{\mathrm{a}}}\right)} \geq \frac{1}{Dr^{a/2}} \qquad (6.26)$$

Equation (6.26) suggests that, for a fixed spinneret diameter, the distance to solidification should approximate the throughput rate to the two-thirds power, and be only weakly dependent on the take-up velocity. Furthermore, the temperature profile and location of the solidification point should be insensitive to rheological constitutive assumptions.

6.5   FINITE-ELEMENT SOLUTIONS

Finite-element solutions for the full two-dimensional velocity and stress fields have been obtained for isothermal spinning in the absence of inertia (creeping flow) for the Newtonian fluid by Fisher *et al.* (1980), and for the Newtonian and Maxwell fluids by Keunings *et al.* (1982). The former used the program developed by Nickell *et al.* (1974), while the latter used both the MIX 1 and MIX 2 programs described by Crochet and Keunings (1982); MIX 2 is equivalent to the formulation of Nickell *et al.* for Newtonian fluids. The boundary conditions are shown in Fig. 6.3. The stress and velocity are assumed to be fully developed upstream of the spinneret exit. The solution method is identical to that for free jet swell, Chapter 3, except that a finite axial force is imposed at a downstream position.

For the Newtonian fluid, the thin filament equations become valid within one spinneret diameter. The position of maximum extrudate swell moves closer to the spinneret with increasing dimensionless force $4F/3\pi\eta d_{0}v_{0}$, and the magnitude of extrudate swell decreases; velocity profile relaxation and stress development are weakly dependent on the dimensionless stress, however. Use of the spinneret area and mean velocity as initial conditions for the thin filament equations should have little effect except perhaps to shift predictions along the spinline by up to one spinneret diameter. There are two limitations on the calculation for the Maxwell fluid. Smooth solutions cannot be obtained for the free jet flow from a capillary for $\lambda v_0/d_0$ greater than about 1/8, as discussed in Chapter 3. The spinning calculations were carried out with $\lambda v_0/d_0 = 1/8$ for various values of dimensionless force, $4F/3\pi\eta d_0 v_0$; smooth solutions were obtained with MIX 1 only up to a value of about 0.75, and up to about 1.50 with MIX 2, and convergence could not be obtained at all beyond 2.0. The asymptotic solution to the thin filament equations for large forces, eqns (6.23) and (6.24), is governed by the parameter $F/A_0 G$, which (to within a factor of 3) is simply the product of these two dimensionless groups. The current upper bound for convergence, based on limited calculations to date, seems to be $F/A_0 G = 3/4$, and the limit for smooth solutions seems to be $F/A_0 G = 3/8$.

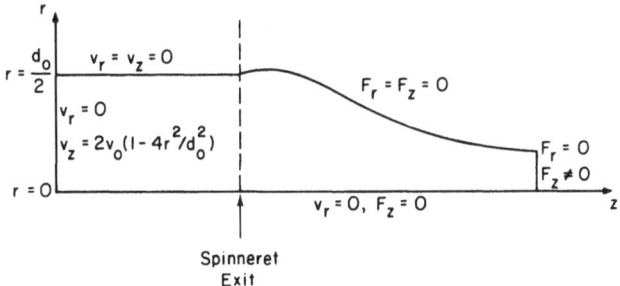

Fig. 6.3.   Boundary conditions for finite-element solution.

The maximum extrudate swell occurs within one-half diameter of the spinneret, and the area increase is less than 20 % for $4F/3\pi\eta v_0 d_0 > 0.50$. The axial velocity becomes uniform to within 5 % at $z/d_0 = 0.9$, which is an only slightly longer rearrangement length than for the Newtonian fluid. The stresses do not become uniform to within 10 % over the cross-section until $z/d_0 = 1.8$, however. The computed ratio of mean radial to axial normal stress $(-t_{xx}^E/t_{zz}^E)$ is plotted versus $F/A_0 G$ for $\lambda v_0/d_0 = 1/8$ in Fig. 6.4. These results indicate that it should generally be adequate to use spinneret area and mean velocity as initial conditions for the thin filament equations, with $t_{xx}^E(0) = 0$, but simulations at small values of $F/A_0 G$ (say, less than 3/8) should be checked for consistency by use of the alternative condition $t_{xx}^E(0)/t_{zz}^E(0) = -0.50$.

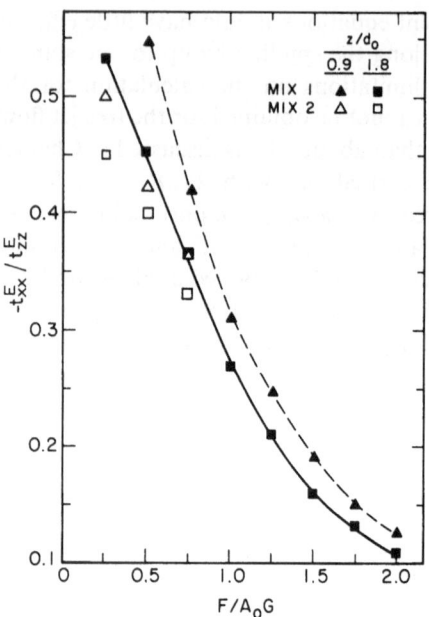

Fig. 6.4.   Ratio of transverse to axial extra-stress at two positions as a function of dimensionless force for a Maxwell fluid (Keunings *et al.*, 1982).

6.6 STEADY-STATE SIMULATION

6.6.1 Measurement

The purpose of a simulation program is to relate filament properties at solidification to polymer properties and processing variables. The program will predict area, velocity, temperature, and stress profiles. Velocity profiles in a single-filament experiment are easily made over most of the filament length, using commercially available laser–Doppler velocimetry, for example; measurement within a few diameters of the spinneret is usually not possible because of geometrical constraints. Area profiles can be measured using still photography, and this is conventionally done for low speed, laboratory studies; resolution of the filament edge is a problem for thin filaments. Contact methods are used to measure the temperature beyond the point where the filament becomes stiff; optical measurements of the mean temperature in the liquid state have been used, but they are very difficult and of questionable reliability. The tension can be measured on the solidified portion using a contact device. Tension measurements for the spinning of solutions, where temperature control is not important, have been made by employing a calibrated displacement transducer on a horizontal capillary just upstream of the point of extrusion. The stress at solidification cannot be measured directly, but it can be inferred (to within a proportionality factor) from the optical birefringence. (The Toyobo simulation program compares birefringence to the stress at 95% of the take-up velocity.)

6.6.2 Computation

Steady-state simulation requires simultaneous solution of eqns (6.5) and (6.6) with $\partial/\partial t$ set to zero. Area or velocity is eliminated, since the steady-state mass balance reduces to $\rho A v = \text{constant}$. The stress T_E must be included, either by direct substitution of eqns (6.10) to (6.12) for an inelastic fluid, or simultaneous, steady-state solution of eqns (6.13) to (6.15) or their equivalent. The computational procedure using the boundary conditions, eqns (6.16) through (6.19), is straightforward. $t_{zz}^E(0)$, or equivalently the force F at $z = 0$, must be assumed. For an inelastic liquid, this is equivalent to assuming the derivative dv/dz at $z = 0$. The differential equations are then integrated until the stopping condition is reached. The stopping condition is normally taken to be the position at which $T = T_f$ (within a preset error); the force is then adjusted, and the calculation repeated, until v is equal to the take-up velocity when the stopping condition is satisfied. The velocity–force function is monotonic, so any elementary iterative scheme is adequate. (George (1982) notes that a bad choice of initial force can cause either the tension or area to become negative before the stopping condition is reached, causing an error stop because the logarithm of a negative number is computed. This can be avoided by monitoring for negative values and making the force adjustment prior to the stopping condition if necessary.) Rapid convergence is obtained using a Newton–Raphson scheme with a finite-difference approximation to the derivative. For situations in which the length is fixed by a chill bath or take-up roll, the computational procedure is unchanged except that the stopping condition is simply $z = z_f$ and the temperature boundary condition is dropped.

There are few practical computational problems in integrating the equations. George (1982) has used a modified Hamming predictor-corrector method from the IBM SSP Library for a simulation treating the fluid as Newtonian. The differential equations become 'stiff' for a viscoelastic liquid when there is a whole distribution of relaxation times, or when the smallest relaxation time becomes small relative to the residence time. Gagon and Denn (1981) found that a fourth-order Runge-Kutta method was adequate for the PET simulation described below, using two relaxation times ($N = 2$) and $\lambda_{max}/\lambda_{min} = 5$, but the program would not run when the minimum relaxation time at spinneret conditions was less than 3×10^{-4} s, and a separate Newtonian code was required. Similar problems can be anticipated with a larger number of terms in the relaxation spectrum and a wider spectrum. An integration scheme for stiff systems, like the Gear algorithm, is preferable for viscoelastic liquids.

Some programs allow position-dependence of process variables that appear as parameters in the system equations. Kase's (1974) MS3 program, for example, which is the forerunner of the Toyobo simulation package, permits the user to specify quench air velocity and temperature at 15 locations, and then interpolates linearly between the specified values. There may be 'communication' problems in transmittal of interpolations of tabulated data to some numerical integration packages, particularly if a variable step size is used, and these should be anticipated and the codes checked carefully.

Non-dimensionalisation of the equations, which is usually desirable, can lead to an unanticipated computational complication. It has become common practice to non-dimensionalise axial position with respect to spinline length, z_f. With this non-dimensionalisation, the dimensionless heat transfer coefficient (the *Stanton number*) depends explicitly on z_f, and both the length and initial force must be adjusted on each iteration. Gagon and Denn (1981) used this formulation, taking as the stopping condition the first of three events: reaching a dimensionless length of unity, the solidification temperature, or the take-up velocity, and adjusting length and force with separate Newton–Raphson schemes. Nine iterations were typically required for convergence for Newtonian, Maxwell, and Phan-Thien/Tanner models.

6.6.3 Sensitivity

George (1982) has reported pilot plant data on spinning of 0·67 intrinsic viscosity PET filaments at speeds ranging from 1000 to 3000 m/min under the conditions given in Table ·6.1, with a spinneret mean velocity of 18·2 m/min. A simulation of these data, using the Phan-Thien/Tanner model (eqns (6.14) and (6.15)) with mean PET rheological properties reported by Gregory (1973), is shown in Fig. 6.5. (Gregory's data were reported at 265 °C and above, and must be extrapolated for calculation to 70 °C.) The heat transfer coefficient is given by eqn (6.9), and the coefficient β in eqn (6.7) was taken to be 0·6. A 'wedge' spectrum with $N = 2$ was used

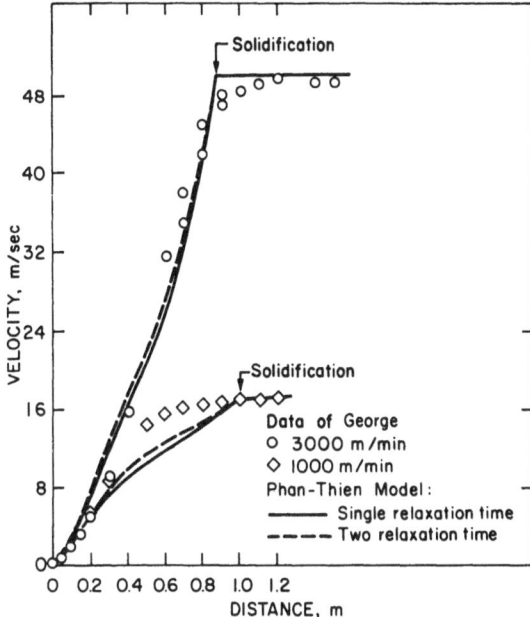

Fig. 6.5. Simulation of data of George (1982) (Gagon and Denn, 1981).

(i.e., $\lambda_1 G_1 = \lambda_2 G_2 = \eta/2$), with $\lambda_2/\lambda_1 = 5$, and a single relaxation time ($N = 1$). λ in eqns (6.14) was set to zero, since PET has a nearly constant viscosity, and ε was taken as 0·015. Following George, who used an extrudate swell correlation, the initial area was taken to be 100% greater than the spinneret area.

The simulation agrees reasonably well with the data, given the fact that the calculation is completely *a priori*. The bounds on the solidification distance at 1000 m/min from eqn (6.26) are $0·73 \leq z_f \leq 1·39$ m, with a computed solidification at 0·98 m. Comparison of the results for $N = 1$ and $N = 2$ suggests that the curvature of the data may require a more detailed spectrum, which is currently unavailable for PET.

Sensitivity to the relaxation time is shown in Fig. 6.6. Passing from Newtonian to slightly viscoelastic (10% of the measured relaxation time) causes a large change at 3000 m/min; a further increase in λ of more than an additional order of magnitude has less effect. This behaviour is a consequence of the relative insensitivity of the temperature profile to kinematics; a small amount of viscoelasticity is sufficient to make the velocity profile approximately linear, and further change can cause only

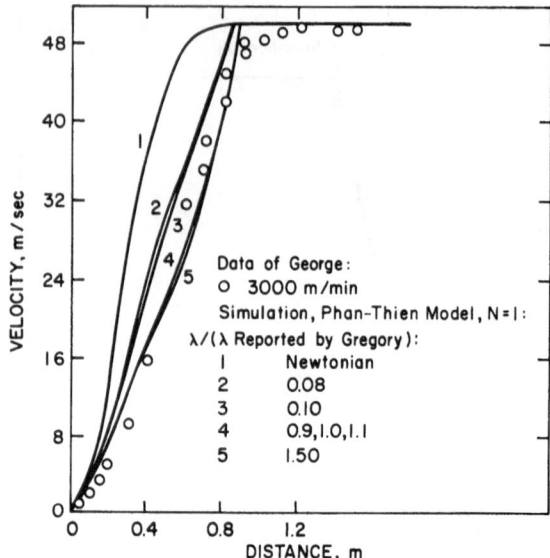

Fig. 6.6. Sensitivity of spinline simulation to changes in magnitude of the relaxation time (Gagon and Denn, 1981).

small variations about the straight line with two fixed ends. The calculation is insensitive to the 'stress saturation' parameter, ε, for $0 \le \varepsilon \le 0.015$, so the observed response is that of a Maxwell fluid. There is much greater sensitivity to ε under other operating conditions.

Sensitivity to the heat transfer coefficient is substantial. Figure 6.7 shows the effect of a 25 % decrease in the coefficient in eqn (6.9). This calculation is shown for the Newtonian fluid ($\lambda = 0$); it is significant that the Newtonian fluid, with a small change in the heat transfer coefficient, fits the data as well as the viscoelastic liquid. Both rheological data on a polymer like PET and heat transfer data have considerable uncertainty. *It is not possible on the basis of the available data alone to choose between a Newtonian and viscoelastic representation of PET for a spinning simulation!* This observation has serious implications regarding extrapolation outside the range where the data were taken.

Finally, the sensitivity to the drag coefficient parameter β is shown in Fig. 6.8. The value $\beta = 0.37$, which is at the lower end of the data but corresponds to the more recent experiments, seems preferable to the mean value of 0.6. Sensitivity to the drag coefficient is less than to the heat transfer coefficient, as would be expected.

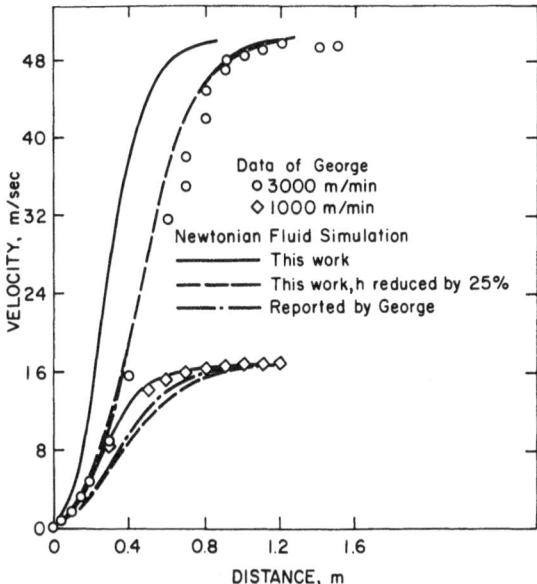

Fig. 6.7.   Sensitivity of a Newtonian fluid spinline simulation to changes in the
heat transfer coefficient (Gagan and Denn, 1981).

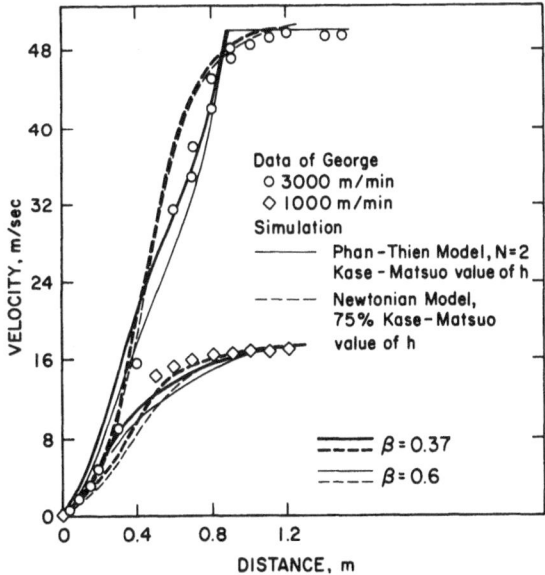

Fig. 6.8.   Sensitivity of spinline simulations to changes in drag coefficient (Gagon
and Denn, 1981).

6.6.4 Multifilament spinning

Heat transfer and air drag characteristics are likely to be altered in a multifilament case, causing property variations across the filament bundle. The description of the Toyobo simulation package (Yasuda *et al.*, 1978), indicates that eqn (6.7) for the drag coefficient is used unchanged, with $\beta = 0.37$, for each filament in the bundle. Equation (6.9) is also used unchanged, but the quench air velocity and temperature are recomputed by means of a macroscopic mass and energy balance over each row of filaments. The mass balance assumes that the volume of cross-flow air is reduced at each row by an amount proportional to the axial velocity and to the square of distance from the spinneret; the details of this calculation are not given. The Toyobo program also performs a force balance normal to the spinning direction in order to calculate the deflection.

The description of the Celanese program (George, 1982) indicates that a single calculation of the change in quench properties is made, rather than a row-by-row calculation. No computational details are given. The air drag on a single filament in a filament bundle is computed by reducing the value given by eqn (6.7), with $\beta = 0.37$, by a factor of (number of filaments/bundle)$^{0.31}$; this factor is based on unpublished in-house data. The Celanese program also uses a dimensional heat transfer correlation based on actual multifilament spinline measurements; the predicted values are of the same order as those in eqn (6.9), but the correlation does not reduce to a unique Nussult number–Reynolds number relation.

The usual analogies between heat and momentum transfer seem to be followed, at least for spinning into quiescent air; see the discussion following eqn (6.8). It would seem, then, that the heat transfer and drag coefficients should be adjusted in a consistent fashion when simulating multifilament spinning.

6.7 VARIATION WITHIN A FILAMENT

Variations within a filament, because of a radial temperature gradient and possible circumferential variations in the heat transfer coefficient, have been treated only approximately. The steady-state microscopic energy equation, without viscous heating and neglecting axial conduction, is written

$$v_r \frac{\partial T}{\partial r}\bigg)_{z,\theta} + v_z \frac{\partial T}{\partial z}\bigg)_{r,\theta} = \frac{\alpha}{\rho\gamma}\left[\frac{1}{r}\frac{\partial}{\partial r}r\frac{\partial T}{\partial r}\bigg)_{z,\theta} + \frac{1}{r^2}\frac{\partial^2 T}{\partial \theta^2}\bigg)_{r,z}\right] \quad (6.27)$$

Numerical solution of this equation will be very difficult, since the filament radius $r = R$ varies with z. If it is assumed that $v_z = v$ is independent of r, then eqn (6.27) can be rewritten in terms of a new independent variable $\xi = r/R(z)$:

$$\left(\frac{\partial T}{\partial z}\right)_{\xi,\theta} = \left(\frac{\pi\alpha}{\dot{M}\gamma}\right)\left[\left(\frac{1}{\xi}\frac{\partial}{\partial\xi}\xi\frac{\partial T}{\partial\xi}\right)_{z,\theta} + \frac{1}{\xi^2}\left(\frac{\partial^2 T}{\partial\theta^2}\right)_{\xi,z}\right] \qquad (6.28)$$

where $\dot{M}$ is the mass flowrate of polymer. Lines of constant ξ are streamlines. The boundary conditions are now given at *fixed* values of ξ:

$$\text{at } \xi = 1, \qquad \frac{\partial T}{\partial\xi} = -\left(\frac{hR}{\alpha_u}\right)(T - T_u) \qquad (6.29)$$

where h is a local value of the heat transfer coefficient, which may depend on θ; if symmetry is assumed, then hd/α_u will be given by eqn (6.8) or eqn (6.9).

Equation (6.28) is cited as the basis for computation in the papers describing the Toyobo program (Yasuda *et al.*, 1978, 1979), but eqn (6.27) is not given as the starting point; both the $v_r\,\partial T/\partial r$ and $\partial^2 T/\partial\theta^2$ terms are missing. The derivation from the correct starting equation is given in an earlier Toyobo paper by Matsuo and Kase (1976), however, and the example given by Yasuda as an application of the simulation package appears to be identical to their example. Stehle and Brückner (1979), in a paper on the simulation of glass spinning, give an equation (containing a misprint) that has the appearance of eqn (6.28), but there appears to be some confusion regarding the distinction between partial differentiation at constant r and at constant ξ.

The computational problems are different in the solution of eqns (6.28) and (6.29), depending on whether or not angular symmetry is assumed. If symmetry is assumed ($\partial/\partial\theta = 0$), then we have nothing more than the conventional diffusion equation, although with a boundary condition that varies with the time-like variable.† Stehle and Brückner discretise the radial direction (details are not given) to obtain a set of ordinary differential equations in the z-direction; this is the 'method of lines'. A Runge-Kutta method is used to solve the ordinary differential equations simultaneously with the one-dimensional momentum equation, using an average temperature to compute the viscosity. The considerable experience in the solution

---

† In fact, hd/α_u does not vary greatly; Matsuo and Kase report a value between 0·1 and 0·2 in one simulation, and a value of order 0·5 would be typical. In that case, the temperature differential across the filament can be closely approximated by the exponential decay and spatial eigenfunction of the dominant term in the exact eigenfunction expansion of the solution.

of the diffusion equation with reaction in the chemical reaction engineering literature, suggests that use of a method of weighted residuals in the radial direction could be used (e.g., Villadsen and Michelsen, 1978), since an accurate solution using only a small number of approximating functions is to be expected. The Nussult number, hd/α_a, will appear explicitly in the approximating function; neglect of the rate of change of the Nussult number in calculating the residual is consistent with the thin filament approximation.

The algorithm used by Matsuo and Kase, and evidently incorporated in the Toyobo simulation package, incorporates data of Eckert and Soehungen (1952) on the circumferential variation of the local heat transfer coefficient. They divide the cross-section into 81 elements bounded by lines of constant angle and arcs of constant radius. Each element is taken to be of uniform temperature, and the rate of heat transfer between each element is taken proportional to α/distance between centres of the elements. The resulting ordinary differential equations for the elements have the form

$$\frac{\mathrm{d}T_j}{(\mathrm{d}z/v)} = \frac{\alpha}{\rho\gamma A_j} \sum_i \frac{p_{ij}}{r_{ij}} (T_i - T_j) \qquad (6.30)$$

where A_j is the cross-sectional area of the element, and p_{ij} and r_{ij} are the perimeters of contact and centre-to-centre distances of adjacent elements, respectively. An explicit marching technique is used, with equal increments in the time-like variable $\tau = \int \mathrm{d}z/v$. It was not demonstrated that this scheme, which was apparently developed prior to 1968, is a proper approximation to the partial differential equation, but Kase (1982) has recently shown that it closely approximates the use of central differences for the spatial derivatives. A more fruitful approach might be to use local polynomial approximations in the r–θ plane in a weighted residuals-based finite-element algorithm.

6.8 DYNAMIC SIMULATION

6.8.1 Simulation program

There appears to be no dynamic simulation program for melt spinning that includes viscoelasticity or high speed effects and simultaneously allows a range of disturbance inputs. Some specialised treatments are discussed subsequently. The only broadly applicable treatment seems to be the MS3 program of Kase (1974), which computes the dynamic response of a Newtonian spinline at low speed to small disturbances. The starting point is eqns (6.4) through (6.6), with only the first term retained on the right-hand side of each of eqns (6.5) and (6.6), with eqns (6.10) and (6.12) to describe the fluid rheology. Neglect of air drag and inertia restricts the application to take-up speeds of less than 1000 m/min.

The equations are rewritten in terms of normalised perturbation variables, denoted by an overbar and defined as follows:

$$\bar{v}(z, t) = \frac{v(z, t) - v_s(z)}{v_s(z)} \tag{6.31a}$$

$$\bar{A}(z, t) = \frac{A(z, t) - A_s(z)}{A_s(z)} \tag{6.31b}$$

$$\bar{T}(z, t) = \frac{T(z, t) - T_s(z)}{T_s(z) - T_{as}(z)} \tag{6.31c}$$

$$\bar{F}(t) = \frac{F(t) - F_s}{F_s} \tag{6.31d}$$

(In comparing this to Kase's original publication, it should be noted that he uses t for temperature, τ for time, x for spatial position, and z for $\int_0^x dx/v_s(x)$.) The equations are linearised about the steady state to obtain a set of linear partial differential equations. The coefficients of the linear equations depend on the steady-state solution, which is stored.

The values of the perturbation variables are specified at $z = 0$:

$$\bar{v}(0, t), \bar{A}(0, t), \bar{T}(0, t) \text{ given} \tag{6.32}$$

Kase's publications have usually considered a disturbance imposed at the take-up, so the value of all perturbations at $z = 0$ is zero. Chang et al. (1982) have considered input variations; it is useful to note that a disturbance in area or velocity at constant flowrate is characterized by $\bar{v}(0, t) = -\bar{A}(0, t)$. The program is coded to allow time-dependent inputs, but this aspect has not been exploited.

At the dynamic solidification point, the temperature equals the solidification temperature and the velocity is taken equal to the take-up velocity. Despite the fact that the solidification point moves, the linearity of the equations can be exploited to formulate a boundary value problem over a fixed spatial domain. The proper dynamic boundary condition, written at the steady-state solidification point, z_{fs} is (Chang et al., 1982; Pearson et al., 1976)

$$\bar{v}(z_{fs}, t) + \frac{(T_f - T_{as}) \, dA_s/dz}{A_s \, dT_s/dz}\bigg|_{z = z_{fs}} \bar{T}(z_{fs}, t) = 0 \tag{6.33}$$

Kase's original formulation uses instead the condition $\bar{v}(z_{fs}, t) = 0$, which we refer to as the 'fixed boundary condition'.

The finite-difference numerical scheme is explicit, using equal increments in time and in the transformed spatial variable $\bar{z} = \int_0^z dz/v_s(z)$. The finite-difference approximations are as follows:

$$\frac{\partial \bar{T}}{\partial t} = \frac{\bar{T}(i,j) - \bar{T}(i,j-1)}{\Delta} \tag{6.34a}$$

$$\frac{\partial \bar{T}}{\partial \bar{z}} = \frac{\bar{T}(i,j) - \bar{T}(i-1,j)}{\Delta} \tag{6.34b}$$

All spatially dependent coefficients are evaluated at the previous value, $i - 1$. This backwards difference formulation is equivalent to the first-order Euler method along the characteristic lines, and it enables explicit solution of the finite-difference approximations over the spatial regime at each successive time step. The linearity of the equations leads to a scheme for solution of the boundary value problem at each time step in three iterations, as follows:

1. Set $\bar{F} = 0$. This converts the problem to an initial value problem, and results in a non-zero value $\bar{v}_1 + C\bar{T}_1$ for the combination of terms in eqn (6.33). (C is the coefficient of $\bar{T}(z_{fs}, t)$.)
2. Set $\bar{F} = 1$, which gives a non-zero value $\bar{v}_2 + C\bar{T}_2$ for the combination of terms in eqn (6.32).
3. The value of $\bar{F}$ that satisfies the zero boundary condition follows from the principle of superposition as

$$\bar{F} = -\frac{\bar{v}_1 + C\bar{T}_1}{(\bar{v}_2 + C\bar{T}_2) - (\bar{v}_1 + C\bar{T}_1)} \tag{6.35}$$

A typical simulation result for a PET filament at 750 m/min is shown in Fig. 6.9, corresponding to a case in which there is a 1 % decrease in initial area with a constant throughput. The dimensionless area perturbation plotted here is the change at the dynamic freeze point relative to conditions at the steady-state solidification point:

$$\bar{A}^* = \frac{A(z_f, t) - A_s(z_{fs})}{A_s(z_{fs})} = \bar{A}(z_{fs}, t) + \bar{v}(z_{fs}, t) \tag{6.36}$$

For the case shown, and for nearly all conditions simulated, the take-up area transient is essentially the same for the correct boundary condition,

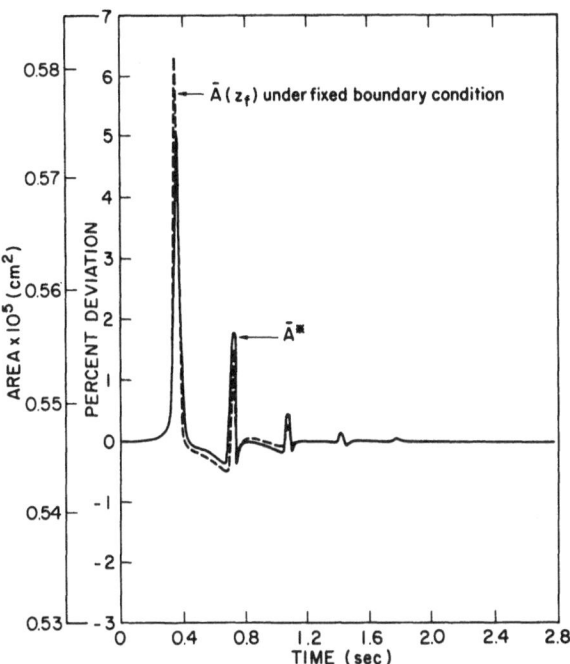

Fig. 6.9. Computed area from MS3 program following a 1 % increase in initial velocity at constant flowrate. Solid line, dynamic boundary condition; broken line, fixed boundary condition (Chang *et al.*, 1982).

eqn (6.33), and the fixed boundary condition. The exception is for a marginally stable spinline with negligible temperature dependence of the viscosity.

6.8.2 Non-linear transients

Ishihara and Kase (1976) have extended the backward difference method described above to the full non-linear equations for isothermal (i.e., water-quenched) low speed spinning; Ishihara's dissertation (1977) contains some details that are not in the papers. The computation remains explicit for a Newtonian fluid, but the force required to solve the boundary value problem at each time step must be found iteratively using Newton's method. If a power-law viscosity is used, then an approximation is made in which one term in the momentum equation that should be evaluated at (i,j) is evaluated instead at $(i-1,j)$; the importance of this approximation is not known.

It follows from the finite-difference formulation in Eqns (6.34) that the number of spatial increments, n, is equal to the number of time increments in one steady-state residence time. Sustained oscillations with a constant amplitude and period are generally obtained after about $3n$ time steps. Most calculations reported by Ishihara and Kase are for $n = 100$; they report a computation time of about 5 min with $n = 100$ on an IBM 360/155 computer for $70n$ time steps, and on an IBM 360/158 computer for $50n$ time steps. Ishihara reports about 5 h on a HITAC-8250 machine for $n = 400$ and $15n$ time steps. A typical calculation is shown in Fig. 6.10. There is a problem of step-size convergence; the computed onset of sustained oscillations occurs at the same draw ratio for $n = 100$ and 400, but there is a difference in the amplitude and period of the subsequent oscillations, as shown in Table 6.2.

Equations (6.14) for a Maxwell fluid ($N = 1$, $\varepsilon = 0$, $\chi = 0$) can be written

$$T_E + \lambda \frac{\partial T_E}{\partial t} + \lambda v \frac{\partial T_E}{\partial z} = \frac{\partial v}{\partial z}\left\{3\lambda G + 2\lambda T_E\left(1 + \frac{3t_{zz}^E}{T_E}\right)\right\} \quad (6.37)$$

Ishihara (1977) has extended his backwards difference scheme for low speed, isothermal spinning to include this equation, with the *ad hoc*

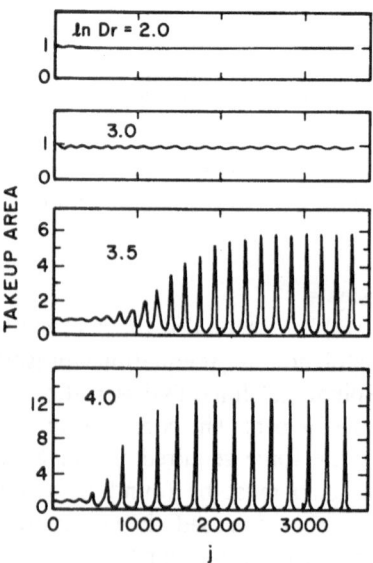

Fig. 6.10. Computed take-up area versus time for an isothermal Newtonian spinline following a 10% increase in take-up speed (Ishihara, 1977).

TABLE 6.2
EFFECT OF NUMBER OF SPATIAL INCREMENTS ON AMPLITUDE
AND PERIOD OF TAKE-UP AREA OSCILLATIONS, $Dr = 60$, LOW
SPEED, ISOTHERMAL, NEWTONIAN FLOW
(Ishihara, 1977)

n	Take-up area/ steady-state value		Dimensionless period of oscillations
	Maximum	Minimum	
100	12·70	0·033 0	0·546 7
200	17·42	0·021 9	0·595 8
300	20·22	0·017 5	0·626 2
400	22·11	0·015 5	0·646 3
500	22·95	0·014 0	0·661 2
600	23·26	0·013 0	0·671 3

additional assumption that $3t_{zz}^E/T_E$ equals a small constant (0·01 in his calculations). This approximation is probably reasonable at high stress levels, as noted in the previous discussion of steady-state simulation. The computation time for the transient following a step change in take-up speed is about 20 % larger than for the Newtonian fluid.

6.8.3 Methods of lines
Methods of lines, in which the transient partial differential equations are discretised spatially and converted to coupled ordinary differential equations in time, have not been widely applied in the published literature to the thin filament equations. Gupta and Ballman (1982) have carried out one such analysis for the linearised low speed, isothermal Newtonian case, using orthogonal collocation to discretise the spatial variable. Collocation can be considered a method of weighted residuals, in which the dependent variables are expressed as sums of the form

$$\bar{v}(z,t) = \sum_{i=0}^{n} \bar{v}_i(t)\phi_i(z) \tag{6.38a}$$

$$\bar{A}(z,t) = \sum_{i=0}^{n} \bar{A}_i(t)\phi_i(z) \tag{6.38b}$$

The ϕ_i are pre-chosen approximating functions; in the application of orthogonal collocation they are Lagrange polynomials of the form

$$\phi_i(z) = \prod_{\substack{m=0 \\ m \neq k}}^{n} \frac{z - \rho_m z_f}{(\rho_k - \rho_m) z_f} \tag{6.39}$$

where the ρ_k are the zeroes of the $(n-1)$th order Legendre polynomial, together with zero and unity. The result is a set of ordinary differential equations for the $\bar{v}_i(t)$, and algebraic equations relating the $\bar{A}_i(t)$ and $\bar{v}_i(t)$.

The procedure is straightforward to apply and gives results comparable to other simulation methods discussed previously and subsequently. Hence, it holds some promise for application to spinning dynamics problems of greater interest. Each additional approximating function corresponds to the inclusion of an additional dynamic mode of the distributed system. The spatial eigenvalues for the starting linear partial differential equations are complex, so an even number of approximating functions should be used. The eigenvalue pairs are widely spaced; thus, the ordinary differential equations will be increasingly stiff as the order of the approximation is increased. Gupta and Ballman found that $n = 7$ represented an acceptable compromise between approximation accuracy and difficulty of numerical integration. The stiffness problem will be compounded by addition of a viscoelastic fluid constitutive equation, which will introduce new time scales.

6.8.4 Frequency response analysis

The dynamic characteristics of a linear system are often summarised in the form of the *frequency response*, which expresses the amplitude and phase of the output following a sinusoidal forcing of unit amplitude. Since any input disturbance can be expressed through a Fourier transform as a linear combination of sinusoids, the output following an arbitrary input disturbance can be computed from the frequency response by superposition, as long as frequency response data are available over the relevant frequency range. Frequency response information is often recorded on a Bode diagram, as in Fig. 6.11, where log amplitude and phase are plotted versus log frequency. The Bode diagram in Fig. 6.11 shows the response of the take-up area of a low speed Newtonian spinline to changes in initial area at different quench temperatures. The peaks in the amplitude indicate the frequency range over which the system is most sensitive to input disturb-

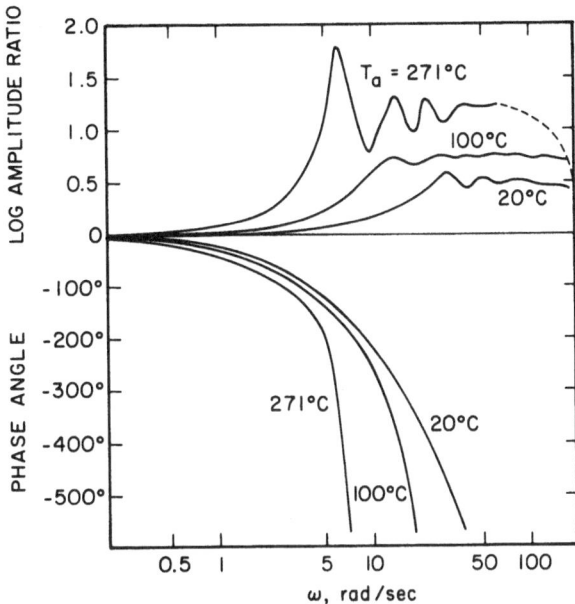

Fig. 6.11. Frequency response of take-up area of a Newtonian spinline to forcing in the initial area, $Dr = 20$ (Kase and Denn, 1978).

ances. The curves shown here demonstrate the stabilising effect of quench air.

The frequency response for low speed, isothermal Newtonian spinning can be reduced to a quadrature (Geyling, 1976; Kase and Denn, 1978; Pearson and Matovich, 1969; Pearson et al., 1976). Frequency response data for other cases can be obtained by integration of the transient following a step change in an input. This approach has been used by Kase (Kase and Araki, 1982; Kase and Denn, 1978) with the MS3 program described above, and it was in this way that Fig. 6.10 was constructed. (When the system is unstable, and the linear transient grows without bound, the overall process must be decomposed into stable elements, as discussed in the references cited here.) Pearson and coworkers (Pearson and Matovich, 1969; Shah and Pearson, 1972) have solved the linear partial differential equations numerically with sinusoidal inputs. This appears to be a less efficient procedure, since larger computational times are required at each frequency than for a single step response.

An attractive alternative approach, which has been applied only in unpublished work by George (cf. Denn, 1979), is to transform the linear equations first, and then to solve in the frequency domain. The transformed equations for isothermal, low speed spinning are as follows:

$$\frac{d\phi}{dz} = \frac{\ln Dr}{Dr^{2z}}\psi - \frac{1}{Dr^{2z}}\tau - \frac{i\omega}{Dr^z}\phi \tag{6.40a}$$

$$\frac{d\psi}{dz} = \tau \tag{6.40b}$$

$$\frac{d\tau}{dz} = (\ln Dr)\tau - Dr^{2z}(\ln Dr)^2\phi - Dr^{2z}\ln Dr\frac{d\phi}{dz} \tag{6.40c}$$

where z is dimensionless length and varies from zero to unity, and time is dimensionless with respect to z_f/v_0. ϕ is the transformed area, defined in terms of eqn (6.31b) as

$$\bar{A}(z, t) = Dr^z \int_{-\infty}^{\infty} \exp{(i\omega t)}\phi(z, \omega)\,d\omega \tag{6.41}$$

Similarly, ψ and τ are transforms of velocity and stress perturbations, respectively. The three functions are complex, and $i^2 = -1$. Boundary conditions on ϕ and ψ at $z = 0$, and on ψ at $z = 1$, are zero unless there is a disturbance, in which case the real part equals unity. A sinusoidal disturbance in input area would require Real $\phi(0) = 1$, Imag $\phi(0) = 0$. The Bode diagram is constructed from the amplitude and phase of the complex number $\phi(1)$ for each ω.

The computational scheme is to choose the real and imaginary parts of $\tau(0)$ and integrate the complex ordinary differential equations (or the corresponding set of 6 real equations) to $z = 1$ to satisfy the boundary condition there. Because of the linearity of the equations, the solution can be constructed from the linearly independent solutions obtained from three independent initial conditions. For example, if the response to an area perturbation is required, then the solutions need to be obtained for initial conditions (Real $\phi(0)$, Real $\tau(0)$, Imag $\tau(0)$) = $(1, 0, 0)$, $(0, 1, 0)$, and $(0, 0, 1)$, respectively, with other initial conditions equal to zero. If these solutions are denoted by subscripts 1, 2, and 3, respectively, then the solution satisfying the boundary condition $\psi(1) = 0$ is

$$\phi(z) = \phi_1(z) + C_2\phi_2(z) + C_3\phi_3(z) \tag{6.42}$$

where the real coefficients C_2 and C_3 are solutions of the linear algebraic equations

$$C_2 \operatorname{Real}\psi_2(1) + C_3 \operatorname{Real}\psi_3(1) = -\operatorname{Real}\psi_1(1) \tag{6.43a}$$

$$C_2 \operatorname{Imag}\psi_2(1) + C_3 \operatorname{Imag}\psi_3(1) = -\operatorname{Imag}\psi_1(1) \tag{6.43b}$$

George's calculations were done with no apparent computational difficulty using a modified Hamming predictor–corrector method from the IBM SSP Library.

6.9 STABILITY

It is important to distinguish carefully between *stability* and *sensitivity*. A system is unstable to infinitesimal disturbances if a steady state cannot be maintained following any arbitrarily small disturbance. The sustained oscillations characterising draw resonance, which occur reproducibly at a critical draw ratio, represent such a process instability. Sensitivity is a reflection of the propagation of disturbances through a stable system, and is analysed using the dynamic simulation methods discussed in the preceding section. Large peaks on the Bode diagram indicate high sensitivity, and it may not be practical to operate such a system even if it is stable in a formal sense. This distinction has not always been recognised in the published literature, and the propagation of input disturbances, probably from quench air, has sometimes been erroneously identified as draw resonance. Spinning instability has been reviewed by Petrie and Denn (1976), Pearson (1976), and Denn and Pearson (1981).

It is possible to identify the onset of an instability using the dynamic simulation methods discussed in the preceding sections. Identification from a transient simulation must be done with care to ensure that it is the system, and not the numerical method, that is exhibiting instability. An instability will be characterised on a Bode diagram by a 360° change in phase beyond some frequency as the critical draw ratio is reached. Kase (Kase, 1981; Kase and Denn, 1978) has shown that stability can be analysed in terms of the frequency response relation (*transfer function*) between velocity and tension at the solidification point, placing this stability analysis in the context of classical feedback control theory. Some analytical results can be obtained in this way (Kase, private communication, 1982).

The traditional way of studying stability is through a linear stability analysis (Denn, 1975). Separation-of-variables solution of the linearised equations is sought in a form

$$\bar{A}(z, t) = Dr^z \sum_{n=1}^{\infty} \exp(l_n t)\phi_n(z) + \exp(l_n^* t)\phi_n^*(z) \qquad (6.44)$$

and similarly for the other variables, where the superscript * denotes a complex conjugate. The resulting equations for low speed, isothermal,

Newtonian spinning are identical to eqns (6.40) with l substituted for $i\omega$, and with zero boundary conditions on ϕ and ψ at $z = 0$, and on ψ at $z = 1$. This is an eigenvalue problem, with solutions existing only for certain values of l. A set of spinning conditions is stable to infinitesimal disturbances if all eigenvalues have negative real parts, since the perturbations defined by eqn (6.44) then decay to zero. The spinning conditions are unstable if any one eigenvalue contains a positive real part. Thus, in principle, all eigenvalues in the infinite set must be examined. In practice, the stability behaviour of this system seems always to be governed by the eigenvalue with the smallest modulus; this is not a general result for non-self adjoint operators, and counter examples exist (cf. Porteous and Denn, 1972).

Several approaches have been used for solving the eigenvalue problem. The first, by Chang and coworkers (1981), following Gelder (1971), is to replace the spatial derivatives by finite differences, leading to a matrix eigenvalue problem that can be solved by any good library program, and then to use Richardson extrapolation to obtain the limiting value as the number of discretisation points tends to infinity. A sequence of such calculations for the eigenvalue of smallest modulus is shown in Table 6.3 for low speed, isothermal Newtonian spinning at $Dr = 20$. The fact that the real part of the eigenvalue tends to zero indicates that this draw ratio corresponds to the transition between stable and unstable flow; a more precise calculation gives the critical value as 20·218.

The orthogonal collocation approach to the transient equations used by Gupta and Ballman (1982), and described above, is quite similar in concept. Here, the discretisation is done first by transforming to a set of linear ordinary differential equations at the spatial collocation points. The

TABLE 6.3
SUCCESSIVE APPROXIMATIONS TO THE EIGENVALUE
FOR LOW SPEED, ISOTHERMAL, NEWTONIAN SPINNING
AT $Dr = 20\cdot2$
(Chang et al., 1981)

Discretisation points	l_{Real}	l_{Imag}
11	−1·0959	−12·594
21	−0·1489	−13·561
31	−0·0547	−13·808
41	−0·0294	−13·895
49	−0·0205	−13·930
Extrapolation, $N \to \infty$	0·0	−14·012

eigenvalues of the system matrix then determine the stability characteristics. Gupta and Ballman found a transition at a draw ratio between 20 and 21 using 7 approximating functions in the collocation expansion. Denn and coworkers (Chang *et al.*, 1981; Fisher and Denn, 1976) have directly integrated the ordinary differential equations, eqns (6.40) (with *l* in place of $i\omega$), and the corresponding equations for non-isothermal flow both with and without viscoelasticity, using a Runge-Kutta method. The equations can be formulated as an initial-value problem by setting Real $\tau(0) = 1$, Imag $\tau(0) = 0$; there is no loss of generality here in taking $\tau(0)$ as real with unit magnitude, since the equations are homogeneous and the phase angle of $\tau(0)$ simply sets the phase of the solution. Iteration to the eigenvalue satisfying the downstream boundary condition is most easily done interactively, and convergence is rapid. The interaction is aided by the observation that the real part of the first eigenvalue is nearly linear in *Dr*. The first two eigenvalues are shown in Table 6.4 for low speed, isothermal, Newtonian spinning. A stability 'map' of the critical draw ratio as a function of $\lambda_{max} v(0)/z_f$ is shown in Fig. 6.12 for low speed isothermal spinning of a Phan-Thien/Tanner fluid with $N = 2$, $G_1 = G_2$, $\lambda_1/\lambda_2 = 5$ (Chang and Denn, 1980). The stabilisation at large values of $\lambda_{max} v(0)//z_f$, and at large draw ratios for some conditions, is in agreement with experiment. The open questions in the application of linear stability theory to spinning have been discussed by Denn and Pearson (1981).

The stability to finite disturbances and the approach to a limit cycle with draw resonance has been studied numerically by Ishihara (1977) and Ishihara and Kase (1975, 1976) using the dynamic simulation methods discussed previously. Fisher and Denn (1975, 1976) have used an approximate method based on an expansion in the spatial eigenfunctions of the linear system, using the method of weighted residuals to obtain a set of

TABLE 6.4
EIGENVALUES FOR LOW SPEED, ISOTHERMAL, NEWTONIAN SPINNING
(Fisher and Denn, 1975)

ln Dr	Dr	First eigenvalue		Second eigenvalue	
		l_{Real}	l_{Imag}	l_{Real}	l_{Imag}
1·0	2·718	−3·810	7·671	−6·395	18·332
2·0	7·389	−2·025	10·845	−4·310	26·075
2·95	19·105	−0·120	13·814	−2·347	33·907
3·006	20·210	0·0	13·989	−2·219	34·380
3·15	23·336	+0·309	14·437	−1·884	35·596

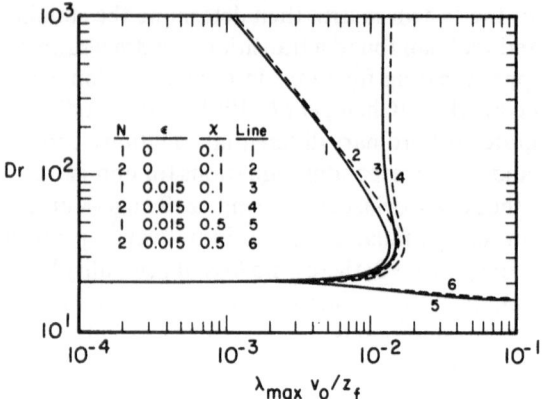

Fig. 6.12. Critical draw ratio for onset of draw resonance, isothermal spinning of a Phan-Thien/Tanner fluid (Chang and Denn, 1980).

ordinary differential equations for the time-dependent coefficients in the expansion. A procedure described by Hyun (1978) is incorrect (Denn, 1980b); this work fails a critical test of any theory of stability to finite disturbances, i.e. that it reduce to the result of the linear theory as the disturbance amplitude becomes vanishingly small.

REFERENCES

Acierno, D., LaMantia, F. P., Marrucci, G., Rizzo, G. and Titomanlio, G. (1976). *J. Non-Newtonian Fluid Mechanics*, **1**, 125, 147.

Chang, J. C. and Denn, M. M. (1980). In *Rheology*, Vol. 3, (Ed. G. Astarita, G. Marrucci and L. Nicolais), Plenum Press, New York, p. 9.

Chang, J. C., Denn, M. M. and Geyling, F. T. (1981). *Ind. Eng. Chem. Fundam.*, **20**, 147.

Chang, J. C., Denn, M. M. and Kase, S. (1982). *Ind. Eng. Chem. Fundam.*, **21**, 13.

Crochet, M. J. and Keunings, R. (1982). *J. Non-Newtonian Fluid Mechanics*, **10**, 85.

Denn, M. M. (1975). *Stability of Reaction and Transport Processes*, Prentice-Hall, New Jersey.

Denn, M. M. (1977). In *The Mechanics of Viscoelastic Fluids*, AMD, Vol. 22, (Ed. R. S. Rivlin), ASME, New York, p. 101.

Denn, M. M. (1979). *Control and Dynamic Systems*, **15**, 147.

Denn, M. M. (1980a). *Ann. Rev. Fluid Mech.*, **12**, 365.

Denn, M. M. (1980b). *AIChE J.*, **26**, 292.

Denn, M. M. and Pearson, J. R. A. (1981). In *Proc. 2nd World Congress Chem. Engr.*, p. vi–354.

Denn, M. M., Petrie, C. J. S. and Avenas, P. (1975). *AIChE J.*, **21**, 795.
Eckert, E. R. G. and Soehungen, E. (1952). *Trans. ASME*, **74**, 343.
Fisher, R. J. and Denn, M. M. (1975). *Chem. Eng. Sci.*, **30**, 1129.
Fisher, R. J. and Denn, M. M. (1976). *AIChE J.*, **22**, 236.
Fisher, R. J. and Denn, M. M. (1977). *AIChE J.*, **23**, 23.
Fisher, R. J., Denn, M. M. and Tanner, R. I. (1980). *Ind. Eng. Chem. Fundam.*, **19**, 195.
Gagon, D. K. and Denn, M. M. (1981). *Polym. Eng. Sci.*, **21**, 844.
Gelder, D. (1971). *Ind. Eng. Chem. Fundam.*, **10**, 534.
George, H. H. (1982). *Polym. Eng. Sci.* (in press).
Geyling, F. T. (1976). *Bell System Tech. J.*, **55**, 1011.
Gould, J. and Smith, F. S. (1980). *J. Textile Inst.*, **72**, 38.
Gregory, D. R. (1973). *Trans. Soc. Rheol.*, **17**, 191.
Gupta, R. K. and Ballman, R. L. (1982). *Chem. Eng. Commun.*, **14**, 23.
Gupta, R. K. and Metzner, A. B. (1982). *J. Rheology*, **26**, 181.
Hyun, J. (1978). *AIChE J.*, **24**, 418.
Ishihara, H. (1977). 'Theoretical Analysis of Draw Resonance in Melt Spinning and its Application to Characterization of the Tensile Rheological Properties of Molten Polymers,' Ph.D. Dissertation, Kyoto University.
Ishihara, H. and Kase, S. (1975). *J. Appl. Polym. Sci.*, **19**, 557.
Ishihara, H. and Kase, S. (1976). *J. Appl. Polym. Sci.*, **20**, 169.
Kase, S. (1974). *J. Appl. Polym. Sci.*, **18**, 3279.
Kase, S. (1982). *J. Appl. Polym. Sci.*, **27**, 2729.
Kase, S. and Araki, M. (1982). *J. Appl. Polym. Sci.* (in press).
Kase, S. and Denn, M. M. (1978). In *Proc. 1978 Joint Automatic Control Conf.*, Vol. 2, Inst. Soc. America, Pittsburgh, p. 71.
Kase, S. and Matsuo, T. (1967). *J. Appl. Polym. Sci.*, **11**, 251.
Keunings, R., Crochet, M. J. and Denn, M. M. (1982). Submitted to *Ind. Eng. Chem. Fundam.*
Malkus, D. S. (1981a). *J. Non-Newtonian Fluid Mechanics*, **8**, 223.
Malkus, D. S. (1981b). In *New Concepts in Finite Element Analysis*, AMD, Vol. 44, (Ed. T. J. R. Hughes, D. Gartling and R. L. Spiker) ASME, New York, p. 29.
Matsui, M. (1976). *Trans. Soc. Rheol.*, **20**, 465.
Matsuo, T. and Kase, S. (1976). *J. Appl. Polym. Sci.*, **20**, 367.
Matsuo, T., Yasuda, H. and Sugiyama, H. (1977). In *Preprints 2nd Int. Symp. Man-Made Fibers*, Vol. 2, p. 206.
Mewis, J. and De Cleyn, G. (1982). *AIChE J.* (in press).
Mewis, J. and Denn, M. M. (1982). *J. Non-Newtonian Fluid Mechanics*, (in press).
Nakamura, K., Watanabe, T., Katayama, K. and Amano, T. (1972). *J. Appl. Polym. Sci.*, **16**, 1077.
Nakamura, K., Watanabe, T., Katayama, K. and Amano, T. (1974). *J. Appl. Polym. Sci.*, **18**, 615.
Nickell, R. E., Tanner, R. I. and Caswell, B. (1974). *J. Fluid Mech.* **65**, 189.
Pearson, J. R. A. (1976). *Ann. Rev. Fluid Mech.*, **8**, 163.
Pearson, J. R. A. and Matovich, M. A. (1969). *Ind. Eng. Chem. Fundam.*, **8**, 605.
Pearson, J. R. A., Shah, Y. T. and Mhaskar, R. D. (1976). *Ind. Eng. Chem. Fundam.*, **15**, 31.
Petrie, C. J. S. (1979). *Elongational Flows*, Pitman, London.

216   M. M. DENN

Petrie, C. J. S. and Dealy, J. M. (1980). In *Rheology*, Vol. 1, (Ed. G. Astarita, G. Marrucci and L. Nicolais), Plenum Press, New York, p. 171.
Petrie, C. J. S. and Denn, M. M. (1976). *AIChE J.*, **22**, 209.
Phan-Thien, N. (1978). *J. Rheology*, **22**, 259.
Phan-Thien, N. and Tanner, R. I. (1977). *J. Non-Newtonian Fluid Mechanics*, **2**, 353.
Porteous, K. C. and Denn, M. M. (1972). *Trans. Soc. Rheol.*, **16**, 295.
Shah, Y. T. and Pearson, J. R. A. (1972). *Ind. Eng. Chem. Fundam.*, **11**, 150.
Stehle, M. and Brückner, R. (1979). *Glastechn. Ber.*, **52**, 82, 105.
Toyobo Co., Ltd. (1980). *Simulation Series for Melt Spinning (SP-Series). General Explanation.*
Villadsen, J. and Michelsen, M. L. (1978). *Solution of Differential Equation Models by Polynomial Approximation*, Prentice-Hall, New Jersey.
White, J. L. (1982). *Polym. Eng. Rev.*, **1**, 297.
Yasuda, H., Ishihara, H. and Yanagawa, H. (1978). *Sen-i Gakkaishi (J. Soc. Fiber Sci. and Tech.*, *Japan)* **34**(1), 20.
Yasuda, H., Sugiyama, H. and Yanagawa, H. (1979). *Sen-i Gakkaishi (J. Soc. Fiber Sci. and Tech.*, *Japan)*, **35**(9), 370.
Ziabicki, A. (1976). *Fundamentals of Fibre Formation*, Wiley, New York.

CHAPTER 7

Film Blowing, Blow Moulding and Thermoforming

C. J. S. PETRIE

Department of Engineering Mathematics,
University of Newcastle upon Tyne, UK

7.1 PROCESSES, MODELS AND ASSUMPTIONS

7.1.1 Introduction

In this chapter we discuss processes involving the stretching of thin sheets of molten or soft solid polymers. The mechanics are modelled on the basis of membrane theory (sheet thickness very small, in particular, small compared with principal radii of curvature of the sheet) and here, as in every published calculation, axial symmetry is assumed. The most significant forces acting on the polymer are boundary tractions and the force arising from any pressure difference across the sheet. Once the problem has grown to the complexity where computer solution of the equations is sought, the incorporation of the less significant forces into the model is straightforward. The influence of gravity may be moderately important and may easily be taken into consideration, as may inertia which is generally insignificant. Surface tension is not likely to be important unless very thin films are being produced, but only requires physical data in order to be incorporated in the model. Air drag likewise requires physical data, probably in the form of a correlation and some empirically determined coefficients.

The cooling of the polymer sheet is of central importance to the processes considered here, and this can only be modelled if we know relevant heat transfer coefficients for forced convection and radiation in the case of film blowing and for free convection, radiation, and polymer–mould contact in the case of blow moulding. Once we start considering the cooling of the

217

218   C. J. S. PETRIE

sheet, we must consider the non-isothermal rheology of the polymer melt. At the very least we have to consider a material that deforms quite readily above its melting point and much less readily below that temperature.

7.1.2 The film-blowing process
The first process we consider, and the one on which most work has been done, is the film-blowing process. Here we consider only the part of the process involving the film formation and stretching, from the extrusion of molten polymer from an annular die to the solidification of the polymer at the freeze-line. Figure 7.1 illustrates the process schematically. We consider the stretching of the tubular film of molten polymer both in the machine direction (i.e. vertically) and in the transverse direction (around the tubular film bubble). This stretching is brought about in the machine direction by

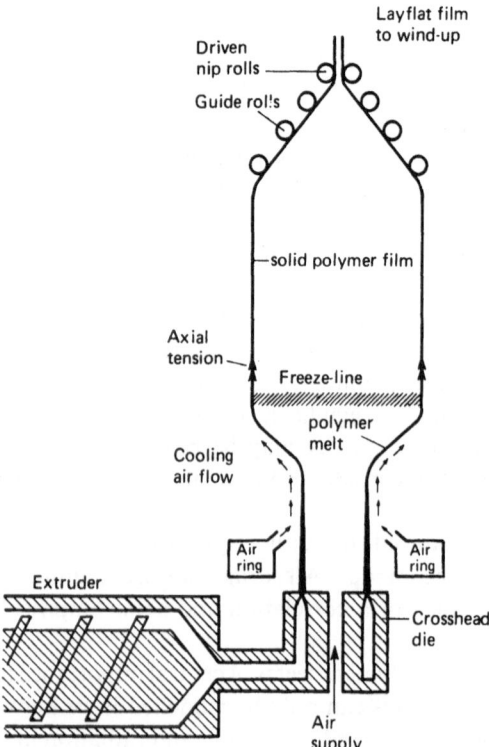

Fig. 7.1. Schematic diagram of the film-blowing process. (Reproduced from Petrie (1975b) by permission of the Society of Plastics Engineers.)

the force applied by the driven nip rolls, and in the transverse direction by the inflation of the bubble by air supplied to its interior. The cooling air, supplied through the air ring is regulated to adjust the height of the freeze-line, and this is important in determining the balance of properties of the film, as well as how stable is the operation of the process.

Details of the modelling of the process are given in a number of papers (e.g. Petrie, 1975a, b) and books (Han, 1976; Middleman, 1977; Pearson, 1982). Essentially, by ignoring details of the flow at the die exit (the region where extrudate swell takes place), and by treating the flow as steady and axisymmetric, with no variations across the film, we end up with a two-point boundary-value problem for a set of ordinary differential equations. If the rheology of the material is modelled by an integral equation, we have integro-differential equations in place of some of the differential equations, and this may be preferable to using rate equations for an elastico-viscous material because then we obtain a higher order system of differential equations with numerical problems both of stiffness and inherent instability.

7.1.3 The mechanics of idealised film blowing

The equations are obtained using local Cartesian coordinates, x_1 in the direction of flow, x_2 in the transverse direction, and x_3 normal to the film. The simplest force balance normal to the film (from standard membrane theory, neglecting gravity, inertia, surface tension and air drag) gives

$$p_i = \frac{ht_{11}}{R_1} + \frac{ht_{22}}{R_2} \tag{7.1}$$

where p_i is the inflation pressure, h is the film thickness, R_1 and R_2 are principal radii of curvature, and t_{11} and t_{22} are principal stresses. A similar balance in the machine direction gives

$$F_1 = 2\pi r h t_{11} \cos \theta - \pi r^2 p_i \tag{7.2}$$

where F_1 is the applied force in the machine direction, r is the radius of the cylindrical bubble, and θ is the angle between the velocity vector (the local x_1 direction) and the machine direction. We take z as our coordinate in the machine direction and then straightforward geometry gives

$$\frac{dr}{dz} = \tan \theta \tag{7.3}$$

$$R_1 = -1 \left/ \left\{ \cos \theta \frac{d\theta}{dz} \right\} \right. \tag{7.4}$$

$$R_2 = r/\cos \theta \tag{7.5}$$

As usual with free-surface flows we take $t_{33} = 0$ (while surface tension is being neglected) and so

$$t_{11} = t_{11}^E - t_{33}^E \tag{7.6}$$

$$t_{22} = t_{22}^E - t_{33}^E \tag{7.7}$$

The kinematics, which we shall need for the rheological equation of state (see Chapter 1, Sections 1.2 and 1.3) in order to obtain $\mathbf{T}^E$, are those of elongational flow, with rate of strain tensor

$$\mathbf{D} = \cos\theta \begin{bmatrix} dv/dz & 0 & 0 \\ 0 & (v/r)(dr/dz) & 0 \\ 0 & 0 & (v/h)(dh/dz) \end{bmatrix} \tag{7.8}$$

where we write v for the velocity (in the x_1-direction), and deformation tensor

$$\mathbf{C}_0^{-1} = \begin{bmatrix} v^2/v_0^2 & 0 & 0 \\ 0 & r^2/r_0^2 & 0 \\ 0 & 0 & h^2/h_0^2 \end{bmatrix} \tag{7.9}$$

relative to the initial state of the material (at $z = 0$, after allowing for the extrudate swell if necessary). For an incompressible material we have the constraint that the trace of $\mathbf{D}$ is zero (in isothermal flow) and more generally, allowing the density to vary with temperature and hence with position in the more realistic non-isothermal situation, we have

$$\dot{M} = 2\pi\rho rhv \tag{7.10}$$

where $\dot{M}$ is the mass flowrate, constant in steady flow.

A constitutive equation, to relate $\mathbf{T}^E$ to $\mathbf{D}$ or $\mathbf{C}^{-1}$ or some function of the history of these quantities completes the set of equations modelling of the dynamics, and since the choice of constitutive equation affects the mathematical structure of the model, detailed discussion will be postponed to later sections of this chapter. Here we note that this has an effect on the number and type of boundary conditions required at the die exit ($z = 0$). The complete model of the dynamics requires knowledge of initial values of r, h and v and the value of θ at the freeze-line ($\theta = 0$ is used—see discussions in the various papers cited, and in Pearson and Petrie (1970)).

7.1.4 More realistic mechanical equations for film blowing

Incorporation of other forces merely requires alteration of eqns (7.1) and (7.2) to

$$p_i + \rho g h \sin \theta = \frac{h(t_{11} - \rho v^2) + 2\sigma}{R_1} + \frac{ht_{22} + 2\sigma}{R_2} \tag{7.11}$$

$$F = 2\pi r[h(t_{11} - \rho v^2) + 2\sigma]\cos \theta - \pi r^2 p_i \tag{7.12}$$

$$\frac{dF}{dz} = 2\pi r(T_{\text{drag}} + \rho g h / \cos \theta) \tag{7.13}$$

Here we have allowed for gravity (the terms involving ρg), inertia (the terms involving ρv^2), surface tension (coefficient σ) and air drag (shear stress T_{drag} acting on the film in the negative x_1 direction).

7.1.5 Modelling heat transfer for film blowing

Completion of the model requires knowledge of the temperature of the film as a function of z so that the density and the rheological properties of the polymer may be correctly incorporated into the computations. One crucial aspect of this is knowledge of the position of the freeze-line, for the geometrical and mechanical boundary conditions involve assumptions about the freeze-line as well as the die exit region. In order to make calculations of the temperature we need to use the energy equation which may be written

$$\dot{M}\frac{d(\gamma T)}{dz} = - \{h(T - T_c) + \varepsilon\sigma(T^4 - T_a^4)\}\frac{2\pi r}{\cos \theta} \tag{7.14}$$

where $T(z)$ is the film temperature, γ its specific heat, T_c the temperature of the cooling air, h the appropriate heat transfer coefficient, ε the emissivity of the film of molten polymer, σ the Stefan–Boltzmann constant and T_a the ambient temperature.

In writing eqn (7.14) we ignore heat transfer into the interior of the tubular bubble. Convection inside the bubble could provide a mechanism for additional cooling of the bubble near the die but any such heat transfer from film to air would need to be balanced by heat transfer back to the film further away from the die because the inflating air is not changed. No detailed estimates of the effect of this have been obtained, and it is taken to be small. The dominant mechanism for cooling the film is external forced convection, and the heat transfer coefficient h must be provided. It has been estimated that radiative cooling may be responsible for up to 20% of the heat transfer, and typically ε is taken to be between 0·4 and 0·6.

Standard 'engineering' correlations for h have been wrongly used in some discussions of the process (see Petrie, 1974, for a discussion of this). The correlation

$$h = K_1(w_{max})^n \tag{7.15}$$

with $n = 1\cdot5$, $K_1 = 4\,W\,m^{-2}\,K^{-1}$ is the best guess from limited data, where w_{max} is the maximum velocity of cooling air at any distance from the die. This in turn is given by a correlation

$$w_{max} = K_2(z)^m \tag{7.16}$$

with the value $m = -0\cdot75$ suggested. The coefficient K_2 is dependent on the volumetric flowrate of cooling air, and on the geometry of the air ring through which the cooling air is directed onto and along the bubble. In a computation one might think of varying K_2 until a particular freeze-line height was obtained.

The freeze-line height will be obtained as the value of z at which the film temperature reaches the value at which solidification takes place. Observation of polyethylene film production indicates that this temperature may be significantly below that at which solidification takes place in slow cooling of the bulk polymer (and could be as low as 90 °C for LDPE). In addition to determining the freeze-line height, solution of the energy equation for the temperature of the film allows temperature-dependent properties to be used in the dynamical equation. The appearance of r and θ in eqn (7.14) means that the solutions of the equations are coupled.

7.1.6 Blow moulding and thermoforming

The inflation stage in blow moulding has many similarities to bubble inflation in film blowing, and the limited amount of computation which has been done uses the same dynamical equations (Petrie and Ito, 1980). A realistic model here would need to drop the axisymmetry requirement, but before doing this, it is possible to make calculations of wall thickness distribution along an article such as a blow-moulded bottle. The new feature that this introduces is the movement of the point of contact of the parison with the mould as inflation proceeds. We shall discuss this, and some related calculations by Wineman, below.

7.2   VISCOUS FILM BLOWING

7.2.1   The isothermal Newtonian model

The first model for which solutions were obtained was that for a Newtonian liquid and isothermal flow (Pearson and Petrie, 1970). The constitutive equation gives

$$T^E = 2\eta D \tag{7.17}$$

with D given by eqn (7.8) and η a constant, and this is substituted in eqns (7.6) and (7.7). It was found more satisfactory to integrate the differential equations from the freeze-line to the die, and a fourth-order Runge-Kutta method proved adequate for the numerical integration.

In order to carry out the integration we require values of parameters p_i, F_1, $\dot{M}$, η, ρ and values of two of the variables v, h, and r at the freeze-line ($z = z_1$), as well as the condition $\theta = 0$. A solution then gives us values of v, h and r at the die (or at least at a point distant z_1 upstream from the freeze-line). Clearly these are known (at least in terms of assumptions about extrudate swell) so two of our specified conditions have to be varied to obtain the correct extrudate dimensions. We may either vary the radius and thickness of the film at the freeze-line, as was done originally, or we may argue that the practical operation of the process is best represented in terms of adjusting the haul-off tension in order to give the correct value of velocity, and adjusting the inflating pressure to give the correct blow ratio. (This is discussed in Petrie, 1975b.)

Whatever procedure we adopt leads to the solution of a boundary-value problem by a shooting method. Experience here suggests that because of the highly non-linear nature of the system, an interactive approach is more likely to succeed than an automated iterative approach. It seems desirable to design a computer program whose input uses direct values of physical constants and actual dimensions even though the calculations are more economically done using dimensionless variables. The program user should not have to calculate these or be required to accept results in terms of them only; non-dimensionalisation can be an internal matter, though there are cases where a knowledge of dimensionless groups would be highly instructive.

Typical input for a program might then be values of ρ, η, r_0, h_0, $\dot{M}$, r_1, h_1, z_1, where r_0, and h_0 are die dimensions and r_1 and h_1 are dimensions of the film to be produced. Refinements to this are not worth considering until we consider a non-isothermal model. In dimensionless form the equations of the process model may be written

$$\bar{r}' = \tan \theta \tag{7.18}$$

$$2\bar{r}^2(\bar{F} + \bar{r}^2\bar{B})\theta' = 3\sin 2\theta + \bar{r}(\bar{F} - 3\bar{r}^2\bar{B}) \tag{7.19}$$

$$2\bar{h}^{-1}\bar{h}' = -\tan \theta/\bar{r} - \sec^2 \theta(\bar{F} + \bar{r}^2\bar{B})/2 \tag{7.20}$$

$$\bar{v} = (\bar{r}\bar{h})^{-1} \tag{7.21}$$

with initial conditions

$$\bar{r}(\bar{z}_1) = \bar{r}_1 \equiv r_1/r_0 \qquad \bar{h}(\bar{z}_1) = \bar{h}_1 \equiv h_1/h_0 \qquad \theta(\bar{z}_1) = 0 \tag{7.22}$$

The dimensionless variables are

$$\bar{r} = r/r_0 \qquad \bar{h} = h/h_0 \qquad \bar{v} = 2\pi\rho r_0 h_0 v/\dot{M} \qquad \bar{z} = z/r_0 \tag{7.23}$$

and constants $\bar{F}$ and $\bar{B}$ are chosen (iteratively) so that the solution satisfies the conditions

$$\bar{r}(0) = 1 \qquad \bar{h}(0) = 1 \tag{7.24}$$

In terms of the physical parameters the dimensionless parameters are

$$\bar{B} = \pi\rho r_0^3 p_i/\eta\dot{M} \qquad \bar{F} = \rho r_0 F_1/\eta\dot{M} - \bar{r}_1^2\bar{B} \tag{7.25}$$

so that in fact η is only required if we wish to calculate the required internal pressure, p_i, and applied force, F_1.

One of the computational problems arises from the fact that solutions to the problem with initial values prescribed at $\bar{z} = \bar{z}_1$ (or at $\bar{z} = 0$) do not exist for a large enough interval (in $\bar{z}$) for all values of the parameters $\bar{B}$ and $\bar{F}$ (see the discussion in Pearson and Petrie, 1970). This causes difficulties particularly if extrapolation to obtain new values of $\bar{B}$ and $\bar{F}$ is done without taking any precautions to restrict the values of these parameters. It is necessary to insist that $\bar{B} > 0$, and $\bar{F} > 0$ is safe (sufficient but not necessary; $\bar{F} > -B$ is necessary but not proved to be sufficient). A very cautious approach which has been used is to vary $\bar{F}$ to satisfy $\bar{v}(0) = 1$ with each $\bar{B}$, and in an outer loop to vary $\bar{B}$ until $\bar{r}(0) = 1$ also. This can be associated with the idea of altering the applied force to get the correct velocity or machine-direction draw (v_1/v_0) and the inflation pressure to get the correct blow ratio (r_1/r_0).

7.2.2 A non-isothermal Newtonian model

A clear improvement in the model is obtained by allowing the viscosity η, to vary with temperature. In the first instance this may be done with a given temperature variation along the bubble, and a linear decrease from die to freeze-line is not too unrealistic. The use of a correlation, cf. eqn (1.8),

$$\eta = \eta_* \exp(-\zeta(T - T_*)) \qquad (7.26)$$

has been reasonably successful (Petrie, 1975a), but

$$\eta = \eta_* \exp(E/RT) \qquad (7.27)$$

(with T in Kelvin) may be more accurate at temperatures close to freezing. The computation turns out to be as easy to carry out as in the isothermal case. Perhaps since the calculations give more realistic results, the computation could be said to be a good deal easier, since initial guesses (based on processing experience) for values of the parameters are nearer to values obtained by the iterative scheme and, as has already been implied, good initial guesses are needed to avoid much numerical experimentation.

One observation, from comparison of theory with experiment (Petrie, 1975a; Wagner, 1976), is that better agreement between the theoretical predictions for bubble shape, $r(z)$, is obtained using the die dimensions without 'corrections' for extrudate swell, except possibly when there is very little machine-direction drawing (i.e. v_1 is close to v_0). Another observation is that calculations with this model overestimate the inflation pressure by a factor of three or four. Conversely, if the melt viscosity were not known, measurement of the inflation pressure could be used to estimate an apparent viscosity, and would give values one-third or one-quarter of that obtained in shear viscometry. [Note that this factor is not the well-known Trouton ratio, of 3 in uniaxial extension or 6 in equal biaxial extension, as this is implicitly included in the model already.]

7.2.3 A non-Newtonian non-isothermal model

A further improvement in the model may be sought by allowing the viscosity to vary with strain rate (Han and Park, 1975). The expression used is

$$\eta = K_0 (I_2)^{(v-1)/2} \exp(E/RT - E/RT_0) \qquad (7.28)$$

and from eqn (7.8), using $\mathrm{tr}(\mathbf{D}) = 0$,

$$I_2 = v^2 \cos^2 \theta [(r'/r)^2 + (h'/h)^2 + (r'h'/rh)] \qquad (7.29)$$

(There is a misprint in Han and Park (1975), eqn (21)—two terms should be squared as in eqn (7.29) here, and as in their earlier paper.) Details of the computation are not available in the published papers, but Han and Park report no particular difficulty in solving the fourth-order system they obtain by a standard Runge-Kutta method. They appear to use a mean axial tension in allowing for the effect of gravity, and they find this tension iteratively. It is not made explicit which boundary values are used and which determined from the computation—in particular how the freeze-line height is specified.

7.3   ELASTIC FILM BLOWING

A rubber-like elastic or neo-Hookean model may be used in place of the Newtonian model. The constitutive equation

$$\mathbf{T}^E = G\mathbf{C}_0^{-1} \tag{7.30}$$

with $\mathbf{C}_0^{-1}$ given by eqn (7.9) and G a constant, may be used in eqns (7.6) and (7.7). However, since this model seeks to describe the 'elastic memory' of the material, a reference state which is not necessarily that of the material at $z = 0$ is worth introducing, and we may replace $\mathbf{C}_0^{-1}$ by $\mathbf{C}_*^{-1}$, so that

$$\mathbf{T}^E = G\mathbf{C}_*^{-1} = G \begin{bmatrix} v^2/v_*^2 & 0 & 0 \\ 0 & r^2/r_*^2 & 0 \\ 0 & 0 & h^2/h_*^2 \end{bmatrix} \tag{7.31}$$

and the reference state is a hypothetical thin-walled tube of radius r_* and thickness h_*, with v_* given by incompressibility, i.e.

$$v_* = r_0 v_0 h_0 / r_* h_* \tag{7.32}$$

Difficulties are experienced using this model (Petrie, 1975a; Wagner, 1976) and similarly with an elastic model for stretching of polypropylene film in the solid phase (Pearson and Gutteridge, 1978). The main problem lies in finding initial guesses for parameters or boundary values (in solving by a shooting method) which give a solution for the whole interval between die and freeze-line. Near to operating conditions which were to be modelled the solutions appeared extremely sensitive to initial guesses, and there is no guarantee that the solutions which were sought do in fact exist. Solutions have been obtained only for relatively small machine-direction draw ratios, $v_1/v_0 = 1 \cdot 6$ by Petrie (1975a) and $v_1/v_0 = 1, 2$ and in one case 4 by Wagner (1976). The polypropylene model is not directly comparable because as well as the different strain measure the modulus varies with temperature. The solution shown for $v_1/v_0 = 8$ is highly unrealistic and led the authors (Pearson and Gutteridge, 1978) to consider a viscoelastic model. The temperature-independence of G in the model used by Wagner and by Petrie for molten polyethylene is claimed to be much more reasonable than the similar approximation for η in the Newtonian model. One of the more arbitrary features of the model, which is embodied in eqn (7.31) is the choice of the reference, or stress-free, state. All three authors cited have discussed this; Wagner uses the state after unconstrained extrudate swell (in the absence of applied forces) as the reference state. The differences in the physical problem studied mean that Pearson and Gutteridge take a different approach here and consider a cold cylindrical film bubble in equilibrium with the applied forces for $z < 0$. At $z = 0$ the film is heated, and a jump in temperature is postulated, with corresponding jumps in v and in $d\theta/dz$, necessary to maintain the equilibrium of forces. In their example this gives an excessively large jump in v so that v decreases from that point to its value at the freeze-line where the prescribed temperature reaches a low enough value for further changes in deformation to be negligible. Two other points may be made. The first is that in Petrie's calculations the value of r_* chosen did not affect the predicted bubble shape appreciably, although it did affect the predicted inflation pressure, p_i, and applied force F_1. Some unpublished work (Thompson, 1975) confirms this conclusion for a restricted range of values of r_* and also shows how the effect of r_* is largely confined to the parts of the bubble nearer the die. The second point can be made by considering the equations for the film, which may be written

$$\tilde{r}' = \tan \theta \tag{7.33}$$

$$\tilde{v}\tilde{r}(\tilde{F} + \tilde{r}^2 \tilde{B})\theta' = (\tilde{r}^2 - \tilde{r}^{-2}\tilde{v}^{-2})\cos \theta - 2\tilde{v}\tilde{r}^2\tilde{B} \tag{7.34}$$

$$\tilde{v}(\tilde{F} + \tilde{r}^2 \tilde{B}) = (\tilde{v}^2 - \tilde{r}^{-2}\tilde{v}^{-2})\cos \theta \tag{7.35}$$

where the dimensionless variables and parameters are

$$\tilde{r} = r/r_* \qquad \tilde{v} = v/v_* \qquad \tilde{z} = z/r_* \tag{7.36}$$

$$\tilde{B} = \pi\rho v_* r_*^2 p_i/\dot{M}G \qquad \tilde{F} = \rho v_* F_1/\dot{M}G - \tilde{r}_1^2\tilde{B} \tag{7.37}$$

It is seen that eqn (7.35) is an algebraic equation, so the order of the system is one less than is the viscous case. If we pose the same problem as in the viscous case, we find that eqn (7.35) gives an additional relation between the parameters when applied at the freeze-line. This can be treated as equivalent to a relation between v_* and r_* (Petrie, 1975a)—with the underlying intuitive argument that the extrudate swell region would allow a matching of this with values appropriate to the material's deformation history.

The boundary conditions we wish to satisfy are the five equations

$$\tilde{r}(0) = r_0/r_* \qquad \tilde{v}(0) = v_0/v_* \qquad (7.38)$$

$$\tilde{r}(\tilde{z}_1) = r_1/r_* \qquad \tilde{v}(\tilde{z}_1) = v_1/v_* \qquad \theta(\tilde{z}_1) = 0 \qquad (7.39)$$

and we have a second-order system of equations to solve which requires the conditions on $\tilde{r}$ and θ from eqn (7.39). If the values of r_* and v_* were prescribed, eqn (7.35), at $\tilde{z} = \tilde{z}_1$ with $\tilde{v}(\tilde{z}_1)$, would then give a relation between $\tilde{F}$ and $\tilde{B}$, and the one degree of freedom left would not allow both conditions of eqn (7.38) to be satisfied. Allowing $\tilde{v}(\tilde{z}_1)$ to vary gives the necessary freedom and, in effect, determines v_* (so that it cannot be prescribed independently). As a practical point it was found convenient to solve a third-order system of equations obtained by replacing eqn (7.35) by its derivative, which gives (by use of eqns (7.33) and (7.34)) an equation for $\tilde{v}'$. This gives $\tilde{v}$ more easily than solving the non-linear algebraic equation, or at least it is easier to program and it reduces the attendant risk of moving between different branches of the solution. Raising the order of the system presents no additional problem, as the extra condition is already prescribed in eqn (7.39).

In contrast to this, Wagner (1976), who takes r_* and v_* to be determined independently of each other and of the flow, finds that one condition cannot be satisfied. His solutions do not satisfy $v(0) = v_0$, and it could be argued that the extrudate swell region is being invoked, as a region where the velocity changes from v_0 at the die to $v(0)$ at the start of the bubble flow being modelled here. This seems rather arbitrary—why should all the adjustment be in v and none in r? However, the earlier conclusion that the value of r_* has relatively little influence suggests that this way of proceeding may be adequate. In his model Wagner introduces temperature variation, but keeps G constant and only allows coupling between the equations through temperature dependence of density which is a relatively small effect. What is done is that $\tilde{r}^{-2}\tilde{v}^{-2}$ in eqns (7.34) and (7.35) is replaced by $\tilde{r}^{-2}\tilde{v}^{-2}\tilde{\rho}(T)^{-2}$, where

$$\tilde{\rho}(T) = \rho(T)/\rho_* = (1 + \beta(T - T_*))^{-1} \qquad (7.40)$$

and ρ in eqn (7.37) is replaced by ρ_*. Typical values of the parameters are LDPE are (Petrie, 1975a)

$$\rho_* = 801 \, \text{kg m}^{-3} \qquad T_* = 388 \, \text{K} \qquad \beta = 0.000\,69 \, \text{K}^{-1} \qquad (7.41)$$

Equation (7.14) is solved, along with the other differential equations, with a given temperature at the die, and the convective heat transfer coefficient is adjusted until the correct temperature at the freeze-line is obtained. (Wagner finds integration from die to freeze-line satisfactory, in contrast to experience described above). In addition Wagner formulates the problem with gravity included, but ultimately decided that this effect was so small that the adjustment in the initial value, $F(0)$ (for differential equations, eqn (7.13), with $T_{drag} = 0$), to get F_1 equal to the observed value was not worth doing. This emphasises a slightly different point, to be discussed below in connection with viscoelastic models, namely that where data on applied force and inflation pressure are available these may be used and attempts made to find a best fit by choice of material parameters.

Finally, the feeling of this author is that, while the elastic models pose some subtle and fascinating mathematical problems, the work described in this section is unlikely to be successful in modelling film-blowing processes. The successful use of viscoelastic models, described in the next section, probably renders tackling the mathematical problems discussed ˌabove unnecessary from a practical point of view. Whether this feeling will be seen to be well-founded, and whether the same can be said about parison inflation and vacuum-forming processes, remains to be discovered.

7.4 VISCOELASTIC FILM BLOWING

There is such a choice both in the constitutive equation used to describe viscoelastic behaviour and in the way the equations modelling the process may be set up and solved, that it seems impossible to do other than describe briefly the four different attempts that have been made. These start from a formulation of the problem for the Oldroyd eight-constant model and include two Maxwell models and a Voigt model. All apart from the first consider the non-isothermal problem, and all apart from this model may claim some success in achieving limited objectives.

7.4.1 Oldroyd–Maxwell models

The full equations for the Oldroyd model are given by Petrie (1973) and since no computation has been done with them they are not reproduced here. (It may be worth noting that eqn (11) of that paper contains two misprints: 'λ_1' should be replaced by '$\lambda_1 u$' and '$(2r'/r - h'/h)L$' by '$(2r'/r + h'/h)L$'.) The sixth-order system of equations obtained is reduced to fifth-order when the retardation time is set to zero (which leaves us with a convected Maxwell model). This results in equations analogous to eqns

(7.18), (7.19), (7.20) and two further first-order differential equations from the constitutive equation, and r, θ, h, t_{11} and t_{22} may be used as our basic set of dependent variables. It appears more convenient to use the isotropic pressure p ($p = t_{33}^E$ since $t_{33} = 0$) instead of t_{11}, and to think of specifying extra initial conditions $t_{33}^E(0)$ and $t_{22}^E(0)$ (whence $t_{22}(0)$ is obtained using eqn (7.7)). Very little successful computation was done with this model, even for the upper convected Maxwell model, and conclusions are necessarily tentative. Numerical integration from freeze-line to die, which was preferred for the Newtonian model, is highly unstable here. This is not really surprising since the process is an integration backwards in time for a material particle and the property of stress relaxation (for time increasing) gives exponential growth of stresses (whose initial values have to be guessed). Numerical integration from die to freeze-line was slightly more successful, but existence of a solution from die to freeze-line depended critically on values of guessed parameters and initial stresses. Solutions obtained were for small freeze-line heights and one is left with the feeling that a good deal of luck (or a great deal of exploratory computation) would be needed to find solutions satisfying any given set of conditions. Experience with the Newtonian model suggests that any effort that might be devoted to this approach must include at least a temperature-dependent viscosity. A constant modulus would then imply a temperature-dependent relaxation time.

7.4.2 A non-isothermal Maxwell model expressed as an integral equation

Wagner (1976) uses a temperature-dependent upper convected Maxwell model expressed in an integral form equivalent to

$$\mathbf{T}^E = \int_{-\infty}^{t} \frac{G}{\lambda(t')} \mathbf{C}_t^{-1}(t') \exp\left\{ - \int_{t'}^{t} \frac{dt''}{\lambda(t'')} \right\} dt' \qquad (7.42)$$

where G is the constant modulus and $\lambda(t)$ is the relaxation time at time t, with dependence on t arising from a specified dependence of λ on temperature and a known or calculated temperature history. The (zero shear rate) viscosity for this model is

$$\eta(T) = \lambda(T)G \qquad (7.43)$$

and the temperature dependence of η and λ is given by eqn (7.27), with $E/R = 7020$ K and, for one typical LDPE at 433 K, $\eta = 8 \cdot 18 \times 10^4$ Pa s (i.e. $\eta_* = 7 \cdot 44 \times 10^{-3}$ Pa s in eqn (7.27)). With the integral form of constitutive equation we require a history of the deformation for $t \leq 0$ (taking $t = 0$ to correspond to $z = 0$). Wagner takes a notional history in which the film deforms as though its state for $t < 0$ were $r = r_*$, $h = h_*$ (and $v = v_* = \dot{M}/2\pi\rho r_* h_*$) at temperature T_0; for $t = 0$ the state is $r = r_0$, $h = h_0$, $v = v_0$. Here r_* and h_* are the values observed if the extruded tube is allowed to relax with no applied forces (unconstrained extrudate swell). This gives

$$\mathbf{T}^E = \int_0^t \frac{G}{\lambda(t')} \mathbf{C}_t^{-1}(t') \exp\left\{ - \int_{t'}^t \frac{dt''}{\lambda(t'')} \right\} dt' + G\mathbf{C}_*^{-1}(t) \exp\left\{ - \int_0^t \frac{dt''}{\lambda(t'')} \right\}$$

(7.44)

where $\mathbf{C}^{-1}$ is given in eqn (7.31).

The numerical solution of the equations obtained from eqns (7.1), (7.2), (7.14) and (7.44) (with use of several other equations, and possibly eqns (7.11), (7.12) and (7.13) in place of eqns (7.1) and (7.2)) is carried out from die to freeze-line. The velocity v is used as a dependent variable rather than the film thickness h, as in eqn (7.20). The integrals in eqn (7.44) are approximated by use of Simpson's rule and a fourth-order Runge-Kutta method is used for the differential equations. In his work Wagner seeks computed solutions which describe experimental results in which the inflation pressure, p_i, and applied force, F_1, are measured. In addition the values of bubble radius, velocity and temperature are measured along the film and so values at die and freeze-line are available for the computation. The computation is structured so that values of $\eta(T_0)$, $\lambda(T_0)$, h (the heat transfer coefficient, eqn (7.14)) and $\theta(0)$ are guessed. Integration from $z = 0$ up to a value of z at which $\theta = 0$ gives values of z, r, v and T which should agree with the observed z_1, r_1, v_1 and T_1. An iterative scheme (described as Newton–Raphson, but using secant-like approximations to derivatives) is used to find the correct values of the four guessed parameters. In the development of this program, the effect of gravity was incorporated. This required a guess for $F(0)$ and solution of the additional differential equation, eqn (7.13). $F(0)$ has to be adjusted so that $F(z_1) = F_1$, the measured value. The change in F along the bubble for the experiments considered was so small that this feature of the program was dropped. One final point concerns $v(0)$. This was not set equal to v_0 ($= \dot{M}/2\pi\rho r_0 h_0$) but chosen so that the axial stress satisfies eqn (7.44) at $z = 0 +$ (i.e. for $t = 0 +$, when it is effectively eqn (7.31)) as well as eqn (7.2).

7.4.3 A non-isothermal Maxwell model expressed as a rate equation

Gupta and Metzner (1980) use a similar constitutive equation (with some novel features) expressed as a rate equation, and adopt the same philosophy of using the measured force, pressure, bubble shape, velocity and temperature to deduce parameters in the constitutive equation. They found that the reverse procedure of trying to make predictions of these measured quantities based on given values of the parameters, was extremely sensitive to initial values of stresses. This is in agreement with the experience reported above (Petrie, 1973). The equation they use is an upper convected Maxwell model 'generalised to account for (the) transient non-isothermal effects':

$$\mathbf{T}^{E}(1 + \lambda B\dot{T}/T) + \lambda\overset{\triangledown}{\mathbf{T}}^{E} = 2\eta\mathbf{D} - \eta(B + 1)(\dot{T}/T)\mathbf{I} \qquad (7.45)$$

where λ and η are the usual Maxwell parameters but are allowed to depend on the temperature and on the second invariant of $\mathbf{D}$ as in the White–Metzner model to which eqn (7.45) reduces if $\dot{T} = 0$ ($\dot{T}$ is the material derivative, $\partial T/\partial t + \mathbf{v}.\nabla T$, of the temperature). B is a constant, and a value of 4 is used to fit some film-blowing data for polystyrene. ($B = 12$ has been suggested for other data; there is not a great deal of data available from which values of B may be deduced.)

The technique used by Gupta (1980) is to use the constitutive equation to express the relaxation time in terms of the stress and rate of strain components and their derivatives, the viscosity function and the parameter B. Values of λ are calculated from the experimental data at chosen points along the bubble. These are compared with those obtained from normal stress measurements in shear at corresponding rates of strain. Quite clearly the problem being addressed here is one of estimating material properties rather than using the analysis of the mechanics of the process to make predictions about bubble shape, kinematics, stresses or temperatures. This does seem to lead to an easier computational problem, but even so Gupta notes that some care was needed to ensure that the calculated λ was positive. The process does involve differentiation of experimental data and so has to be carried out with great care. Finally we note that Metzner (in an as yet unpublished lecture) reported that while predictions of normal stress differences obtained from the axial stress component (t_{11}) are acceptable (within 20% of predictions from shear flow data in the worst case), the predictions from the transverse stress component (t_{33}) are extremely poor. It may be appropriate to contrast this with the predictions of strain-rate components using the non-isothermal Newtonian theory (Petrie, 1975a) the results of which do not agree well with experimental data (Farber and Dealy, 1974); however, agreement is better for the transverse components.

7.4.4 A non-isothermal Voigt model

The only other viscoelastic model used in film-blowing studies is the Voigt model adopted by Pearson and Gutteridge (1978). This can be regarded most simply as having a stress given by a linear combination of elastic and viscous stresses (the elastic component being obtained from their unusual elastic model). Using dimensionless variables defined in eqn (7.36) the stress components are

$$t_{11} = G(T)(\tilde{v} - \tilde{v}^{-3}\tilde{r}^{-2}) + G(T)\tilde{\Lambda}(T)\tilde{v}\cos\theta$$

$$\times \left\{2\tilde{v}^{-3}\tilde{r}^{-3}\frac{\mathrm{d}\tilde{r}}{\mathrm{d}\tilde{z}} + (1 + 3\tilde{v}^{-4}\tilde{r}^{-2})\frac{\mathrm{d}\tilde{v}}{\mathrm{d}\tilde{z}}\right\} \quad (7.46)$$

$$t_{22} = G(T)(\tilde{r} - \tilde{v}^{-2}\tilde{r}^{-3}) + G(T)\tilde{\Lambda}(T)\tilde{v}\cos\theta$$

$$\times \left\{2\tilde{v}^{-3}\tilde{r}^{-3}\frac{\mathrm{d}\tilde{v}}{\mathrm{d}\tilde{z}} + (1 + 3\tilde{v}^{-2}\tilde{r}^{-4})\frac{\mathrm{d}\tilde{r}}{\mathrm{d}\tilde{z}}\right\} \quad (7.47)$$

Here $G(T)$ is the temperature-dependent modulus, $\Lambda(T)$ is the temperature-dependent retardation time and $\tilde{\Lambda}(T) = \Lambda(T)v_*/r_*$. Putting these into the dynamical equations, eqns (7.1) and (7.2), gives a third-order system similar in structure to the Newtonian model.

The boundary conditions prescribed are values of r, v and θ at $z = 0$ and $z = z_1$. Three of these give solutions of the systems of differential equations, and the other three are used to determine (iteratively) values of z_1 and the parameters p_i and F_1. The material parameters G and Λ are given functions of temperature for the calculation, and the temperature profile $T(z)$ is also given (imposed by the heating and cooling arrangements for the process). The basic difference from the Newtonian model is that z_1 is determined by the condition $\theta(0) = 0$, which could not be imposed for the Newtonian model (even if it appeared to be physically correct). Integration from the freeze-line to the die was employed for this model, and a fourth-order Runge-Kutta procedure was satisfactory. Integration from die to freeze-line (with values of parameters found to be correct) was found liable to instability and, typically, the numerical solution broke down ($\theta \to \infty$). An attempts to use an integral formulation of eqns (7.46) and (7.47) (for the strains in terms of the stresses) was abandoned; it was only successful with good initial guesses for the stress functions (Gutteridge, 1977). The iterative determination of F_1 and p_i is done very simply. The integration is stopped when $\theta = 0$ and this determines z_1 for the guessed F_1 and p_i. The values of $v(0)$ and $r(0)$ are compared with the correct values and empirically determined corrections are made to F_1 and p_i (proportional to the errors in $r(0)$ and $v(0)$ respectively). It should perhaps be remarked that, as carried out by Gutteridge, the calculation does not involve p_i explicitly, but rather the ratios F_1/p_i and $G(T)/p_i$, and the latter is adjusted in a way that implies changes to $T(z)$ and p_i.

7.4.5 Discussion
The present state of knowledge can be described simply as inadequate. It appears that solutions of viscoelastic model equations for the film-blowing dynamics is possible, but difficult. Use of an integral form of constitutive equations seems, on the basis of one set of computations (Wagner, 1976), to be the most promising approach. The experience of Gutteridge (1977) shows that a Voigt model leads more easily to solutions than do the elastic models discussed above. It remains important to distinguish this analytical and predictive use of models from the 'rheometrical' use, determining material functions from the kinematics and dynamics.

7.5   BLOW MOULDING AND THERMOFORMING

Some simplified processes, which serve as prototypes for a variety of moulding and forming operations will be discussed. The basic extrusion blow-moulding process consists of four stages:

(i)   a cylindrical tube of molten polymer (the parison) is extruded;
(ii)   the mould is closed on the parison, clamping it at both ends and separating it from the material still in the extrusion die;
(iii) air is blown into the parison to inflate it and force it into contact with the mould;
(iv)   the moulded article is allowed to cool, and after sufficient time the mould is opened and the moulded article ejected.

In thermoforming we start with a sheet of softened polymer instead of the cylindrical parison, and a combination of blowing or suction and mechanical assistance may be used to force the sheet into a mould. A typical simple article to be formed might be a beaker or a flowerpot, while blow moulding is commonly used for bottles.

In modelling these processes a useful first objective is the prediction of the wall thickness of the moulded or formed article, and in particular prediction of how the wall thickness varies along and around the article. We can identify three problems in the mechanics which influence this; first the parison formation (extrusion), secondly the inflation of a cylindrical tube clamped at both ends, or of a sheet clamped around its edge, and thirdly the contacting of an inflating sheet of material with the mould.

7.5.1 Parison formation in blow moulding
No full mechanical analysis has been attempted for the parison formation stage, and there is little. to be said in the context of modelling and

computation. The dominating factor is the extrudate swell and in particular the fact that this is time-dependent. There is the additional feature of parison sag; as the parison is extruded the weight of material already extruded draws the parison down, and so the top of the parison will be thinner than the bottom. The cooling of the parison as it is extruded will affect both these processes. The mechanics of extrudate swell for this unsteady flow have not been studied theoretically or computationally, even for an isothermal flow.

An alternative to a full mechanical analysis is the use of some empirically obtained functions and even this is frustrated by a lack of data, as Kamal *et al.* (1981) report. They ignore the cooling of the parison and present a method of calculation of parison diameter based on knowledge of the transient diameter and thickness swell, given by

$$B_1(t) = \frac{\text{Mean extrudate diameter at time } t \text{ after extrusion}}{\text{Mean die diameter}}$$

$$B_2(t) = \frac{\text{Parison thickness at time } t \text{ after extrusion}}{\text{Die gap}}$$

If parison sag is to be considered the tensile creep compliance $J(t)$ is required. In view of the lack of data we shall not discuss this further here.

7.5.2 Parison and sheet inflation (blow moulding and thermoforming)

The inflation stage in blow moulding, before the inflating parison contacts the mould, is governed by the same equations, eqns (7.1) to (7.7), as film blowing. For an axisymmetric situation (which implies an idealisation of end effects where the mould closes onto the parison and cuts it) with a parison whose initial thickness is everywhere the same, identical solutions would be obtained with an elastic constitutive equation. The same difficulties were experienced in finding solutions here (Petrie and Ito, 1980) as in the elastic film-blowing model discussed above.

A simpler approach has been adopted by Kamal *et al.* (1980). If the mechanics are ignored, or the assumption made that the parison radius is large, one may assume that a longitudinal section of the parison will be the arc of a circle (of decreasing radius as inflation proceeds). This approach is based on the work of Fukase *et al.* (1978) which will be discussed below in connection with the contact of parison and mould. The simplest model here does not give results very close to experimental observation. Kamal *et al.* therefore make unspecified modifications to the model to take account of the redistribution of material near the pinched end of the parison. This gives better agreement with experiment. These authors also modify the model to take account of the elliptical cross-sectional shape caused by the pinching-off at the end of the parison. This is intended to predict thickness variation around the circumference, but again details are not completely specified, and agreement with experiment is not particularly good.

Williams (1970) has performed an analysis based on a large strain approximation to the rubber–elastic constitutive equation. This takes a form equivalent to eqn (7.31) with h/h_* small so that t_{33}^E is small and instead of eqns (7.6) and (7.7), $t_{11} = t_{11}^E$ and $t_{22} = t_{22}^E$ are used. This particular constitutive equation allows a spherical shape, if this is obtained during thermoforming from a circular disc, to satisfy the dynamical equations. This gives us a greatly simplified problem, and it has to be noted that the model represents a material in which stresses equal to the modulus G, are present in an unstretched sheet. Clearly this is a situation for which the approximation is not expected to work and Williams offers suggestions for correction of the solution. The approximation to the deformation is used to calculate corrections to the stresses by including the terms in t_{33}^E omitted from eqns (7.6) and (7.7). Averaged values of these (first) corrections give better estimates of inflation pressure, and Williams suggests an iterative scheme of successive correction in turn to the stress and to the deformation, but this has not been implemented.

Wineman (1976) considers the inflation of a circular sheet of a non-linear viscoelastic solid, using material properties for a vulcanised styrene–butadiene rubber. The technique is discussed in detail and a number of important comments are made—on extrapolating beyond the range of validity of a constitutive equation, on stability and bifurcation of solutions, and on computational difficulties. The formulation of the problem suffers from the disadvantage that lengthy algebraic manipulation, the details of which are affected by the form of constitutive equation, is required before the problem is ready for computational solution. In a subsequent paper Wineman (1978) reformulates the problem in a way which allows for much easier use of a variety of constitutive equations. He solves the same problem for an elastico-viscous liquid described by a Kaye–BKZ constitutive equation with data describing a polyisobutylene. A third paper (Wineman, 1979) applies the same technique to the simultaneous inflation and elongation of a circular cylindrical sheet (a tubular membrane). The geometrical difference does lead to different equations in the two cases, but the structure of the two problems is so similar that we need only discuss details for one, and the inflation and elongation of a tubular membrane bears the closest resemblance to the film-blowing problems we have discussed above.

The geometrical situation is that a cylindrical membrane has its ends bonded to discs of radius a which are moved apart so that the distance between them is given by the function $2L(t)$, with $L(0) = L_0$. An inflation pressure $p_i(t)$ is applied, and the cylinder becomes barrel-shaped. The material particle with polar coordinates (a, θ, z) at $t = 0$ moves to $(r(z, t), \theta, y(z, t))$ at time t. The stretch ratios are λ_1 in the longitudinal (meridional) direction and λ_2 in the transverse (circumferential) direction, so $\lambda_2 = r/a$ and $\lambda_1^2 = (\partial r/\partial z)^2 + (\partial y/\partial z)^2$. The additional variable $x = \lambda_1^{-1}\partial r/\partial z$ is introduced and dimensionless stresses

$$\hat{T}_1 = T_{11}^E/\lambda_1 S_0 \qquad \hat{T}_2 = T_{22}^E/\lambda_2 S_0 \qquad (7.48)$$

are defined (note the scaling by λ_1 and λ_2; S_0 is a characteristic stress). With lengths scaled by $L_0(\hat{z} = z/L_0$, etc.) and $l = L_0/a$, the system of equations obtained is

$$\frac{\partial \hat{T}_1}{\partial \hat{z}} = \hat{T}_2 \times l \qquad (7.49)$$

$$\frac{\partial \lambda_2}{\partial \hat{z}} = \lambda_1 \times l \qquad (7.50)$$

$$\frac{\partial x}{\partial \hat{z}} = l((1 - x^2)\hat{T}_2 - p_i\lambda_1\lambda_2\sqrt{(1 - x^2)})/\hat{T}_1 \qquad (7.51)$$

where all the dependent variables and the parameter p_i are functions of time, t. The constitutive equation is written in the form

$$\hat{T}_k(t) = F_k(\lambda_1(t), \lambda_2(t), t) + \int_0^t G_k\left(\frac{\lambda_1(t)}{\lambda_1(s)}, \frac{\lambda_2(t)}{\lambda_2(s)}, t - s\right)ds \qquad (7.52)$$

for $k = 1$ and 2.

With origin $(y = z = 0)$ midway between the two ends of the tubular membrane, boundary conditions are

$$\lambda_2(1, t) = 1 \qquad (7.53)$$

since the ends have constant radius a,

$$x(0, t) = 0 \qquad (7.54)$$

from symmetry, and

$$\int_0^1 \lambda_1 \sqrt{(1 - x^2)} \, d\hat{z} = \hat{L}(t) \qquad (7.55)$$

If we think of eqn (7.52) as giving values of $\hat{T}_2(t)$ and (by inversion) $\lambda_1(t)$ in terms of $\hat{T}_1(t)$ and $\lambda_2(t)$ we can treat eqns (7.49), (7.50) and (7.51) as a third-order system of ordinary differential equations for $\hat{T}_1$, λ_2, and x as functions of $\hat{z}$ for a particular time t. With eqns (7.53), (7.54) and (7.55) we have a two-point boundary-value problem, which Wineman solves by a shooting method. Obviously we need values of $\lambda_1(s)$ and $\lambda_2(s)$ for past times s for each material particle in order to evaluate the integrals in eqns (7.52).

The details of computation are important both in regard to the solution of the initial value problem and the associated evaluation of the integrals in eqns (7.52), which are discussed by Wineman (1978), and the two-dimensional shooting method used to satisfy the boundary conditions (Wineman, 1979). In order to discuss the problem we consider a set of times $t_1, t_2, t_3, \ldots, t_n$ (with $t_1 = 0$, $t_n = t$ which is the 'present' time) and a set of values of $\hat{z} = z_1, z_2, z_3, \ldots, z_m$ (with $z_1 = 0$, $z_m = 1$). To start the computation, initial values are required: $\lambda_1 = \lambda_2 = 1$ and $x = 0$, for all z, for an unstretched membrane. The relations $\hat{T}_1 = \hat{T}_2 = 0$ are deduced from these for the form of eqn (7.52) used by Wineman. These conditions assume that the inflation pressure, $p_i(t)$, and length, $L(t)$, are continuous functions of t starting from values 0 and L_0 respectively at $t = 0$. The zero initial stresses are what we expect from a reasonable constitutive equation. Then in proceeding from time t_{n-1} to time t_n, values of $\lambda_1(0, t_n)$ and $\lambda_2(0, t_n)$ are guessed. From these, using discretised forms of eqns (7.52), $\hat{T}_1$ and $\hat{T}_2$ may be calculated and x is given by eqn (7.54). Then eqns (7.49) to (7.51) can be solved for an increment in z to give $\hat{T}_1$, λ_2 and x at (z_2, t_n). $\hat{T}_2(z_2, t_n)$ is obtained from the second of eqns (7.52) and the first of eqns (7.52) has to be solved for $\lambda_1(z_2, t_n)$ by iteration. This process is repeated for $z_3, z_4, \ldots, z_m$.

Then conditions (7.53) and (7.55) have to be satisfied and these are, in effect, two non-linear equations to be solved for $\lambda_1(0, t_n)$ and $\lambda_2(0, t_n)$. An iterative process is described by Wineman (1979) which is a discretisation of the Newton–Raphson method for two unknowns. The initial guesses for $\lambda_1(0, t_n)$ and $\lambda_2(0, t_n)$ are obtained by quadratic extrapolation from values at $t_{n-1}, t_{n-2}, t_{n-3}$ for $n > 3$. Linear extrapolation is used for $n = 3$, and for $n = 2$, values are supplied by the user of the computer program. Two further observations may be made. Wineman (1978, pp. 259–60) notes that the evaluation of the integrals in eqns (7.52) need not be done for each iteration of the Newton–Raphson process if the constitutive equation is of a suitable form, so that we may write

$$\int_0^t G_k \left(\frac{\lambda_1(t)}{\lambda_1(s)}, \frac{\lambda_2(t)}{\lambda_2(s)}, t - s \right) ds$$

$$= \sum_{j=1}^{J} F_{kj}(\lambda_1(t), \lambda_2(t)) \int_0^t M_{kj}(\lambda_1(s), \lambda_2(s), t - s) \, ds \qquad (7.56)$$

This requires the time-consuming calculation of integrals once only for each value of z_i at each t_n. Secondly it seems to this author that the integral in eqn (7.55) would most conveniently be evaluated by adding an extra differential equation,

$$\frac{\partial \zeta}{\partial \hat{z}} = \lambda_1 \sqrt{(1 - x^2)} \qquad (7.57)$$

to the set, eqns (7.49) to (7.51), with boundary conditions

$$\zeta(0, t) = 0 \qquad \zeta(1, t) = \hat{L}(t) \qquad (7.58)$$

the second of which clearly replaces eqn (7.55).

For further details and careful discussion of many aspects of this work, the papers of Wineman (1978, 1979) are recommended.

7.5.3 Contact with the mould

The contacting of the inflating sheet or parison with the mould is dealt with by assuming that there is no further deformation of the material once it comes in contact with the mould. This may be explained on the basis of friction (Williams, 1970) or solidification of the material (Petrie, 1975c) and has been discovered or rediscovered by these and other authors (Fukase et al., 1978; Muller, 1973).

The latter authors assume that the geometry of the inflation is given and that the membrane assumes a spherical or cylindrical shape (depending on the details of the process being considered). In general this assumption will not be strictly true except for truly two-dimensional situations (Petrie, 1975c) and a complete solution of the problem would require the proper analysis of the inflation as discussed in the previous section. Without this the results obtained, notably by Fukase et al. (1978), are usable engineering approximations whose reliability may vary with different geometries. The computation involved is straightforward, generally involving the evaluation of algebraic expressions. It should be noted that the model assumes a thin membrane, and so will not be reliable near corners where the radius of curvature is as small as the thickness. The only computational point is that some situations arise where solutions of a differential equation gives a very complicated algebraic expression, and when numerical solutions are sought it may be more convenient and faster to solve the differential equation numerically rather than using its algebraic solution.

Finally some preliminary work on combining the full analysis of the membrane inflation and contacting stages may be mentioned. This was done (Petrie and Ito, 1980) with an elastic constitutive equation, so that each stage in the inflation can be analysed as a static (or quasi-static) problem. The equations are the same as eqns (7.33) to (7.35) if $\tilde{v}$ is replaced by λ_1, the stretch ratio in the longitudinal direction. Also, $\tilde{r}$ is interpreted as λ_2, the transverse stretch ratio, and the thickness of the membrane is reduced in the ratio $1/\lambda_1\lambda_2$. The main problem is not one of computation so much as of formulation of the boundary-value problem. We omit details here and merely note that, once again, it is found that numerical solution of the two-point boundary-value problem requires good initial guesses if a shooting method is to be used.

ACKNOWLEDGEMENTS

The author thanks those who have provided material from their dissertations, and those who have supplied unpublished work and comments on the issues involved, including W. Ast, R. Farber, R. K. Gupta, P. A. Gutteridge, M. R. Kamal, R. R. Muller, M. H. Wagner and A. S. Wineman.

REFERENCES

Farber, R. and Dealy, J. M. (1974). Strain history of melt in film blowing, *Polym. Eng. Sci.*, **14**, 435–40.

Fukase, H., Iwaaki, A. and Kunio, T. (1978). A method of calculating the wall thickness distribution in blow moulded articles, *SPE Tech. Papers*, **24**, 650–2.

Gupta, R. K. (1980). A new non-isothermal rheological constitutive equation and its application to industrial film blowing processes, Ph.D. Thesis, University of Delaware.

Gupta, R. K. and Metzner, A. B. (1980). Non-isothermal flow of viscoelastic fluids, In *Rheology*, Vol. 3 (Ed. G. Astarita, G. Marrucci and L. Nicolais), Plenum Press, New York, pp. 3–8.

Gutteridge, P. A. (1977). A theoretical analysis of the film-blowing and related processes for elastic and viscoelastic materials, Ph.D. Thesis, Imperial College, London.

Han, C. D. (1976). *Rheology in Polymer Processing*, Academic Press, New York.

Han, C. D. and Park, J. Y. (1975). Studies on blown film extrusion II. Analysis of the deformation and heat transfer processes, *J. Appl. Polym. Sci.*, 19, 3277–90.

Kamal, M. R., Tan, V. and Kaylon, D. (1981). Measurement and calculation of parison dimensions and bottle thickness distribution during blow moulding, *Polym. Eng. Sci.*, 21, 331–8.

Middleman, S. (1977). *Fundamentals of Polymer Processing*, McGraw-Hill, New York.

Muller, R. R. (1973). Vacuumdieptrekken, Bachelor's Dissertation, Twente University of Technology.

Pearson, J. R. A. (1982). *Mechanics of Polymer Processing*, Hemisphere, New York.

Pearson, J. R. A. and Gutteridge, P. A. (1978). Stretching flows for thin film production I. Bubble blowing in the solid phase, *J. Non-Newtonian Fluid Mechanics*, 4, 57–72.

Pearson, J. R. A. and Petrie, C. J. S. (1970). The flow of a tubular film. Part 2. Interpretation of the model and discussion of solutions, *J. Fluid Mechanics*, 42, 609–25.

Petrie, C. J. S. (1973). Memory effects in a non-uniform flow: a study of the behaviour of a tubular film of viscoelastic fluid, *Rheol. Acta*, 12, 92–9.

Petrie, C. J. S. (1974). Mathematical modelling of heat transfer in film blowing: a case study, *Plastics and Polymers*, 42, 259–64.

Petrie, C. J. S. (1975a). A comparison of theoretical predictions with published experimental measurements on the blown film process, *AIChE J.*, 21, 275–82.

Petrie, C. J. S. (1975b). Mathematical modelling and the systems approach in plastics processing: the blown film process, *Polym. Eng. Sci.*, 15, 708–24.

Petrie, C. J. S. (1975c). The analysis of polymer processing operations involving free surface elongational flows. In *Polymer Rheology and Plastics Processing* (Ed. P. L. Clegg, F. N. Cogswell, D. E. Marshall and S. G. Maskell), Plastics and Rubber Institute, London, pp. 307–19.

Petrie, C. J. S. and Ito, K. (1980). Prediction of wall thickness of blow moulded containers, *Plastics and Rubber: Processing*, 5, 68–72.

Thompson, J. W. (1975). An elastic model for the film-blowing process and problems in its numerical solution, M.Sc. Dissertation, University of Newcastle upon Tyne.

Wagner, M. H. (1976). Ein rheologisch-thermodynamisches ProzeBmodell des Folienblasverfahrens, Dr.-Ing. Dissertation, Universitat Stuttgart.

Williams, J. G. (1970). A method of calculation for thermoforming plastics sheets, *J. Strain Anal.*, 5, 49–57.

Wineman, A. S. (1976). Large axisymmetric inflation of a nonlinear viscoelastic membrane by lateral pressure, *Trans. Soc. Rheol.*, 20, 203–25.

Wineman, A. S. (1978). On axisymmetric deformations of nonlinear viscoelastic membranes, *J. Non-Newtonian Fluid Mechanics*, 4, 249–60.

Wineman, A. S. (1979). On the simultaneous elongation and inflation of a tubular membrane of BKZ fluid, *J. Non-Newtonian Fluid Mechanics*, 6, 111–25.

CHAPTER 8

Coating Flows

S. F. KISTLER and L. E. SCRIVEN

Department of Chemical Engineering and Materials Science,
University of Minnesota, Minneapolis, USA

8.1 INTRODUCTION TO THE THEORY OF COATING FLOWS

Coating flows are small-scale viscous free surface flows by which a film of liquid, Newtonian or non-Newtonian, is continuously deposited on a flexible or rigid moving substrate. The liquid displaces gas, usually air, from the substrate; where this occurs is known as a wetting line. The preferred flows are steady and deposit a uniformly thin film of liquid devoid of imperfections of any sort. Coating flows share essential features with mould-filling flows (cf. Chapter 5), extrudate swell flows (cf. Chapter 3), film-blowing flows (cf. Chapter 7), and fibre-spinning flows (cf. Chapter 6). Coating applications are legion and industrial coating operations come in great variety: blade-, slot-, slide-, extrusion-, curtain-, and roll-coating are examples, some of which are used as illustrations in Section 8.5.

The objective of the computer-aided methods summarised in this chapter is an accurate fluid mechanical theory of coating flows. There are various complications: see Fig. 8.1. Gas–liquid interfaces appear to intersect solids in seemingly singular contact lines, the physics of which is not yet fully understood. Contact lines come in two varieties: the static attachment or separation line where a fluid interface intersects a die surface with respect to which it is at rest; and the dynamic wetting line where the liquid at the interface continually encounters new substrate to be coated. Immediately downstream of the latter the newly deposited liquid is generally accelerated in boundary-layer fashion by shear-stress-transmitted momentum from the substrate, whilst deeper into the liquid film and at least in the immediate

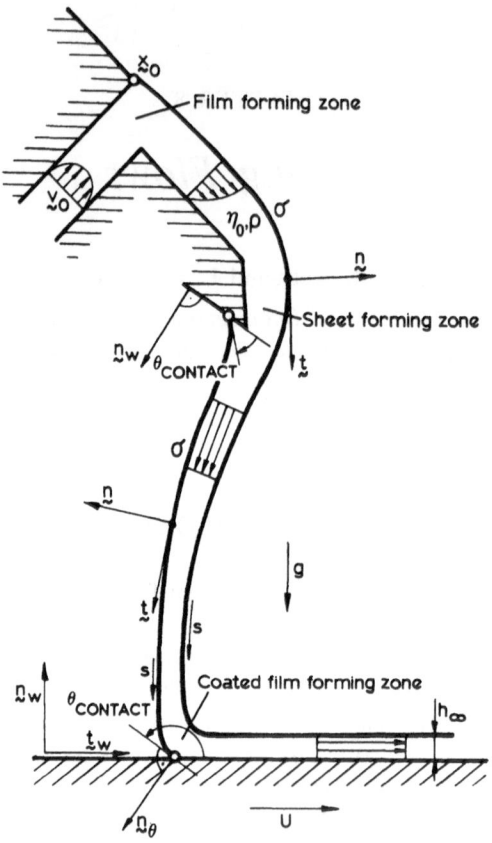

Fig. 8.1.   Schematic of a coating flow with two free surfaces, two separating
contact lines and a dynamic contact line.

vicinity upstream of the wetting line the liquid is accelerated in large
measure by extensional viscous-stress-transmitted momentum from the
liquid already accelerated by shear stress. Upstream of a static separation
line, liquid adheres to or slips very little along the solid wall and so there is
appreciable shear stress; downstream the liquid contacts gas that exerts
negligible shear stress, so that liquid accelerates rapidly along the free
surface. In this situation too there is a sharply varying mixture of shear and
extension in two-dimensional flow. Further complicating theoretical
analysis is the shape of the solid boundaries, which rarely coincide with a
coordinate surface of any convenient coordinate system, and of the free
surfaces between liquid and gas. The location of the free surfaces is

unknown *a priori*. They can be highly curved and where they are the flow typically rearranges drastically over a short distance. Short zones of two-dimensionally rearranging flow we call *forming zones*. Deformations in them are usually combinations of shear and extension that vary rapidly in space. In the forming zones boundary layers and internal layers not infrequently appear, but apart from them the standard approximations from fluid mechanics rarely apply because viscous, pressure, and capillary forces, together often with inertial and gravity forces, all contend with one another in a two-dimensional flow. The extrusion swell problem of Chapter 3 provides a particularly simple case of this.

Upstream and downstream, however, the forming zones merge asymptotically with rectilinear or near-rectilinear (in some cases circular or near-circular) inflow and outflow zones that are predominantly shearing or extensional flows. In these zones, which we call *developing zones*, the flow can be approximated by various conventional asymptotic techniques. Strategies for recombining approximations to the basic equations, in order to arrive at appropriate equations for interfacial shape that lead to acceptable estimates of flow in developing zones, are described by Higgins *et al.* (1977), Higgins and Scriven (1979), Higgins (1980) and Kistler (1983). In limiting cases in which the shape of the free boundary comes close enough to that of a simple coordinate surface, coating flows can also be analysed by perturbation methods (Bixler and Scriven, 1983a; Higgins, 1980, 1982; Ruschak, 1974; Ruschak and Scriven, 1976).

The complications in the forming zone defy traditional mathematical analysis, relying as it does on standard shapes of domains and ostensibly infinite sets of basis functions defined on those domains. But those complications yield to modern computer-aided functional analysis by the subdomain/Galerkin approach in which finite-element basis functions are employed (cf. Chapters 2, 3 and 7). These consist of low-order polynomials, each being non-zero only within a few of the relatively small subdomains into which the flow domain is tessellated. The coefficients of these functions are values of the dependent variables on nodes of the tessellation, which is extremely convenient, and the analysis ultimately reduces to solving a set of algebraic equations for these coefficients.

Representations of velocity and pressure fields with finite-element basis functions have limited differentiability, in contrast to the typically analytic or meromorphic basis functions of traditional analysis. Yet no more differentiability is required than the minimum called for by the physics of the flow. For analysis of coating flows, subdomain basis functions are particularly useful and superior to simulation by finite-difference equations for several reasons. Irregular flow domains and free surfaces are handled effectively by spine parametrisations of the surfaces, by coordinated tessellation into subdomains, and by isoparametric mapping of all subdomains onto standard squares—a great array of local transformations to fixed Cartesian coordinates. Complicated boundary conditions, in particular those along the free surfaces, are neatly imposed by retaining them in their physically natural form. Boundary layers and analogous internal zones are readily resolved by reducing the size of the subdomains or elements according to the local steepness of gradients of velocity, strain rate, stress, free surface elevation, etc.—a parallel to the use of stretched coordinates in conventional analysis. When the mathematical formulation omits the physics that would remove a singularity, certain inaccuracies in its neighbourhood are accommodated so that a global solution can be found without full details of the singularity, as in the case of contact lines discussed below. The global solution, approximated as it is by a finite number of continuous basis functions, is a complete functional representation of velocity and pressure, from which pathlines, strain rate, vorticity, stress and other quantities of interest are readily deduced. Furthermore, the Jacobian matrix that arises in Newton iteration, the technique of choice for solving the non-linear equation set to which the Galerkin/finite-element approach leads, is valuable for analysis of the sensitivity, stability, and controllability of the flow field (Brown et al., 1980).

Means have been developed in recent years for making efficient and reliably accurate theoretical predictions of coating flows. The effort was founded in the theory of static meniscus shapes (Brown, 1979; Huh, 1969; Orr, 1976; Rivas, 1972) and asymptotic analyses of viscous free surface film flows (Ruschak, 1974). The first published finite-element analysis of such a flow pertained to die swell in extrusion (Nickell et al., 1974, described in Chapter 3). Flows in forming zones of coating operations of increasing complexity have been analysed, as described below, and both the finite-element formulation and the associated computational techniques have been systematically developed to the state that the complications summarised above are regularly treated in analyses of coating flows. This chapter recounts that state of development as far as it relates to isothermal, steady, and two-dimensional flow of a Newtonian liquid.

Section 8.2 summarises the conservation equations and boundary conditions of viscous free surface flow, and emphasises the conditions at free boundaries and near contact lines. Section 8.3 covers details of the subdomain Galerkin/finite-element approach. Stressed are a very flexible parametrisation of free surfaces, the effective use of the isoparametric mappings, and the weak form of the conservation equation, with natural boundary conditions imposed in the most natural way. Section 8.4 describes the algorithms for computing the coefficients in the basis function expansions of the dependent variables. Discussed in detail is the implementation of Newton's method which allows the shape and location of the free boundaries to be calculated simultaneously with the velocity and pressure fields in a single iteration procedure. Section 8.5 illustrates the predictive capabilities of the approach by reviewing the finite-element analyses of coating flows that are now available. Featured is how each flow was broken down into forming zones that require fully two-dimensional theory entailing relatively costly computations, and developing zones where various cost-saving approximations are adequate. Section 8.6 summarises the current status of coating flow theory and computation, and of research on developments that are needed; see also Kistler and Scriven (1983).

8.2 GOVERNING EQUATIONS AND BOUNDARY CONDITIONS FOR STEADY COATING FLOWS

For steady, isothermal flow of incompressible liquid, the dimensionless form of the equation expressing local conservation of linear momentum is

$$Re\,\nabla \cdot \mathbf{vv} = \nabla \cdot \mathbf{T} + St\,\mathbf{f} \tag{8.1}$$

Here $\mathbf{v}$ is the dimensionless velocity. For a Newtonian liquid the total stress $\mathbf{T}$ is $-p\mathbf{I} + [\nabla\mathbf{v} + (\nabla\mathbf{v})^T]$. $\mathbf{f}$ is the unit vector in the direction of the force of gravity. Length is made dimensionless with L, a length scale appropriate to the system; velocity with U, a characteristic velocity such as substrate speed; and pressure and stress with $\eta_0 U/L$, where η_0 is the viscosity of the liquid. $Re \equiv \rho UL/\eta_0$ is the Reynolds number and $St = \rho gL^2/\eta_0 U$ is a Stokes number characterising the ratio of viscous forces to gravity (ρ is liquid density and g is gravitational acceleration). The Navier–Stokes equation, eqn (8.1), is retained in the divergence form in which it comes from the global principle of momentum conservation.

At the free surface between liquid and any gas that to good approximation is inviscid and inertialess, the momentum principle leads to the force balance†

$$\mathbf{n} \cdot \mathbf{T} = \frac{1}{Ca}\frac{d\mathbf{t}}{ds} - \mathbf{n}p_a \tag{8.2}$$

† This is equivalent to eqn (1.13) when p_a is included in $\mathbf{T}$, $C_f = 0$ and $d\mathbf{t}/ds$ is recognised as $H\hat{\mathbf{n}}$.

where $Ca \equiv \eta_0 U/\sigma$ is the capillary number, σ being the surface tension; p_a is the ambient gas pressure; $\mathbf{t}$ is the unit tangent that points in the direction of increasing arc length s along the free surface; $\mathbf{n}$ is the outward pointing unit normal (Fig. 8.1). Boundary condition, eqn (8.2), relates the total normal stress $\mathbf{nn}{:}\mathbf{T}$ on the liquid side to the curvature $d\mathbf{t}/ds$ by which surface tension generates a normal resultant, and requires the tangential or shear stress $\mathbf{tn}{:}\mathbf{T}$ on the liquid side to vanish. At the interface between two viscous liquid layers A and B the force balance is

$$\mathbf{n}_A \cdot \left(\mathbf{T}_A - \frac{\eta_B}{\eta_A} \mathbf{T}_B \right) = \frac{1}{Ca_{AB}} \frac{d\mathbf{t}_{AB}}{ds_{AB}} \qquad (8.2a)$$

or equivalently

$$\mathbf{n}_B \cdot \left(\mathbf{T}_B - \frac{\eta_A}{\eta_B} \mathbf{T}_A \right) = \frac{1}{Ca_{BA}} \frac{d\mathbf{t}_{AB}}{ds_{AB}} \qquad (8.2b)$$

Here $\mathbf{n}_A$ and $\mathbf{n}_B = -\mathbf{n}_A$ are the respective unit outward normals and $\mathbf{t}_{AB}$ is the unit tangent pointing in the direction of increasing arc length s_{AB} along the interface. A viscosity ratio η_B/η_A or η_A/η_B appears when $\mathbf{T}_A$ and $\mathbf{T}_B$, the total stress states on either side of the interface, are made dimensionless with η_A/UL and η_B/UL respectively. The capillary number $Ca_{AB} \equiv \eta_A U/\sigma_{AB}$ or $Ca_{BA} \equiv \eta_B U/\sigma_{AB}$ depends on σ_{AB}, the interfacial tension, if there is any. Boundary condition (8.2a) or (8.2b) requires the shear stress to be continuous across the interface and relates the total normal stress difference to the curvature $d\mathbf{t}_{AB}/ds_{AB}$. If the layers are miscible there is no interfacial tension whatsoever and so the normal stress must then also be continuous across the interface.

The continuity equation expresses local conservation of mass within the flow:

$$\nabla \cdot \mathbf{v} = 0 \qquad (8.3)$$

That no mass is supposed to penetrate a free surface or interface is expressed by the kinematic boundary conditions:

$$\mathbf{n} \cdot \mathbf{v} = 0 \quad \text{and} \quad \mathbf{n}_A \cdot \mathbf{v}_A = \mathbf{n}_B \cdot \mathbf{v}_B = 0 \qquad (8.4)$$

The absence of slip at an interface requires continuity of tangential velocity:

$$\mathbf{t}_{AB} \cdot \mathbf{v}_A = \mathbf{t}_{AB} \cdot \mathbf{v}_B \qquad (8.4a)$$

From eqns (8.4) and (8.4a) it follows immediately that at an interface

$$\mathbf{v}_A = \mathbf{v}_B \qquad (8.4b)$$

At solid boundaries the no-slip hypothesis provides an accurate boundary condition except very near contact lines. Ordinarily liquid adheres to stationary confining walls, so that at them

$$\mathbf{v} = \mathbf{0} \qquad (8.5)$$

and liquid sticks to a moving substrate, so that at a substrate having local velocity $\mathbf{v}_{\text{solid}}$ the boundary condition is

$$\mathbf{v} = \mathbf{v}_{\text{solid}} \qquad (8.6)$$

In the neighbourhood of contact lines where fluid interfaces intersect solid boundaries conventional fluid mechanical analysis, based on the hypothesis of steady, two-dimensional Newtonian flow, constant liquid properties and, most significantly, no slip, indicates that stresses would become indefinitely high (Huh and Scriven, 1971). Currently the method of choice for avoiding this apparent stress singularity, which is physically inadmissible, is to replace the no-slip condition near the contact line with some type of slip condition (Silliman and Scriven, 1980). A simple expedient is to specify—or to allow to enter through discretisation—a slip velocity distribution of the form

$$\mathbf{v} - \mathbf{v}_{\text{solid}} = \mathbf{v}_{\text{slip}}(\mathbf{x}_{\text{solid}}) \qquad (8.7)$$

over one or more elements adjacent to the contact line. More physical is Navier's boundary condition that makes the flux of momentum tangential to the wall proportional to the velocity discontinuity or local slip velocity:

$$\beta^{-1}\mathbf{t}_w \cdot (\mathbf{v} - \mathbf{v}_{\text{solid}}) = \mathbf{t}_w \mathbf{n}_w : \mathbf{T} \qquad (8.8)$$

The momentum transfer coefficient is written as $1/\beta$, where β is the slip coefficient. In eqn (8.8) $\mathbf{t}_w$ and $\mathbf{n}_w$ are the unit tangent and normal to the solid surface (cf. Fig. 8.1). Such conditions as eqns (8.7) and (8.8) all involve a length parameter, e.g. $\beta\eta_0$, that measures the extent of the region where slip is significant. At present this parameter remains an empirical input, although various mechanisms can be invoked to justify a transfer coefficient for tangential momentum $1/\beta$. The microfluid mechanics of flows near contact lines is an active research area (Miyamoto and Scriven, 1982; Scriven, 1982; Teletzke et al., 1982).

There are two types of contact lines, static or separating and dynamic or wetting. A prototype flow with the former is the slot extruder, where the free surfaces of the issuing liquid sheet separate from the die lips—the die-swell problem. Actually, this was the first viscous free surface flow problem solved by the finite-element technique (Nickell et al., 1974). Since then it has drawn the attention of various investigators (cf. Chapter 3) and has become a standard problem to test methods for incorporating free surfaces into finite-element simulators of viscous fluid flow (Chang *et al.*, 1979; Ruschak, 1980; Silliman and Scriven, 1980). Richardson (1970) analysed the closely related 'stick–slip' problem and found at the separating contact line a stress singularity inversely proportional to the square root of the distance from the line. Like the leading edge singularity in boundary layer flow, this singularity is integrable: the total shear force on the solid wall is finite. Studying the effect of replacing the usual no-slip boundary condition by a slip condition, eqn (8.8), in die-swell flow, Silliman and Scriven (1978, 1980) found that slip at the wall alleviates the apparent stress singularity. When slip is slight, all the other features of the flow field are matched well by the corresponding flow field with no slip at all. But when the dimensionless slip parameter $\beta\eta_0/b$ exceeds 0·001 (b is the half-width of the die slot) slip reduces die swell substantially, even of Newtonian liquids. Predictions by finite-element calculations employing no slip, however, seem to compare favourably with experiments (Chang *et al.*, 1979; Kistler *et al.*, 1981). Thus near a static contact line no-slip appears to be an adequate boundary condition on liquid motion. A boundary condition on the free surface itself is also required. Either it remains pinned, for instance along a corner or compositional discontinuity, in which case the location of the apparent contact line must be prescribed,

$$\mathbf{x} = \mathbf{x}_{contact} \tag{8.9}$$

or it is free to slide along the wall, in which case the apparent contact angle—the angle between the normal to the wall $\mathbf{n}_w$ and that to the visible free surface at the apparent contact line $\mathbf{n}$—has to be specified:

$$\mathbf{n}_w \cdot \mathbf{n} = \cos\theta_{contact} \tag{8.10}$$

Actually it may remain pinned until the contact angle exceeds a critical value, in which case a combination of eqns (8.9) and (8.10) may be appropriate. The inhibiting effect of sharp edges on spreading of a liquid drop has been studied experimentally (Oliver *et al.*, 1977) and Gibbs' inequality condition for equilibrium of an interface that pins to an edge was confirmed. When there is flow, however, it is unclear whether a similar inequality is the condition for a contact line to remain pinned at a sharp edge or compositional discontinuity, nor is it clear whether the contact angle as the separating contact line moves away from the discontinuity is the static one.

The situation near a dynamic contact line, where the liquid at the free surface continually encounters new substrate to be coated, is quite different. A relatively simple prototype flow incorporating a dynamic wetting line is fluid displacement in a capillary tube or between two plates, which shares essential features with mould-filling flows (cf. Chapter 5). Asymptotic limits of this problem have been analysed by Hocking (1976, 1977) and Huh and Mason (1977), and of related problems by Huh and Scriven (1971) and Dussan (1976). All the analytical results show that the stress singularity that stems from use of the no-slip condition is not integrable, i.e. conventional fluid mechanical analysis predicts unbounded force about the wetting line. Such force is physically impossible, and so one or more of the conventional premises must be wrong. All models used so far to remove the singularity assume that the velocity of the fluid immediately at the contact line and the velocity of the contact line are the same, i.e. the contact line always consists of the same material points. In finite-difference simulations and finite-element analysis this assumption,

$$v_{\text{contact line}} = 0 \qquad (8.11)$$

leads to inherent numerical slip which provides a remedy for the singularity, but unless it is controlled it diminishes as the discretisation is refined at the wetting line. By prescribing a slip coefficient distribution in eqn (8.8) or specifying a slip velocity distribution, eqn (8.7), Silliman (1979) investigated the effect of slip near the moving contact line in capillary displacement flow. Lowndes (1980) used a mesh progressively refined toward the contact line to study the effect of minute slippage of fluid along the wall. Both studies confirmed that the meniscus profile and therefore the macroscopic flow field are sensitive to the slip coefficient or equivalently to the characteristic length of the slip region adjacent to the contact line. The sensitivity is of course greatest nearby. At present the appropriate value of slip coefficient can only be conjectured and then adjusted to improve the computed predictions of measurable flow characteristics.

At a dynamic contact line a specification of the dynamic contact angle θ_{contact} (cf. Fig. 8.1), or more likely the *apparent* dynamic contact angle, is usually required, as in eqn (8.10). At present its value too must be conjectured unless an empirical correlation of pertinent observations is available. Only in the limiting case of a high enough capillary number that the curvature term in eqn (8.2) can be neglected is the dynamic contact angle a dependent variable rather than a parameter to be specified, as illustrated by an example in Section 8.5.

With slip length and apparent dynamic contact angle as input parameters, the governing equations and boundary conditions for steady coating flows are complete, apart from upstream and downstream conditions, which are discussed next. The sensitivity of the finite-element solution to those parameters of momentum transfer in the region close to the wetting line has to be tested by computational experiments in which they are varied. Actually to resolve the flow in that region requires consideration of the microaerodynamics of the gas being displaced, the disjoining or conjoining pressures stemming from long-range inter-molecular forces at submicron scales, and three-dimensional and transient phenomena. All are still under investigation, as noted above.

Far enough upstream and downstream of the forming zones in many, if not most, coating operations the flow is well approximated by simply described rectilinear (sometimes circular) regimes. If inflow and outflow boundaries are placed there, the appropriate Dirichlet or Neumann boundary conditions are easy to specify. Typically the velocity distribution at an inflow plane is plug-like or parabolic,

$$\mathbf{v} = \mathbf{v}_0(\mathbf{x}_{\text{inflow}}) \tag{8.12}$$

and if a free surface reaches there, the position of the meniscus is known,

$$\mathbf{x} = \mathbf{x}_0 \tag{8.13}$$

These are essential or Dirichlet boundary conditions. At the outflow plane it is preferable to specify natural or Neumann conditions, in order to avoid conflicts with overall mass and momentum conservation. Typically the derivatives of velocity and the slopes of interfaces (i.e. unit tangents) are known from analysis of the fully developed regime downstream:

$$\mathbf{n} \cdot \nabla \mathbf{v} = \mathbf{m}(\mathbf{x}_{\text{outflow}}) \tag{8.14}$$

$$\mathbf{t} = \mathbf{t}_{\text{outflow}} \tag{8.15}$$

where $\mathbf{m}$ is a momentum flux distribution over the outflow plane and $\mathbf{t}_{\text{outflow}}$ is the unit tangent to the free surface there. In view of the divergence form of the Navier–Stokes equation, eqn (8.1), the traction condition

$$\mathbf{n} \cdot \mathbf{T} = \mathbf{m}(\mathbf{x}_{\text{outflow}}) \tag{8.16}$$

or, even more generally, a condition on the total momentum flux

$$\mathbf{n} \cdot (-Re\mathbf{vv} + \mathbf{T}) = \mathbf{m}(\mathbf{x}_{\text{outflow}}) \tag{8.17}$$

may be preferred over eqn (8.14). Solutions for which either the asymptotic kinematics or the asymptotic fluxes are imposed at a finite distance upstream or downstream, have to be tested by changing the position of inflow and outflow planes. This procedure succeeds particularly well at inflow planes when the flow is confined and rectilinear. In that case exit effects due to a forming zone further downstream influence the confined flow upstream only within a distance of a few gap widths (Wilson, 1969). Moreover, exit effects are even less pronounced with increasing inertia due to downstream convection. Where there are free surfaces, however, the flow development lengths may be considerably increased and the inflow or outflow planes may have to be placed inordinately far upstream or downstream in order to make the solution insensitive to their location. Examples come from the standing capillary waves that in some cases extend considerable distances above forming zones (Ruschak, 1978) and the slow levelling of liquid films under certain combinations of inertia and surface tension effects (Higgins, 1982).

When an inflow or outflow boundary has to be placed too far upstream or downstream in order to make accurate predictions with essential or natural boundary conditions, a less expensive alternative is to construct a vector boundary condition of the third kind, a Robin condition that relates momentum flux to velocity, and slope of a free surface to its location. Thus with the fully developed state denoted by the subscript ∞, either the traction force $\mathbf{n} \cdot (\mathbf{T} - \mathbf{T}_\infty)$ as in

$$\mathbf{n} \cdot (\mathbf{T} - \mathbf{T}_\infty) = \mathbf{A} \cdot (\mathbf{v} - \mathbf{v}_\infty) \tag{8.18}$$

or the velocity gradient $\mathbf{n} \cdot (\nabla \mathbf{v} - \nabla \mathbf{v}_\infty)$ as in

$$\mathbf{n} \cdot (\nabla \mathbf{v} - \nabla \mathbf{v}_\infty) = \mathbf{B} \cdot (\mathbf{v} - \mathbf{v}_\infty) \tag{8.19}$$

is set proportional to the deviation of velocity $\mathbf{v}$ from the fully developed flow $\mathbf{v}_\infty$. Similarly with a film flow, the proportionality of the free surface slope to the deviation of the local film thickness h from the final thickness h_∞ can be described by

$$\tan [\cos^{-1}(\mathbf{t} \cdot \mathbf{e})] = c(h - h_\infty) \tag{8.20}$$

where $\mathbf{e}$ is the unit vector in the flow direction of the ultimate rectilinear flow regime. The forms of these conditions as well as the values of the proportionality coefficients $\mathbf{A}$ or $\mathbf{B}$ and c emerge from the solution of the appropriate flow problem linearised for small departures from the fully developed regime that is approached asymptotically. Higgins (1982) illustrates the procedure for the downstream development of two-dimensional, steady liquid film flow on a moving horizontal substrate. Boundary conditions (8.18) and (8.20) allow matching the finite-element solution in the forming zone to the asymptotic solution for film flow levelling with distance downstream in the developing zone. In this way the finite-element domain can be shortened from what is needed for comparable accuracy when natural boundary conditions appropriate to fully developed flow rate are imposed (Bixler and Scriven, 1982b).

In some cases, particularly when a liquid film flows as an unsupported sheet or curtain, there is no unidirectional regime around which the equation set can be linearised in order to derive boundary conditions. Nevertheless the flow may approach an evolving non-linear asymptotic regime, where variations in the streamwise velocity and in stress are small across every cross-section of the flowing film, as in the case of the freely falling curtain mentioned below, or where possibly another simple form of velocity distribution in the cross-sections adequately models reality. There are various strategies for recombining the basic equations, eqns (8.1) and (8.3), and boundary conditions into equations of interfacial shape that are useful for such extended developing zones (Higgins and Scriven, 1979; Kistler, 1983). Introducing an approximate velocity distribution and averaging the component momentum equations by integrating over the cross-section of the flow may yield film profile equations that adequately describe the asymptotic regime.† Such equations are referred to as film profile equations in this chapter. They generally are non-linear ordinary differential equations for the location of free surfaces and can be solved by the one-dimensional Galerkin/finite-element method simultaneously with the main flow problem (Kistler and Scriven, 1981; Kistler *et al.*, 1981). This strategy can greatly shorten the finite-element domain in which the full two-dimensional flow problem has to be solved, for it accounts for those downstream features that have an upstream influence. Matching the one-dimensional approximation with the main flow is based on the physical requirements of continuity of mass flux, total momentum flux and interface position, and requires the weak form of the conservation equations as discussed in more detail in Section 8.3.

8.3   FINITE-ELEMENT FORMULATION

Equations (8.1) and (8.3) with associated boundary conditions are solved by a combination of the subdomain and Galerkin methods of weighted residuals, in which the very convenient finite-element basis functions are

---

† This is effectively the approximation used in fibre flow in Chapter 6 and in film flow in Chapter 7.

used. The mathematical foundations of this finite-element technique are well-established (Strang and Fix, 1973). While its application to fluid flow in irregular domains is well worked out (cf. Chapter 2), the presence of one or more free surfaces complicates analysis and raises difficulties that were not fully resolved until recently. This section summarises the most powerful and efficient versions of the finite-element technique currently being used to analyse viscous free surface flows.

The equation set is non-linear owing to the convective momentum transport term in eqn (8.1), if it is important (i.e. if Re is at least of order unity), but always owing to the presence of free surfaces, the locations of which are unknown *a priori* and at which the non-linear boundary conditions, eqns (8.2) and (8.4), apply. Calculating a prediction of a steady flow state requires therefore an iteration procedure that converges to that state, which in turn requires a suitable first approximation to the location of free surfaces as well as to the flow field. After the initial estimate and subsequently after each iteration step the flow domain, which can be quite irregularly shaped, has to be divided into geometrically simple sub-domains, or elements. A simple yet flexible way of parametrising the free surfaces together with an algorithm for appropriately tessellating the computational domain into elements is needed. Saito and Scriven (1981) combined polar parametrisations of highly curved meniscus sections with Cartesian parametrisations of less curved sections.

What is presented here is a much more general scheme which is related in some respect to Ruschak's (1980) boundary-location method. That method requires, however, that free surfaces be approximated by piecewise straight line segments. Here free surfaces and interfaces are parametrised by their locations along conveniently placed, independent spines. Unless it needs to be curved each spine is defined by a base point x_B^i and a direction vector e^i as in Fig. 8.2 which shows an example of a coating flow with two free surfaces. Each free surface is described by its spinal distances from the base points. In Fig. 8.2, for instance, h_T^i and h_B^i stand for the locations of the top and bottom boundaries on the ith spine. Additional spinal distances h_I^i parametrise the location of interfaces, if there are any. Irregular solid boundaries can be represented by taking h_B^i or h_T^i as data, whereas for free surfaces these are unknown coefficients to be evaluated simultaneously with those that come from expansions of velocity and pressure fields in the basis functions. The base points x_B^i and the direction vectors e^i can be chosen as needed to accommodate highly irregular configurations without the free surface representation becoming singular. The spines can be coordinate curves drawn from one or more coordinate systems, when this is

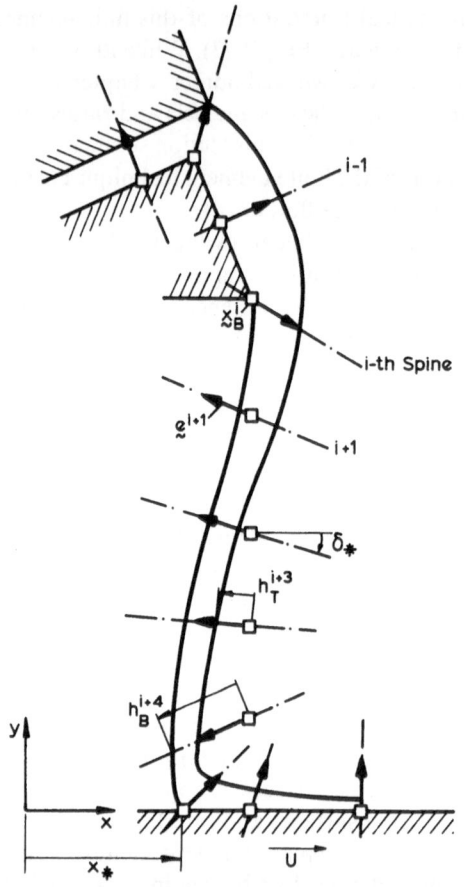

Fig. 8.2.   Free boundary parametrisation scheme.

the most effective way to construct them. If a base curve can be designed so
that normals to it are convenient spines, generating them in this way may be
most effective; but they can also be located and oriented without
restriction. No matter what the choice, the representation of free surfaces
or interfaces is independent of any reference surface. This allows parameters in the coordinates of the base points x_B^i and the directions e^i of the
spines to become additional unknown coefficients so that spines can be
adjusted adaptively and automatically to features of the flow during the
iteration procedure. In the analyses of curtain and slide coating featured in
Section 8.5, for instance, one spine is made to follow the position x_* of the

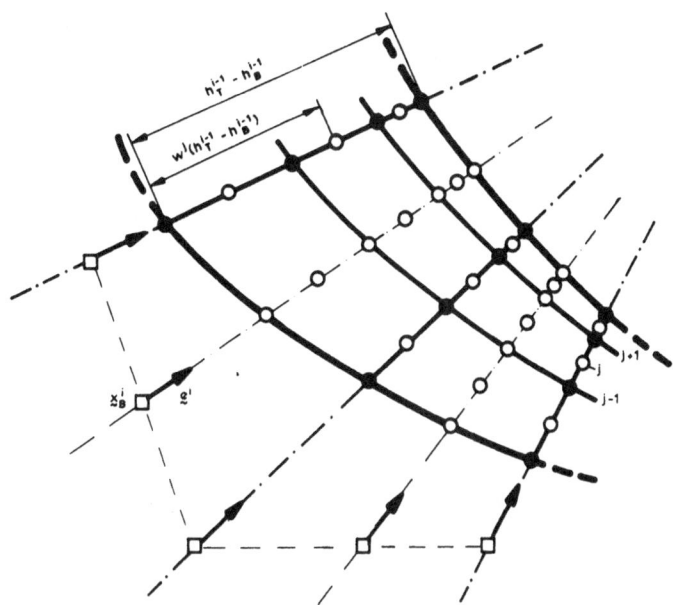

Fig. 8.3. Tessellation of computational domain into finite elements. ●, velocity
and pressure nodes; ○, velocity nodes.

dynamic contact line (see Fig. 8.2) and the others to adjust correspondingly. In another example the spine in the outflow boundary is made
to change its orientation δ_* in order to remain perpendicular to the
midsurface of the liquid sheet so that the conditions of matching to an
asymptotic downstream regime can be imposed.

The domain determined by the free surface locations is tessellated into
quadrilateral elements many of which have two curved sides that reflect the
shape of the free surfaces and two straight sides formed by spines, as shown
in Fig. 8.3. The vertices and midpoints of the elements are called nodes, and
those not on the boundaries lie on spines at positions given by prescribed
proportions w^j of the local thickness of the fluid layer. In the notation of
Fig. 8.3 the location $\mathbf{x}^k$ of node $k(i,j)$ lying on spine i is related to the
positions h_B^i and h_T^i of free surfaces on that spine by

$$\mathbf{x}^k = \mathbf{x}_B^i(\mathbf{x}_*) + [h_B^i + w^j(h_T^i - h_B^i)]\mathbf{e}^i(\delta_*) \qquad (8.21)$$

Here $\mathbf{x}_*$ is the entire set of parameters on which the coordinates of the base
points depend, and δ_* is the entire set of parameters on which the

orientations of the spines depend.† Equation (8.21) links the location of the nodes directly to the position of the boundaries, and there is no possibility of free boundaries overtaking elements during iteration. If there is more than one layer, an expression of the form of eqn (8.21) determines the nodal coordinates in each layer. If the computational domain does not have the shape of a single strip of the kind shown in Fig. 8.2, as for instance for the film-splitting flows encountered in roll-coating, separate families of spines, each leading to a representation of the form of eqn (8.21), are readily patched together. For the algorithms referred to in particular in this chapter the size of elements in the streamwise direction, i.e. the distance between neighbouring spines, is usually refined in regions of steepest gradients on the basis of physical understanding of the fluid mechanics that can be brought to bear beforehand. The size of the elements across the flow, i.e. the nodal spacing along a particular spine, is also refined, for instance progressively toward a contact line, but the total number of elements across the liquid layer is kept constant. It is feasible to vary the number according to the complexities of the flow field; triangular elements are the most obvious choice to manage the transition from one number of quadrilateral elements to another, but quadrilateral elements can also be used (e.g. Jackson and Finlayson, 1982).

Some care is needed placing spines and selecting proportionality factors w^j, to ensure that the mid-side nodes and the centre node of an element are close enough to the centre between adjacent corners and between mid-side nodes, respectively, so that the determinant of the Jacobian of the isoparametric transformation defined below by eqn (8.53) does not vanish. A straightforward way to guarantee this is to choose freely only the spines that form edges of elements, and then to fix the mid-element spines by

$$x_B^i = \tfrac{1}{2}(x_B^{i+1} + x_B^{i-1})$$

$$e^i = (e^{i+1} + e^{i-1})/|e^{i+1} + e^{i-1}|$$

and to require the prescribed proportions w^j of mid-nodes to satisfy

$$w^j = \tfrac{1}{2}(w^{j+1} + w^{j-1})$$

The mathematical theory of the finite-element method rests on the approximation of the dependent variables locally over each element by basis functions that are low-order polynomials. It is generally recognised that in the velocity–pressure finite-element formulation of the Navier–

† Throughout this chapter **v**, **T**, etc., denote physical vectors and tensors, whereas v, **J**, etc., denote column vectors and matrices, i.e. sets of numbers.

Stokes equations 'mixed interpolation' should be used, i.e. the interpolating functions for pressure should be a polynomial of lower degree than that for velocities (e.g. Huyakorn *et al.*, 1978). In the current formulation for viscous free surface flow quadrilateral elements are employed, with nine-node biquadratic basis functions ϕ^k for the velocity field $\mathbf{v}$ and four-node bilinear ones ψ^l for the pressure field p:

$$\mathbf{v} = \sum_{k=1}^{K} \mathbf{v}^k \phi^k(\xi, \eta) \qquad p = \sum_{l=1}^{L} p^l \psi^l(\xi, \eta) \qquad (8.22)$$

The $\mathbf{v}^k = \mathbf{i}u^k + \mathbf{j}v^k$ are the values of the velocity at the nodes, the total number of velocity nodes being K. The p^l are the nodal values of the pressure, the total number of pressure nodes, i.e. those at the element corners, being L. Incidentally, recent experience indicates that the three-node linear basis functions for the pressure field (which are discontinuous at the finite-element boundaries) introduced by Sani *et al.* (1981) may be superior also for free surface flows (Kheshgi and Scriven, 1982), for they appear to afford improved accuracy and computational efficiency. That experience also suggests that a 'penalty' formulation (cf. Hughes *et al.*, 1979) may offer advantages over 'mixed interpolation' (see also Section 8.6).

The polynomial basis functions in eqn (8.22) are constructed in fixed Cartesian coordinates ξ and η on the standard (ξ, η) square shown in Fig. 8.4. This square is mapped into each of the deformed quadrilateral elements in the flow domain by the isoparametric transformation (Strang and Fix, 1973; Zienkiewicz, 1977)

$$\mathbf{x} = \sum_{k=1}^{K} \mathbf{x}^k(\boldsymbol{h})\phi^k(\xi, \eta) \qquad (8.23)$$

where the dependence of the nodal locations $\mathbf{x}^k$ on the position of the free boundaries is given by eqn (8.21). Here $\boldsymbol{h}$ stands for the entire set of coefficients of the free boundary parametrisation and of the spine locations and orientations:

$$\boldsymbol{h}^{\mathrm{T}} = [\boldsymbol{h}_{\mathrm{B}}^{\mathrm{T}}, \boldsymbol{h}_{\mathrm{T}}^{\mathrm{T}}, \boldsymbol{h}_{\mathrm{I}}^{\mathrm{T}}, \ldots, \mathbf{x}_{*}^{\mathrm{T}}, \delta_{*}^{\mathrm{T}}] \qquad (8.24)$$

Given the location of the nodes, eqn (8.23) determines the shape of the entire domain. It can also be thought of as a set of local mappings onto standard squares that have the same topological relationship as have the quadrilateral elements.

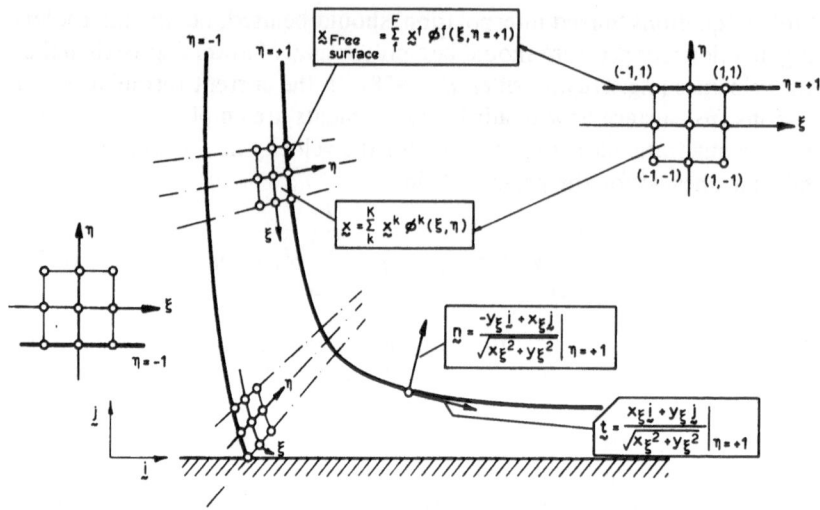

Fig. 8.4.   Isoparametric transformation of elements adjacent to free surface.

At free surfaces, only the basis functions ϕ^f associated with free surface nodes are non-zero. Moreover, they collapse into one-dimensional quadratic functions because free surfaces are necessarily curves of either $\eta = +1$ or $\eta = -1$. The result is a consistent expansion of the free surface shape in terms of the position of the free surface nodes alone:

$$x_{\text{free surface}} = \sum_{f=1}^{F} x^f(h)\phi^f(\xi, \eta = \pm 1) \qquad (8.25)$$

There is an analogous expansion for an interface between two viscous layers. The free surface representation, eqn (8.25), is akin to local coordinate patches well-known in differential geometry (e.g. Lipschutz, 1969). It allows computation of derived quantities at the free surface directly from the isoparametric transformation rather than relying on a reference surface as had been done in earlier formulations (Orr, 1976; Orr and Scriven, 1978; Saito and Scriven, 1981).

The unit tangent and normal vectors to a free surface or interface take on a simple and computationally convenient form,

$$\mathbf{t} = \frac{x_\xi \mathbf{i} + y_\xi \mathbf{j}}{\sqrt{x_\xi^2 + y_\xi^2}}\bigg|_{\eta = \pm 1} \qquad \mathbf{n} = \frac{-y_\xi \mathbf{i} + x_\xi \mathbf{j}}{\sqrt{x_\xi^2 + y_\xi^2}}\bigg|_{\eta = \pm 1} \qquad (8.26)$$

since the Cartesian components x_ξ and y_ξ are obtained directly from eqn (8.23):

$$\mathbf{x}_\xi\big|_{\eta=\pm1} = \sum_{f=1}^{F} \mathbf{x}^f(h) \left.\frac{\partial \phi^f(\xi,\eta)}{\partial \xi}\right|_{\eta=\pm1} \qquad (8.27)$$

Along a free boundary the derivative of a scalar such as a trial function ϕ is simply given by

$$\frac{\mathrm{d}\phi}{\mathrm{d}s} = \mathbf{t}\cdot\nabla\phi = \left.\frac{\partial\phi}{\partial\xi}\right|_{\eta=\pm1} \frac{\mathrm{d}\xi}{\mathrm{d}s} \qquad (8.28)$$

Boundary integrals are evaluated along the free surface or interface by means of the one-dimensional coordinate transformation,

$$\frac{\mathrm{d}s}{\mathrm{d}\xi} = \sqrt{x_\xi^2 + y_\xi^2}\big|_{\eta=\pm1} \qquad (8.29)$$

In the Galerkin/finite-element method, residuals weighted by the basis functions are employed to reduce the original continuous flow problem to a discrete analogue which takes the form of a large set of mostly non-linear algebraic equations. If the Navier–Stokes equation, eqn (8.1), is left in its basic divergence form, then applying the divergence theorem to its weighted residual leads to a boundary integral of the total momentum flux $\mathbf{n}\cdot[-Re\mathbf{vv}+\mathbf{T}]$:

$$\mathbf{R}_M^k \equiv \int_{\Omega_\text{domain}} [\nabla\phi^k\cdot(-Re\mathbf{vv}+\mathbf{T}) - \phi^k St\mathbf{f}]\,\mathrm{d}V$$

$$- \int_{\partial\Omega_\text{boundary}} \phi^k\mathbf{n}\cdot(-Re\mathbf{vv}+\mathbf{T})\,\mathrm{d}s = 0 \qquad (8.30)$$

Weighted by each ϕ^k in turn the vector equation, eqn (8.30), provides $2K$ scalar algebraic equations, the same number as there are velocity coefficients in the expansion, eqn (8.22). Similarly, the vanishing of the ψ^l-weighted residuals of the continuity equation, eqn (8.3),

$$R_C^l \equiv \int_{\Omega_\text{domain}} \psi^l \nabla\cdot\mathbf{v}\,\mathrm{d}V = 0 \qquad (8.31)$$

yields the same number of equations L as there are pressure coefficients in eqn (8.22). The residuals of the kinematic boundary condition, eqn (8.4),

262   S. F. KISTLER AND L. E. SCRIVEN

weighted by the one-dimensional specialisations of the biquadratic basis
function $\phi^f(\xi, \eta = \pm 1)$ along a free surface or interface

$$R_K^f \equiv \int_{\partial\Omega_{\text{free boundary}}} \phi^f(\xi, \eta = \pm 1)\mathbf{n}.\mathbf{v}\,\mathrm{d}s = 0 \qquad (8.32)$$

furnish the additional equations needed to cover the coefficients h for free
surface and interface position and for spine location and orientation
defined in Fig. 8.2. After the set of algebraic equations that results from
eqns (8.30)–(8.32) is completed with the appropriate boundary conditions,
the velocity and pressure fields and free surface and interface locations can
all be calculated simultaneously.

The momentum equation requires boundary conditions on every portion
of the boundary. A feature of the finite-element method that makes it
particularly well suited to the solution of viscous free surface flow problems
is the simple way in which natural boundary conditions that specify the
momentum flux, or components of it, enter into the set of finite-element
equations, eqns (8.30)–(8.32). It is sufficient to specify the fluxes or flux
components in the boundary integral of the weighted residual, eqn (8.30).
Separately constructed approximations to such boundary conditions are
not necessary as they are in most finite-difference simulations. In contrast,
essential boundary conditions specifying the velocity replace those residuals, eqn (8.30), that are formed with the weighting functions associated
with the velocity at boundary nodes. If only one component of the velocity
is specified, only the corresponding component of eqn (8.30) is replaced.
Therefore no-slip boundary conditions, eqns (8.5) and (8.6), at solid walls,
slip velocity distributions of the form of eqn (8.7) close to wetting lines, and
inlet velocity profiles given by eqn (8.12) are imposed by discarding the
momentum residuals, eqn (8.30), formed from those trial functions ϕ^k that
are unity at nodes where the velocity is known, and by entering instead the
essential conditions directly into the algebraic equation set. Thus in the
Galerkin weak form, eqn (8.30), of the momentum equation, which can be
summarised as

$$\mathbf{R}_M^k = \mathbf{R}_{\text{domain}}^k + \mathbf{R}_{\text{boundary}}^k \qquad (8.33)$$

only those portions of the boundary where natural conditions apply have to
be considered for possible contributions to the boundary term. Contri-
butions do come from the free surfaces ($\mathbf{R}_{\text{free surface}}^k$), from the outflow plane
($\mathbf{R}_{\text{outflow}}^k$), from the zone close to the dynamic wetting line, where the slip

boundary condition, eqn (8.8), applies ($\mathbf{R}^k_{\text{slip}}$), and possibly from internal interfaces ($\mathbf{R}^k_{\text{interface}}$). Indeed, the weighted boundary integral can be split:

$$\mathbf{R}^k_{\text{boundary}} = \mathbf{R}^k_{\text{free surface}} + \mathbf{R}^k_{\text{outflow}} + \mathbf{R}^k_{\text{slip}} + \mathbf{R}^k_{\text{interface}} \qquad (8.34)$$

The kth boundary residual, i.e. the one weighted by ϕ^k, has at most two contributions, for very few basis functions are non-zero on more than one part of the boundary. The individual terms in eqn (8.34) are now discussed. At free surfaces the convective momentum flux through the boundary is zero by virtue of eqn (8.4), and in two-dimensional flow the traction $\mathbf{n} \cdot \mathbf{T}$ is $Ca^{-1} \, dt/ds$ according to eqn (8.2), provided the ambient pressure p_a is absorbed in p. This sets the pressure datum; it is not necessary to specify the pressure at a node within the domain of a viscous free surface flow. After eqn (8.2) is substituted in the free surface part of the boundary term, eqn (8.34), the free surface integral can be integrated by parts, i.e. the surface divergence theorem can be applied, to eliminate the free surface curvature dt/ds in favour of the unit tangent $\mathbf{t}$ (Ruschak, 1980):

$$\begin{aligned}
\mathbf{R}^k_{\text{free surface}} &= -\frac{1}{Ca} \int_{\partial\Omega_{\text{free surface}}} \phi^k \frac{d\mathbf{t}}{ds} \, ds \\
&= +\frac{1}{Ca} \int_{\partial\Omega_{\text{free surface}}} \mathbf{t} \frac{d\phi^k}{ds} \, ds - \frac{1}{Ca} [\phi^k \mathbf{t}_1 - \phi^k \mathbf{t}_0] \qquad (8.35)
\end{aligned}$$

The end-point terms in eqn (8.35) can be interpreted physically as shell forces due to surface tension. The directions of these forces are given by the natural boundary conditions, eqn (8.15), which can be imposed by specifying the unit tangents $\mathbf{t}_0$ and $\mathbf{t}_1$ in eqn (8.35). This procedure works where a free surface intersects an inflow or outflow plane. At contact lines, however, where the contact angle is to be specified, the only basis function ϕ^k that is non-zero is associated with the velocity $\mathbf{v}^k$ there; that velocity is prescribed by the essential condition, eqn (8.11). The contact angle condition, eqn (8.10), can be imposed in its natural form by discarding the kinematic boundary residual, eqn (8.32), that is weighted by the basis function $\phi^k(\xi, \eta = \pm 1)$ belonging to the contact line node, and replacing it by the normal component of the momentum residual formed from the basis function ϕ^k at the contact line node (cf. Kistler and Scriven, 1983):

$$\mathbf{n}_\theta \cdot \mathbf{R}^k = \mathbf{n}_\theta \cdot \int_{\Omega_{\text{domain}}} [\nabla\phi^k \cdot (-Re\mathbf{v}\mathbf{v} + \mathbf{T}) - \phi^k St\mathbf{f}] \, dV$$

$$+ \mathbf{n}_\theta \cdot \frac{1}{Ca} \int_{\partial\Omega_{\text{free surface}}} \mathbf{t} \frac{d\phi^k}{ds} \, ds + \mathbf{n}_\theta \cdot \mathbf{R}^k_{\text{slip}} = 0 \qquad (8.36)$$

There is no end-point term here because the unit vector $\mathbf{n}_\theta$ by definition satisfies $\mathbf{n}_\theta \cdot \mathbf{n}_w = \cos\theta_{\text{contact}}$. When the position rather than the slope of a free surface is prescribed at the end point, the essential condition, eqn (8.13) or eqn (8.9), can be condensed to a scalar equation because nodes are restricted to move along a spine. This equation is then entered directly into the equation set in place of the kinematic boundary integral, eqn (8.32), associated with the end-point node, the location of which is known.

The boundary integral remaining in eqn (8.35) can always be evaluated in terms of the computationally convenient isoparametric coordinate ζ along the free surface. Having the tangent vector $\mathbf{t}$ in the form of eqn (8.26) and derivatives of the basis functions computed according to eqn (8.28) is crucial to the present formulation, for evaluating the normal resultant of capillarity amounts to differencing two tangential forces that may be almost parallel and both large in magnitude. If there is more than one free surface, each contributes to the momentum residual, eqn (8.30), with a term of the form of eqn (8.35), but of course none of the basis functions ϕ^k is non-zero at more than one free surface.

Besides considerably simplifying the way natural boundary conditions enter the finite-element equations, applying the divergence theorem lowers the order of the highest derivatives and requires therefore a degree less of differentiability of the basis functions. In principle, this allows approximating the velocity and the free surface by piecewise linear functions and the pressure by piecewise constant functions, an approach taken successfully by Ruschak (1980).

For the outflow plane, the momentum flux that is to be specified in the boundary integral can be calculated in different ways. If it is deduced directly from the flow in the fully developed regime, the boundary residual according to eqn (8.17) becomes

$$\mathbf{R}^k_{\text{outflow}} = -\int_{\partial\Omega_{\text{outflow}}} \phi^k \mathbf{m}(\mathbf{x}_{\text{outflow}}) \, ds \qquad (8.37)$$

If the linearised solution for small departures from the asymptotic regime downstream can be found, it is computationally more efficient to employ the vector boundary condition, eqn (8.18), which leads to the boundary integral

$$\mathbf{R}^k_{\text{outflow}} = -\int_{\partial\Omega_{\text{outflow}}} \phi^k(-\mathbf{n} \cdot Re\mathbf{vv} + \mathbf{A} \cdot (\mathbf{v} - \mathbf{v}_\infty) + \mathbf{n} \cdot \mathbf{T}_\infty) \, ds \qquad (8.38)$$

If, however, the flow evolves non-linearly downstream, neither of these strategies works. An alternative to solving the full set of equations for the entire distance downstream to a place where the appropriate boundary conditions can be applied is to employ the approximate film profile equations described in Section 8.2, which can be solved by various methods. If the Runge-Kutta method or another shooting technique is employed, as by Gifford (1982) to analyse isothermal fibre formation, then it is necessary to iterate back and forth between the part of the domain on which the finite-element equations are solved and the part solved by shooting. The finite-element method, however, is perfectly well suited to the solution of non-linear ordinary differential equations, and the algebraic weighted residual equations that arise from the one-dimensional Galerkin/ finite-element procedure can be solved simultaneously with eqns (8.30) to (8.32) from the main flow problem; that is, all the equations can be handled in the same Newton iteration. This strategy keeps the flow problem a boundary value problem, yet greatly reduces the number of unknowns by limiting finite-element analysis of the full two-dimensional flow problem to forming zones. Applications have been made to curtain- and sheet-drawing flows (Kistler *et al.*, 1981; Kistler and Scriven, 1981), and are illustrated by Fig. 8.5. The thickness $H(s)$ and, in the case of non-symmetrical sheets, the mid-surface profile inclination $\theta(s)$ of the free liquid film beyond the main flow domain are expanded in one-dimensional Hermite cubic basis functions $\chi^m(s)$ and linear ones $\zeta^n(s)$, s being the arc length along the profile of the mid-surface:

$$H = \sum_{m=1}^{M} H^m \chi^m(s) \qquad \theta = \sum_{n=1}^{N} \theta^n \zeta^n(s) \qquad (8.39)$$

Continuity of the mass flux $\mathbf{n} \cdot \mathbf{v}$ across the matching plane is guaranteed by choosing the appropriate form of velocity distribution to arrive at the film profile equations and by imposing continuity of the free surface location as an essential condition. This condition replaces the residual that is weighted by χ^1, i.e. the residual associated with the first node of the asymptotic domain. In the case of an asymmetrical sheet the spine in the matching plane is required to change its orientation δ_* in order to remain perpendicular to the mid-surface profile, so that

$$\delta_* = \theta^1 - \frac{\pi}{2} \qquad (8.40)$$

becomes an essential condition. In order to impose that condition the residual which is weighted by ζ^1 is discarded. The matching conditions that

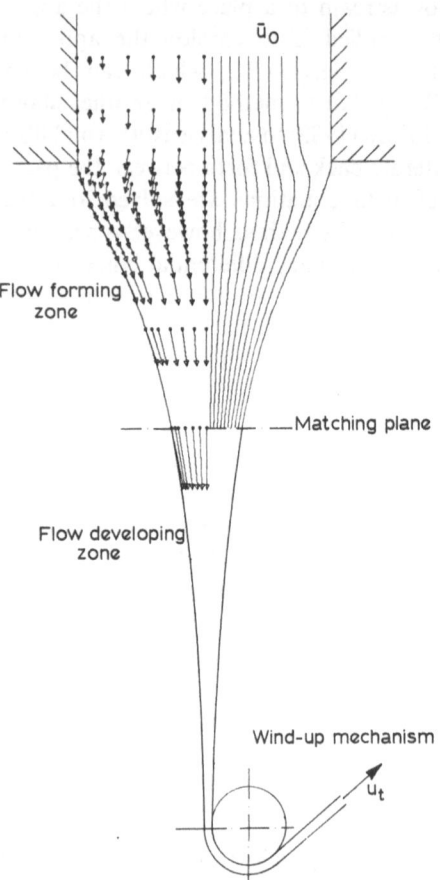

Fig. 8.5. Multiple flow zones in drawn viscous sheet. $Re = 10^{-4}$, $Lr \equiv L/D_0 = 5$, $Dr \equiv u_t/\bar{u}_0 = 30$, where L is length of sheet, D_0 is slot thickness, u_t is take-up velocity, and $\bar{u}_0$ is average velocity in slot. (Note: horizontal scale is exaggerated by a factor of 2.)

require continuity of the total momentum flux $\mathbf{n} \cdot (-Re\mathbf{vv} + \mathbf{T})$ and of the tangential tension $(1/Ca)\mathbf{t}$ fit naturally into the finite-element formulation: the traction $\mathbf{n}(\delta_*) \cdot \mathbf{T}_a$ and the force $(1/Ca)\mathbf{t}_a$, which have to be specified as boundary conditions for the two-dimensional flow problem in the forming zone, are evaluated at the matching plane with the aid of the expansions, eqn (8.39), and the approximate velocity distribution that underlies the film profile equation; the convective momentum flux $Re\mathbf{n}(\delta_*) \cdot \mathbf{vv}$, however, is evaluated with the aid of the velocity expansion, eqn (8.22). The momentum fluxes and the surface tension force are then substituted into the weighted boundary integral at the matching plane:

$$\mathbf{R}^k_{\text{outflow}} = -\int_{\partial\Omega_{\text{outflow}}} \phi^k \mathbf{n}(\delta_*) \cdot (-Re\mathbf{vv} + \mathbf{T}_a)\,\mathrm{d}s \qquad (8.41)$$

and into the end-point term $(1/Ca)\mathbf{t}_a$ in eqn (8.35).

Near a contact line where a slip boundary condition, eqn (8.8), applies, the local shear stress in the boundary integral is replaced by $\beta^{-1}\mathbf{t}_w \cdot (\mathbf{v} - \mathbf{v}_{\text{solid}})$, which leads to a contribution to the momentum residual:

$$\mathbf{R}^k_{\text{slip}} = -\mathbf{t}_w \int_{\partial\Omega_{\text{slip region}}} \phi^k \beta^{-1}\mathbf{t}_w \cdot (\mathbf{v} - \mathbf{v}_{\text{solid}})\,\mathrm{d}s \qquad (8.42)$$

The normal component of $\mathbf{R}_{\text{slip}}$ becomes irrelevant, because it must be replaced by the essential condition $\mathbf{n}_w \cdot \mathbf{v} = 0$. Similarly, at a symmetry plane in the flow the tangential component of the boundary residual is zero due to the vanishing shear stress $\mathbf{tn}:\mathbf{T}$, whereas the normal component drops out of the algebraic equation set because the no-flux requirement $\mathbf{n} \cdot \mathbf{v} = 0$ is essential.

When a coating flow is stratified, a Galerkin weighted residual, eqn (8.30), can be written for each layer separately. It suffices to consider just two layers, namely A above B, for the extension to further layers is then obvious. The separate momentum residuals are

$$\mathbf{R}^k_{M,A} = \int_{\Omega_A} [\nabla\phi^k \cdot (-Re_A\mathbf{vv} + \mathbf{T}_A) - \phi^k St_A\mathbf{f}]\,\mathrm{d}V$$

$$-\int_{\partial\Omega_A} \phi^k \mathbf{n}_A \cdot (-Re_A\mathbf{vv} + \mathbf{T}_A)\,\mathrm{d}s = 0 \qquad (8.43a)$$

$$\mathbf{R}^k_{M,B} = \int_{\Omega_B} [\nabla\phi^k \cdot (-Re_B\mathbf{vv} + \mathbf{T}_B) - \phi^k St_B\mathbf{f}]\,\mathrm{d}V$$

$$-\int_{\partial\Omega_B} \phi^k \mathbf{n}_B \cdot (-Re_B\mathbf{vv} + \mathbf{T}_B)\,\mathrm{d}s = 0 \qquad (8.43b)$$

Here the Reynolds numbers are $Re_A \equiv \rho_A UL/\eta_A$ and $Re_B \equiv \rho_B UL/\eta_B$, the

Stokes numbers are $St_A \equiv \rho_A g L^2 / \eta_A U$ and $St_B \equiv \rho_B g L^2 / \eta_B U$, and the total stresses T_A and T_B are made dimensionless with $\eta_A U/L$ and $\eta_B U/L$, respectively. When the weighting function ϕ^k in eqns (8.43) does not belong to a node in the interface between layers, the corresponding residual equation enters in the same way as in the single-layer case described above; capillary numbers must however be defined for each layer that has a free surface: $Ca_A \equiv \eta_A U/\sigma_A$, σ_A being the surface tension of A, and $Ca_B \equiv \eta_B U/\sigma_B$, σ_B being the surface tension of B. Moreover it may be necessary to consider each layer separately in order to formulate the outflow boundary term in eqns (8.37), (8.38), or (8.41). When ϕ^k in eqns (8.43) belongs to a node in the interface between layers, both residual equations in eqns (8.43) arise. Combining the pair into their viscosity-weighted sum and invoking eqn (8.4) leads to

$$
R_M^k = \int_{\Omega_A} [\nabla\phi^k . (-Re_A \mathbf{vv} + T_A) - \phi^k St_A \mathbf{f}] \, \mathrm{d}V
$$

$$
+ \frac{\eta_B}{\eta_A} \int_{\Omega_B} [\nabla\phi^k . (-Re_B \mathbf{vv} + T_B) - \phi^k St_B \mathbf{f}] \, \mathrm{d}V
$$

$$
- \int_{\partial\Omega_{\text{interface}}} \phi^k \left(\mathbf{n}_A . T_A + \frac{\eta_B}{\eta_A} \mathbf{n}_B . T_B \right) \mathrm{d}s_{AB} + R_{\text{outflow, A}}^k + \frac{\eta_B}{\eta_A} R_{\text{outflow, B}}^k = 0
$$

$$(8.44)$$

This accommodates the force balance, eqn (8.2a), naturally at the interface; indeed, eqn (8.2a) can be used to simplify eqn (8.44), because it is equivalent to

$$
\mathbf{n}_A . T_A + (\eta_B/\eta_A)\mathbf{n}_B . T_B = \frac{1}{Ca_{AB}} \frac{\mathrm{d}t_{AB}}{\mathrm{d}s_{AB}} \tag{8.45}
$$

Consequently the interface contribution can be integrated as the surface contribution is in eqn (8.35); thus in eqn (8.44)

$$
R_{\text{interface}}^k = \frac{1}{Ca_{AB}} \int_{\partial\Omega_{\text{interface}}} t_{AB} \frac{\mathrm{d}\phi^k}{\mathrm{d}s_{AB}} \, \mathrm{d}s - \frac{1}{Ca_{AB}} [\phi^k t_{AB,1} - \phi^k t_{AB,0}] \tag{8.46}
$$

Usually in practice the liquid layers are miscible and there is no interfacial tension between them. Then eqn (8.2a) no longer leads to a contribution $R_{\text{interface}}^k$ to the momentum residual, eqn (8.44). Because pressure, unlike velocity, need not be continuous across the interface, the residuals of the continuity equation, eqn (8.31), that are weighted with basis functions ψ^I belonging to interfacial nodes form distinct residual equations for the two sides of the interface. There are exactly enough of these equations for the number of independent algebraic equations for coefficients of basis functions to just match the number of those coefficients.

8.4 EVALUATION OF EXPANSION COEFFICIENTS

Even at vanishing Reynolds number a free surface generally makes the system of equations governing viscous flow non-linear. Hence whether a steady flow state can be predicted, if indeed it exists, depends not only on the accuracy of the mathematical formulation just presented, but also on understanding the physics of the flow in order to construct a first approximation to a solution from which the chosen iteration procedure converges to a steady state. In most previous attempts to solve free surface flows governed by the full Navier–Stokes equation, free boundary locations have been found by successive approximation techniques, sometimes called Picard methods (Isaacson and Keller, 1966; Rheinboldt, 1974): given a position of the free boundary, the corresponding flow field is computed—iteratively when the Reynolds number is not zero—with one of the boundary conditions at the free surfaces omitted, usually eqn (8.4) or the normal component of eqn (8.2). That boundary condition together with the current approximation to the flow field is then used to generate an updated position of the free boundary, and the cycle is repeated until convergence is reached. Although fairly simple to implement, this method of locating the free boundary has the telling disadvantage of failing to converge in about half of all cases, if the same boundary condition is always chosen for iteration. This was demonstrated by Silliman and Scriven (1980), who settled the controversy about iterating on the kinematic condition (Nickell *et al.*, 1974) or on the normal stress condition (Orr and Scriven, 1978) by showing that at low capillary numbers the normal stress scheme converges whereas the kinematic scheme does not, and conversely at high capillary numbers. Moreover, at intermediate capillary numbers, which are of most practical interest in many coating operations, there is a range of overlap where both schemes succeed but both become slow to converge and sometimes do so only with the aid of artful under-relaxation (Silliman and Scriven, 1980). These features of the convergence behaviour make successive approximation schemes cost-ineffective and unattractive, if not inapplicable to flows with highly irregular free surface geometries.

Only recently have iteration schemes based on the full Newton method been introduced to the finite-element analysis of viscous free surface flows (Ruschak, 1980; Saito and Scriven, 1981). The complete set of non-linear equations is solved simultaneously for the position and shape of the free boundaries *and* for the velocity and pressure fields. Compared to successive approximation techniques, the rate of convergence is strikingly improved: it becomes quadratic as the iteration proceeds, in accord with the asymptotic theory of Newton's method (Isaacson and Keller, 1966). Consequently the convergence behaviour now has an important role in systematic investigations, and insisting on a second order or quadratic rate of convergence affords a useful test of the correctness of the Jacobian and iteration procedure. Besides its rapid convergence which is independent of flow parameters, another great attraction of Newton's method is the wealth of information contained in the Jacobian matrix of the converged solution, information valuable not only for continuation methods to find solutions as parameters are changed, but also for analysing the stability of the solutions found (Bixler, 1982; Ruschak, 1982a) and for computing asymptotic estimates of the effect of small changes in flow parameters, boundary conditions or boundary shape by sensitivity analysis (Brown *et al.*, 1980).

When the weighted residual equations, eqns (8.30), (8.31) and (8.32), are written in terms of a vector of finite-element coefficients α and a vector of weighted residuals $R(\alpha)$, the Newton iteration process finds the updated coefficients α_{n+1} from the last computed set α_n by solving the linear system of equations

$$J[\alpha_{n+1} - \alpha_n] \equiv J\,\Delta\alpha_n = -R(\alpha_n) \qquad J \equiv \frac{\partial R}{\partial \alpha}(\alpha_n) \qquad (8.47)$$

The vector of finite-element coefficients α can be subdivided† into $u^T = [u^1, \ldots, u^k]$, $v^T = [v^1, \ldots, v^k]$, $p^T = [p^1, \ldots, p^L]$ and h, which was previously defined in eqn (8.24), so that $\alpha^T = [u^T, v^T, p^T, h^T]$. Similarly, the residual vector R is made up of $R_x^T \equiv \mathbf{i} \cdot [R_M^1, \ldots, R_M^K]$, $R_y^T \equiv \mathbf{j} \cdot [R_M^1, \ldots, R_M^K]$ (where $\mathbf{i}$ and $\mathbf{j}$ are the unit vectors in the x and y directions), $R_C^T \equiv [R_C^1, \ldots, R_C^L]$ and $R_K^T \equiv [R_K^1, \ldots]$ in such a way that $R^T = [R_x^T, R_y^T, R_C^T, R_K^T]$. This notation brings out the structure of the Jacobian matrix J in eqn (8.47) when it is decomposed in the way shown in eqn (8.48).

The actual numbering of the unknowns in the computer algorithm to solve eqn (8.48) by Gauss elimination of course does not correspond to the ordering shown in eqn (8.48); the latter would destroy entirely the useful

---

† Here v^i has become (u^i, v^i).

$$
\begin{bmatrix}
\dfrac{\partial R_x}{\partial u} & \dfrac{\partial R_x}{\partial v} & \dfrac{\partial R_x}{\partial p} & \begin{matrix}0\\[2pt]\dfrac{\partial R_x}{\partial h}\end{matrix} \\
0 \quad\vdots\quad 1 & 0 & 0 & 0 \\
\dfrac{\partial R_y}{\partial u} & \dfrac{\partial R_y}{\partial v} & \dfrac{\partial R_y}{\partial p} & \begin{matrix}0\\[2pt]\dfrac{\partial R_y}{\partial h}\end{matrix} \\
0 \quad 0 \quad\vdots\quad 1 & 0 & 0 & 0 \\
\dfrac{\partial R_C}{\partial u} & \dfrac{\partial R_C}{\partial v} & 0 & \begin{matrix}0\\[2pt]\dfrac{\partial R_C}{\partial h}\end{matrix} \\
0 \;\vdots\; \dfrac{\partial R_K}{\partial u} \;\; 0 & \;\vdots\; \dfrac{\partial R_K}{\partial v} & 0 & \dfrac{\partial R_K}{\partial h} \\
0 & 0 & 0 & 0 \;\vdots\; 1
\end{bmatrix}
\begin{bmatrix}
u_{n+1}-u_n \\
\\
v_{n+1}-v_n \\
\\
p_{n+1}-p_n \\
\\
h_{n+1}-h_n
\end{bmatrix}
=
\begin{bmatrix}
R_x \\
0 \\
R_y \\
0 \\
R_C \\
R_K \\
0
\end{bmatrix}
$$

$$(8.48)$$

bandedness of the Jacobian matrix. This global matrix is assembled from contributions at element level. If u^e, v^e, p^e and h^e denote the subset of coefficients at the nodes of a particular element, and if R_x^e, R_y^e, R_C^e, R_K^e are the contributions to the residuals arising from integration over that element or its edges, the Jacobian matrix at element level is

$$
J_e =
\begin{bmatrix}
\dfrac{\partial R_x^e}{\partial u^e} & \dfrac{\partial R_x^e}{\partial v^e} & \dfrac{\partial R_x^e}{\partial p^e} & \dfrac{\partial R_x^e}{\partial h^e} \\[10pt]
\dfrac{\partial R_y^e}{\partial u^e} & \dfrac{\partial R_y^e}{\partial v^e} & \dfrac{\partial R_y^e}{\partial p^e} & \dfrac{\partial R_y^e}{\partial h^e} \\[10pt]
\dfrac{\partial R_C^e}{\partial u^e} & \dfrac{\partial R_C^e}{\partial v^e} & 0 & \dfrac{\partial R_C^e}{\partial h^e} \\[10pt]
\dfrac{\partial R_K^e}{\partial u^e} & \dfrac{\partial R_K^e}{\partial v^e} & 0 & \dfrac{\partial R_K^e}{\partial h^e}
\end{bmatrix}
\begin{matrix}\updownarrow 9 \\[12pt] \updownarrow 9 \\[12pt] \updownarrow 4 \\[12pt] \updownarrow \geq 3\end{matrix}
$$

$$\longleftarrow 9 \longrightarrow | \longleftarrow 9 \longrightarrow | \longleftarrow 4 \longrightarrow | \longleftarrow \geq 3 \longrightarrow$$

$$(8.49)$$

for elements adjoining free surfaces or interfaces. For elements that change their shape isoparametrically as the free boundary locations are updated during the iteration procedure, but do not directly adjoin a free boundary or interface, the residuals of the momentum and continuity equations are sensitive to the free boundary parameters h. Those elements, however, do not contribute to the weighted residuals of the kinematic boundary condition; consequently for such elements the last row of submatrices in eqn (8.49) is absent. On assembly of element-level Jacobian matrices this leads to the zero entries in the penultimate row of submatrices in the global Jacobian matrix, eqn (8.48). Elements that lie entirely between fixed boundaries are unaffected by changes in free boundary location. Therefore the element-level Jacobian matrix lacks both the last row and the last column of submatrices. On assembly this leads to zero entries in the last column of submatrices in the global matrix, eqn (8.48). Moreover, in the linearised system of finite-element equations, eqn (8.48), essential conditions on velocity and free boundary position replace some of the residual equations.

If film profile equations arising from an asymptotic regime upstream or downstream are solved in the same Newton iteration as the momentum and continuity equations and the boundary mapping of the main flow domain, the system, eqn (8.48), has to be augmented by the one-dimensional weighted residuals of the film profile equations R_A and by those matching conditions that are stated as essential conditions. The structure of the matrix problem to be solved at each Newton iteration step is then

$$
\begin{bmatrix}
 & 0 & \\
\dfrac{\partial R}{\partial \alpha} & \cdots & 0 \\
 & \dfrac{\partial R}{\partial \gamma} & \\
0 & A & 0 \\
0 & \dfrac{\partial R_A}{\partial \gamma} &
\end{bmatrix}
\begin{bmatrix}
\alpha_{n+1} - \alpha_\eta \\
\\
\\
\gamma_{n+1} - \gamma_\eta
\end{bmatrix}
=
\begin{bmatrix}
R \\
\\
b \\
R_A
\end{bmatrix}
\qquad (8.50)
$$

Here α is the same vector of finite-element coefficients and R is the same vector of weighted residuals of the main flow problem as in eqn (8.47). γ is the set of coefficients in the expansions of the downstream film profile in one-dimensional basis functions. For non-symmetrical, falling liquid sheets, for example, the expansions are those of eqn (8.39), and the set of coefficients is $\gamma^T = (H^T, \theta^T) = (H^1, \ldots, H^M, \theta^1, \ldots, \theta^N)$. A and b in eqn (8.50) stand for the matching conditions that are imposed as essential conditions, whereas those imposed as natural conditions as in eqn (8.41) lead to the sensitivities $\partial R/\partial \gamma$. These sensitivities also arise when the positions or orientations of spines in the main flow domain are adjusted according to characteristic features of the solution γ for the asymptotic regime downstream.

To apply full Newton iteration to free boundary problems requires accurate calculation of the components of $\partial R/\partial h$, i.e. the sensitivities of all the residuals throughout the flow domain to the location and shape of the free boundaries. The residuals, which are made up of area integrals or line integrals

$$I(\beta, h) = \int_{A(h)} F(\mathbf{x}, \beta, h)\, dx\, dy$$

$$L(\beta, h) = \int_{S(h)} G(\mathbf{x}, \beta, h)\, ds$$

(8.51)

where β is the vector of finite-element coefficients for the flow field such that $\alpha^T = (\beta^T, h^T)$, depend on h through not only the integrands, but also the limits of integration. This long-standing obstacle was overcome by realising that the sensitivities to free boundary coefficients can readily be calculated in the isoparametrically mapped domain (Saito and Scriven, 1981), where the domain of integration is a fixed square A_0 or straight line segment ξ_0:

$$I(\beta, h) = \int_{A_0} F(\mathbf{x}(\xi, \eta; h), \beta, h)|\mathscr{J}|\, d\xi\, d\eta$$

$$L(\beta, h) = \int_{\xi_0} G(\mathbf{x}(\xi, \eta = \pm 1; h), \beta, h)\frac{ds}{d\xi}\, d\xi$$

(8.52)

Here $|\mathscr{J}|$ is the determinant of the Jacobian of transformation

$$\mathscr{J} \equiv \frac{\partial(x, y)}{\partial(\xi, \eta)} \equiv \frac{\partial \mathbf{x}}{\partial \boldsymbol{\xi}} = \begin{bmatrix} \partial x/\partial\xi & \partial x/\partial\eta \\ \partial y/\partial\xi & \partial y/\partial\eta \end{bmatrix}$$

(8.53)

The derivatives needed are then simply

$$\frac{\partial I}{\partial h} = \int_{A_0} \frac{\partial}{\partial h}\{\bar{F}(\xi, \eta, \beta, h)|\mathscr{J}|\}\, d\xi\, d\eta$$

$$\frac{\partial L}{\partial h} = \int_{\xi_0} \frac{\partial}{\partial h}\{\bar{G}(\xi, \eta = \pm 1, \beta, h)\sqrt{x_\xi^2 + y_\xi^2}|_{\eta = \pm 1}\}\, d\xi$$

(8.54)

and are easily enough worked out analytically. The integrands in eqn (8.54) depend on h through the Jacobians of transformation and also through derivatives $\partial/\partial x$ and $\partial/\partial y$ when these are present. These operators are transformed by

$$\frac{\partial}{\partial x} = \frac{\partial \xi}{\partial x} \frac{\partial}{\partial \xi} + \frac{\partial \eta}{\partial x} \frac{\partial}{\partial \eta} = \frac{1}{|\mathcal{J}|} \left(\frac{\partial y}{\partial \eta} \frac{\partial}{\partial \xi} - \frac{\partial y}{\partial \xi} \frac{\partial}{\partial \eta} \right)$$

$$\frac{\partial}{\partial y} = \frac{\partial \xi}{\partial y} \frac{\partial}{\partial \xi} + \frac{\partial \eta}{\partial y} \frac{\partial}{\partial \eta} = \frac{1}{|\mathcal{J}|} \left(-\frac{\partial x}{\partial \eta} \frac{\partial}{\partial \xi} + \frac{\partial x}{\partial \xi} \frac{\partial}{\partial \eta} \right) \tag{8.55}$$

where the derivative $\partial/\partial\xi$ and $\partial/\partial\eta$ are independent of h, whereas $\partial x/\partial\xi$ and $\partial x/\partial\eta$ follow directly from the isoparametric transformation, eqn (8.23),

$$\frac{\partial \mathbf{x}}{\partial \xi} = \sum_k^K \mathbf{x}^k(h) \frac{\partial \phi^k}{\partial \xi} \qquad \frac{\partial \mathbf{x}}{\partial \eta} = \sum_k^K \mathbf{x}^k(h) \frac{\partial \phi^k}{\partial \eta} \tag{8.56}$$

and the derivatives with respect to h are then

$$\frac{\partial}{\partial h} \left(\frac{\partial \mathbf{x}}{\partial \xi} \right) = \sum_k^K \frac{\partial \mathbf{x}^k(h)}{\partial h} \frac{\partial \phi^k}{\partial \xi} \qquad \frac{\partial}{\partial h} \left(\frac{\partial \mathbf{x}}{\partial \eta} \right) = \sum_k^K \frac{\partial \mathbf{x}^k(h)}{\partial h} \frac{\partial \phi^k}{\partial \eta} \tag{8.57}$$

The derivatives $\partial \mathbf{x}^k(h)/\partial h$ in turn are readily worked out from the simple dependence, eqn (8.21), of the nodal locations $\mathbf{x}^k$ on the free boundary coefficients h.

Newton's method ordinarily converges quadratically as the solution is approached, as noted above. If there are no errors in the Jacobian matrix, eqn (8.48), i.e. if all the derivatives are calculated correctly, the rate of convergence does indeed become quadratic as the iteration proceeds (cf. Saito and Scriven, 1981). Thus by allowing iteration to proceed until machine round-off error intrudes, it is possible to test the correctness of the Jacobian. The particular convergence criterion chosen does not matter; both the maximum change among the unknown coefficients, $|\alpha_{n+1}^i - \alpha_n^i|_{\max}$, and the largest among the weighted residuals, $|R^i(\alpha_n)|_{\max}$, are useful.

When a viscous free surface flow problem is solved by the Galerkin/finite-element method a significant fraction of the computer time is spent in assembling and solving the set of repeatedly linearised algebraic equations. In particular a major drawback of the Newton method is that the asymmetric coefficient matrix in eqn (8.48) must be computed and factored at each iteration step. That is why a new group of algorithms emerged recently that are known as quasi-Newton methods (Dennis and Moré, 1977): in them the coefficient matrix is updated incompletely or approximately, but in a simple manner, after each iteration step rather than recomputed entirely (full Newton) or left unchanged (modified Newton). A quasi-Newton method was recently shown to be effective in finite-element analysis of fluid flow with fixed boundaries (Engelman *et al.*, 1981), but quasi-Newton methods have not yet been adequately tested on problems with free boundaries.

In the algorithms summarised in this chapter, the full Newton system, eqn (8.48) or eqn (8.50), is assembled and solved by the frontal technique developed by Hood (1976, 1977). A very condensed flow chart of the code is shown in Fig. 8.6. The finite elements are scanned in sequence; at each one the subroutine ABFIND is called to calculate entries in the element-level Jacobian matrix, eqn (8.49), which is called ESTIFM; and the nodal equations of that element are assembled on a submatrix EQ of the global matrix, eqn (8.48) or eqn (8.50). When an equation is complete, i.e. when the dependent variable does not appear in any further equations to be assembled, the equation is eliminated and placed in out-of-core storage until it is needed at the final stage for back-substitution. The subroutine ABFIND contains essentially the entire finite-element formulation described in Section 8.3 and is the central piece in a computer algorithm to solve viscous free surface flows. It is thus the chief subroutine the programmer must provide in order to use Hood's frontal routine. Evaluating the integrals in the weighted residuals, eqns (8.30) to (8.32), takes a numerical integration procedure; here the Gauss quadrature formula is used (cf. for example, Zienkiewicz, 1977, pp. 195 ff) with 3 points to evaluate area integrals along edges of an element and with $3 \times 3 = 9$ points to evaluate area integrals over an element.

The frontal technique puts no stringent constraints on node numbering, which controls the ordering of equations. This flexibility is particularly valuable for viscous free surface flow problems. It is straightforward to introduce nodes that do not coincide with a particular point in space and can be assigned to many elements which may be located in different parts of the tessellation. For instance it is highly advantageous to associate the free surface coefficients h_B^i and h_T^i (see Fig. 8.2) with additional nodes that do not coincide with velocity or pressure nodes, but which appear in every element where the shape is affected by a change of the free surface coefficient in question. Another example is those parameters that control the position or orientation of spines and that can be made additional unknown coefficients which are then associated with a separate node shared

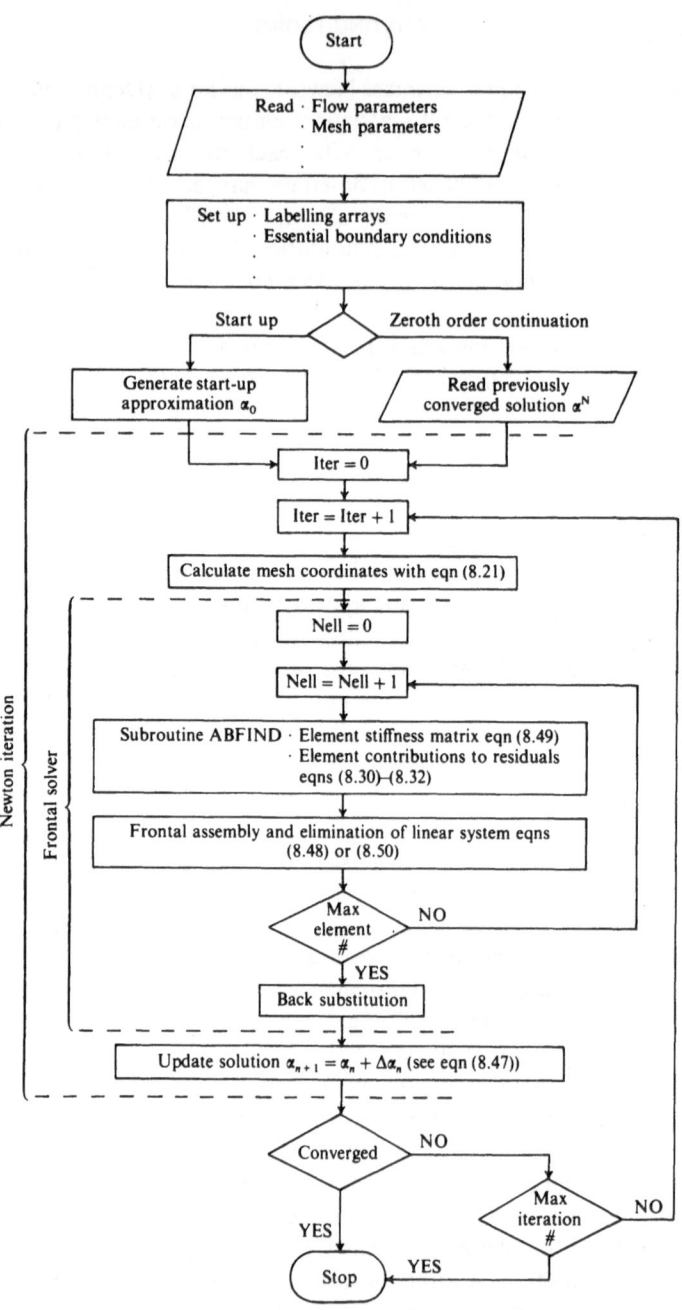

Fig. 8.6.   Condensed flow chart of algorithm to solve viscous free surface flow by finite-element method with zeroth-order continuation (first-order continuation requires computation of continuation vector $\boldsymbol{d}_m$ with eqn (8.59) and of improved initial estimate $\boldsymbol{\alpha}^{N+1}$ with eqn (8.58)).

by all the elements that are deformed as those parameters change—under the control of some flow feature such as a wetting line to which the spines are linked. And when an asymptotic regime is matched, as in eqn (8.41), so that it is found simultaneously with the main flow, the nodes in the first element of one-dimensional basis functions belong also to the elements in the outflow plane of the main flow domain because the momentum fluxes in the boundary integral, eqn (8.41), are calculated from the asymptotic solution.

Computational costs depend strongly on the size of the submatrix EQ that is kept in the core memory of the computer (cf. Hood, 1976). The size required depends on the front width and the number of equations needed for adequate choices of pivots in the Gaussian elimination. The front width in turn depends on the aspect ratio of the finite-element flow domain. For coating flows the domain is usually long in the main flow direction but narrow across the liquid layer, so that a few elements suffice in that direction. This high aspect ratio is quite advantageous because in the mixed formulation of viscous flow problems, the Jacobian matrix acquires zero diagonal entries from the continuity equations but rarely has any other pivots that are near zero. Hence pivoting can be confined to the equations of each element and the least size of EQ allowed by front width can be chosen. Moreover, the general partial pivoting strategy in Hood's (1976) program can be changed to diagonal pivoting, in order to increase execution speed (Silliman, 1979). The effectiveness of the frontal solver for hydrodynamic problems can be raised even further by modifications such as those suggested by Walters (1980). Altogether the frontal solution technique has proven effective for solving the matrix equations, eqn (8.48) or eqn (8.50), to which a viscous free surface flow problem is reduced by the finite-element method.

Crucial to solving iteratively the non-linear equation set is the initial approximation α_0, for if it does not fall in the domain of convergence of the Newton iteration procedure, eqn (8.47), no solution will be found. Most crucial of all is finding a start-up approximation, i.e. the initial approximation α_0 when no solutions of related cases have been previously calculated. If the flow is confined to a large extent or if it is almost rectilinear, the task is fairly straightforward: a limiting shape, such as a straight free surface in the case of die-swell calculations, or an approximate solution obtained by asymptotic techniques, usually suffices. If, however, the free surfaces are highly irregular or are so free to move that the location of the entire computational domain is unknown *a priori*, the search for a successful start-up approximation can be and usually is much more difficult. What is needed is experimental evidence about the free surface shapes to be expected at a set of values of flow parameters, or at least some approximate models of crucial features of the flow, in order that spines can be placed appropriately and start-up approximations suggested that will lead to convergence. For the analysis of curtain coating, for instance, the position of the dynamic wetting line can be predicted approximately with the aid of a simple boundary layer model, and this turns out to give a completely successful start-up approximation for the flow in the impinge-ment region at a particular set of flow parameters. For viscous free surface flows, the domain of convergence of Newton iteration generally depends much more on position and shape of the free boundaries than on the flow field within the flow domain; this accords with the non-linearity caused by the free boundaries usually being stronger than that of curvilinear flow at modest Reynolds numbers. Hence it is often effective in start-up to temporarily fix the free boundaries during the first iteration steps in order to generate an improved approximation to the velocity and pressure fields before allowing boundary adjustments. Moreover the domain of con-vergence from the start-up approximation can sometimes be enlarged by under-relaxing the iteration process, i.e. by applying only fractions of the updating changes called for by the Newton formula, eqn (8.47); this is most likely to be useful when the changes in free boundary positions tend to over-shoot.

Once the first converged solution is obtained the calculation of related solutions for nearby parameter values or boundary configurations is much easier. Often it suffices to use the solution for one set of parameter values to begin Newton iteration at a not-too-different set of parameters, a scheme sometimes called zeroth-order continuation. Alternatively first-order con-tinuation can be used to get an improved initial estimate to the new solution $\alpha(\pi^{N+1})$ for parameter values π^{N+1} given the previously converged solution $\alpha(\pi^N)$ for parameter values π^N, where $\pi^T = (Re, Ca, St, \ldots, \pi^M)$ is the set of M parameters relevant to the problem:

$$\alpha(\pi^{N+1}) = \alpha(\pi^N + \Delta\pi) = \alpha(\pi^N) + \sum_{m=1}^{M} d_m \Delta\pi_m \qquad (8.58)$$

In this multiple Taylor-series expansion the sensitivity vectors are $d_m \equiv \partial\alpha/\partial\pi_m$ and can be found with little additional computation by solving

$$Jd_m = s_m \qquad (8.59)$$

where the Jacobian matrix J is already available from the Newton iteration used to obtain $\alpha(\pi^N)$ and where the sensitivity vectors of residuals $s_m \equiv \partial R/\partial \pi_m$ are easily calculated (Brown et al., 1980). Comparisons of zeroth-order and first-order continuation depend on the size of parameter steps chosen. To succeed at all, any continuation scheme must keep initial approximations within the domain of convergence, which itself depends on the parameters; beyond that the efficacy of the scheme depends on the accuracy of the initialisations it yields. First-order continuation methods with automatic parameter step-size control can be devised (Khesghi, 1982); these are analogous to time step-size control in the solution of initial value problems (Gresho et al., 1979). The optimal strategy for computing sequences of finite-element solutions for viscous free surface flows along solution branches in parameter space, and arriving at particularly desired parameter values is a topic of current investigation.

The examples in the following section typically took 4 to 7 Newton iteration steps from a successful start-up approximation to reduce to less than 10^{-4} the largest change $|\alpha^i_{n+1} - \alpha^i_n|_{max}$ in the basis function coefficients α (which are of order unity excepting pressure coefficients in some cases). The search for a successful start-up approximation in some cases, however, took many steps, particularly when the Jacobian matrix subsequently proved to have been incorrect. In contrast, 3 to 5 steps typically sufficed when a neighbouring solution was available and zeroth-order continuation was used. Most of the calculations were made with a CDC CYBER 74 computer and the FTN compiler. The number of elements ranged from 30 to 70, which led to around 400 to 1000 unknown finite-element coefficients. The amount of core memory required was about 70 K octal words, and the amount of out-of-core memory for the frontal routine, originally designed for disc memory, was up to 200 K octal words. This last number, however, depends strongly on the front width, which in turn depends on the configuration of the computational domain; memory requirements might be quite different for other linear matrix equation solvers. Anywhere from 5 to 12 s of central processor time were needed for one iteration step depending on the number of unknowns and front width.

8.5   ILLUSTRATIVE RESULTS

Rimming flow (Fig. 8.7(a)), in a rotating horizontal cylinder partially filled with liquid, contains two of the essential elements of coating flows: a free surface and a liquid film accelerated by a moving wall or substrate. It lacks,

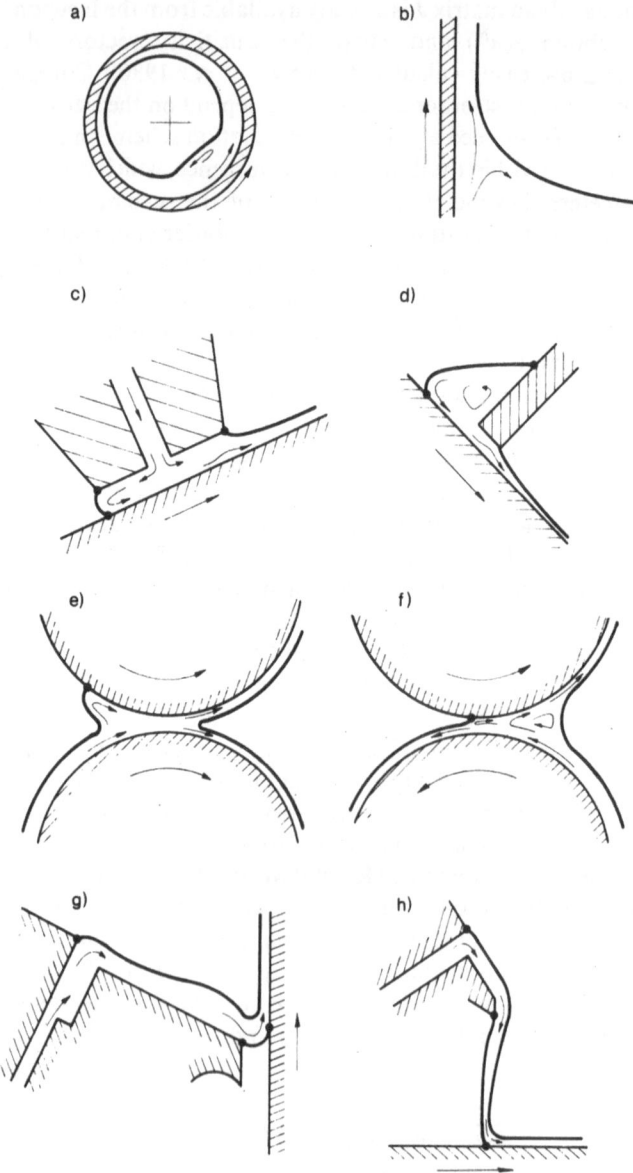

Fig. 8.7. Selected coating operations. a, Rimming flow; b, dip-coating; c, bead-and extrusion-coating; d, knife-coating; e, forward roll-coating; f, reverse roll-coating; g, slide-coating; h, curtain-coating.

however, the complications of contact lines and inflow and outflow. The early version of the finite-element method applied to this problem by Orr and Scriven (1978) was based on a successive iteration procedure that uses the normal component of the stress boundary condition at the free surface and converges only if surface tension effects are appreciable (Silliman and Scriven, 1980). The results for circumferential variation of the liquid film thickness and flow field, which agree well with asymptotic solutions (Ruschak and Scriven, 1976) for various limiting regimes, illustrate the power of the method in providing physical insight into balances of viscous, inertial, pressure, gravitational, and capillary forces that vary through the flow domain and shift as parameters are changed.

Dip-coating (Fig. 8.7(b)) and its axisymmetric counterpart, free-wire-coating, are other examples that lack contact lines and the complications they bring. Numerous studies have been based on the first analytical treatment of a coating flow, the asymptotic analysis by Landau and Levich (1942), later refined by Ruschak (1974), to predict approximately the shape of the free surface and the final thickness of the film entrained. Recently Tanguy and Choplin (1981) applied the Galerkin/finite-element technique to the problem of wire-coating, but their algorithm to locate iteratively the position of the free surface makes less than full use of the possibilities inherent in the finite-element method as outlined above. Frederiksen and Watts (1981) considered the start-up of liquid entrainment by a plate moving vertically out of a bath of liquid of finite depth; they traced liquid film formation by solving the complete time-dependent Navier–Stokes system with the finite-element method.

The bead- or extrusion-coater depicted in Fig. 8.7(c) is an industrially important case that incorporates static separation lines and dynamic wetting lines. A finite-element analysis of the entire bead region (Silliman, 1979) reveals that there are forming zones near the entry, in the turnaround zone near the upstream meniscus, and in the drawdown zone beneath the downstream meniscus. It also shows that between these zones the flow rapidly approaches fully developed shearing motions, the development lengths of which are of the order of one gap width. Thus when the gap is long enough for there to be intervening zones of fully developed flow, each forming zone can be treated separately. And it is of course preferable whenever possible to subdivide a complicated coating flow into simpler, essential parts; to analyse and calculate predictions for each part in turn; to work out the mathematical analysis of the asymptotic regimes in between, as necessary; and finally to combine the results so that the entire flow of practical interest can be understood in physical terms—to say nothing of design, operation, control, and optimisation applications *per se*. This strategy allows concentrating computational resources economically on the flow formation zones in order to produce reliably accurate predictions of coating flows.

Before attempting the entire bead-coater Silliman (1979) attacked different forming zones separately. To analyse the upstream meniscus region requires specifying the vacuum applied. As Ruschak's (1976) analysis of a limiting case and further results by Higgins and Scriven (1980) indicate, a coating bead can be maintained only if a differential pressure in a certain range is maintained between the gas spaces adjacent to the coating bead upstream and downstream—a fact recorded long ago in the patent literature (Beguin, 1954). Computationally it is more convenient to replace the pressure difference, which cannot be set arbitrarily, by the liquid volume enclosed between the upstream meniscus and a reference plane, the volume being set as a constant and the applied vacuum then emerging as part of the solution. The menisci computed by Silliman (1979), all for low capillary numbers, are nearly arcs of a circle, and therefore are almost entirely determined by the applied vacuum and the prescribed dynamic contact angle (cf. Higgins and Scriven, 1980).

The earlier analysis of the downstream section of the bead-coater, or of a slot- or knife-coater as shown in Fig. 8.7(d), was extended by Saito and Scriven (1981) by combining polar and Cartesian coordinate para-metrisations of free surface shape to accommodate cases where the meniscus enters the slot. Calculated free surfaces and fields of velocity vectors are shown in Fig. 8.8. With finite-element analysis the flow parameters can be varied over wide ranges in order to study their effect on flow features such as the separation angle where the meniscus attaches to

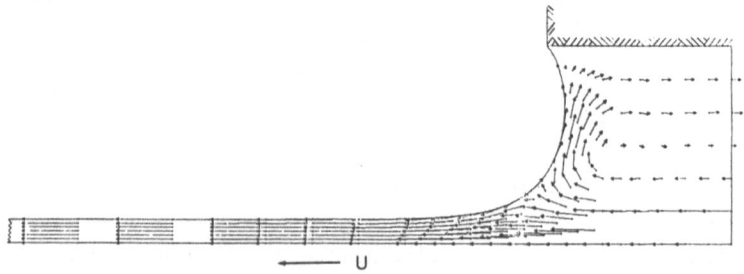

Fig. 8.8.   Slot coating flow with invading meniscus. $Re \equiv \rho U d / \eta_0 = 50$, $Ca = 0 \cdot 125$, $Q = q/Ud = 0 \cdot 13$, where d is slot clearance, and q is flowrate per unit width (Saito and Scriven, 1981).

the corner of the slot, the extent of the recirculating flow, the location of the stagnation line on the meniscus, and the length of the zone of flow relaxation into the asymptotic, fully developed regime downstream. The relaxation sufficiently far downstream is predicted to be exponential by Higgins (1982) and the finite-element results are in excellent agreement with the asymptotic theory (Saito and Scriven, 1981, Fig. 9). Bixler (1982) used the asymptotic analysis to derive vector boundary conditions of the third kind of the form of eqns (8.18) and (8.20), which allow shrinking of the computational domain of slot-coating analysis. This turned out to be particularly advantageous for solving the large asymmetric generalised eigenvalue problem that arises from the finite-element analysis of stability with respect to barring and ribbing disturbances downstream of a slot- or knife-coater (Bixler, 1982).

A close relative of the flow in slot-coating is the separating flow in the gap region of counter-rotating rolls depicted in Fig. 8.7(e). Symmetrical film splitting between a partially submerged pair is the simplest representative of a large variety of flows between rotating rolls. Williamson (1972) neglected inertial effects and solved the resulting biharmonic equation for the stream function by means of finite-difference simulation rather than finite-element analysis, approximating the entire free surface of interest by a single, sixth degree polynomial with two adjustable parameters. Ruschak (1982b) matched asymptotic expansions, employing the finite-element method to solve for the first term of an inner expansion that describes the two-dimensional flow in the vicinity of the interface and sets the appropriate boundary conditions on an outer expansion for the nip region, the first term of which is given by the lubrication approximation. Other examples where the detailed finite-element solution of a forming zone can be viewed as providing appropriate boundary conditions for some approximate film profile equations are discussed below. Coyle et al. (1982a) developed a generally applicable finite-element analysis of flow between rolls with film splitting, in which the free surface parametrisation described in Section 8.3 is crucial: a set of spines that control the tessellation into elements in the splitting region and in the film flow along the rolls is made to shift with the location of the film splitting stagnation line, in order that the nodes remain concentrated where the flow varies most rapidly, the mesh in the gap region between the rolls being stretched or compressed accordingly. The sequence of computed meniscus shapes in Fig. 8.9 shows that the position of the free surface is quite sensitive to the competition between viscous stresses and the curvature-dependent pressure, as expressed by the capillary number; the same applies for an even more striking feature of the flow field, the slowly

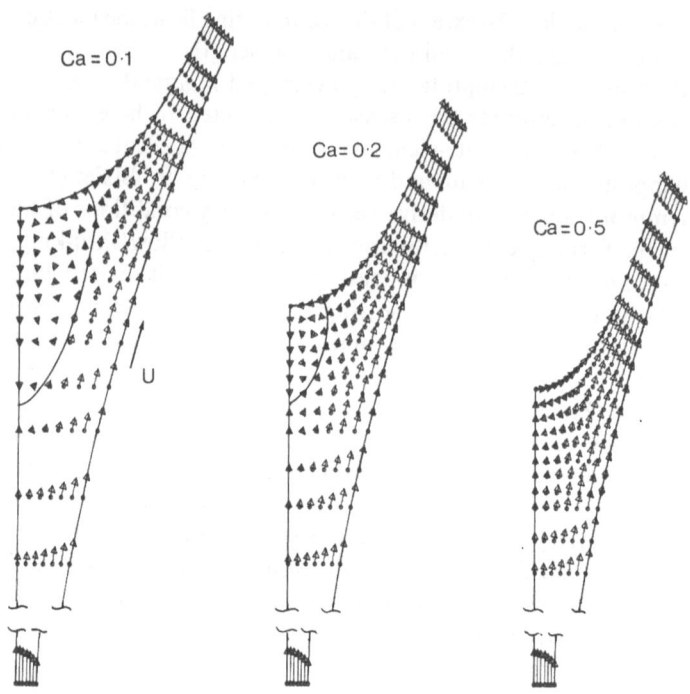

Fig. 8.9.   Symmetrical film splitting of flow between rolls. $Re = 0$, $D/h_0 = 100$, where D is roll diameter and h_0 is gap clearance (Coyle et al., 1982).

circulating eddies just under the stagnation line at which the film splits. The analysis was very recently extended to reverse roll-coating shown in Fig. 8.7(f) with its dynamic contact line (Coyle et al., 1982b), an industrially important means of metering flow rate and coating a uniform, thin film. Eddies are also found there, as illustrated by Fig. 8.10. These examples indicate especially well how computer-aided analysis as outlined in this chapter furnishes valuable insights into the operation of a coating process, inasmuch as the slowly circulating eddies might evade detection but might be responsible for such troubles as intermittent or occasional appearance of prematurely polymerised, dried, or solidified coating liquids, or clumps of particles in coating suspensions.

In slide-coating, a common precision operation, and in certain configurations of curtain-coating, liquid is metered through a narrow distribution slot to form a film that flows down an inclined plane for transfer to a moving web. Various designs of the slot exit have been proposed in the

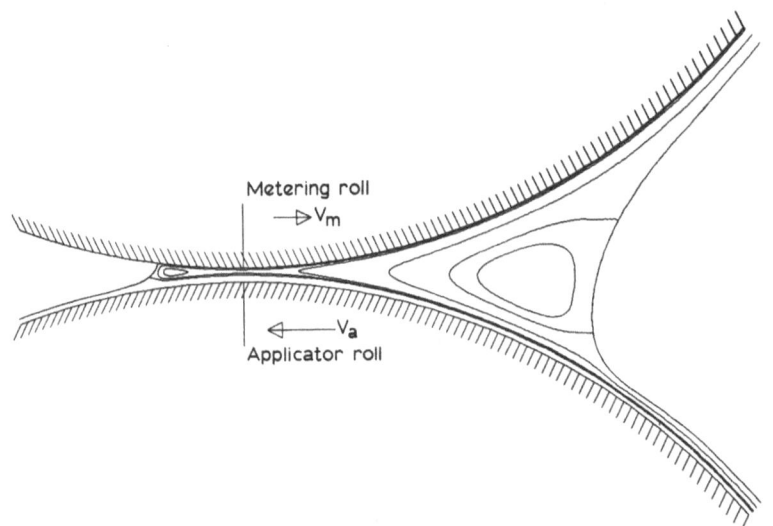

Fig. 8.10.   Flow in reverse roll-coating. $Re = 0$, $Ca \equiv \eta_0 V_a/\sigma = 0\cdot1$, $V_m/V_a = 0\cdot5$, $D/h_0 = 100$, where D is roll diameter and h_0 is gap clearance (Coyle *et al.*, 1982b).

patent literature since Padday (1961) first suggested that a slot enlargement of the kind shown in Fig. 8.7(g) might be advantageous. Finite-element analysis can be used to clarify the effect of die design on the flow field. For instance, Fig. 8.11 shows how the height of the step at a slot exit affects free surface shape at various flowrates. Of particular interest for optimisation of coating die configurations are the separation angle at the static contact line and whether or not there is recirculation in the adjacent region of slow flow. The use of the spines is worth noting: the slot exit and the film downstream were spanned by spines that vary smoothly from being orthogonal to the slot wall to being orthogonal to the inclined plane (cf. also Fig. 8.2).

Figure 8.12 is a theoretical prediction of stream lines in the coating bead zone where the liquid film flowing down the inclined slide is transferred over a narrow gap to an upward-moving substrate (Kobayashi *et al.*, 1982). One boundary condition used at the substrate was limited slip in the element adjoining the dynamic wetting line, prescribed in the form of a slip velocity distribution, eqn (8.7), from zero speed at the wetting line to substrate velocity at the next node or next element. In addition a dynamic contact angle had to be specified, as by eqn (8.36); its magnitude was found to affect

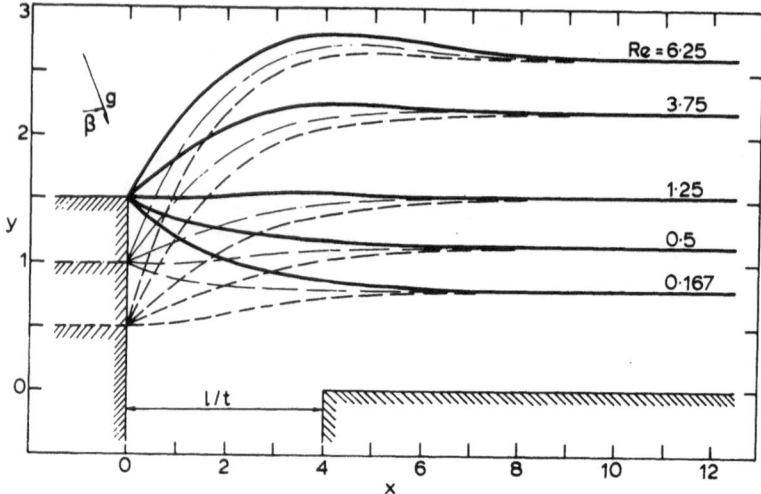

Fig. 8.11. Progression of free surface profiles in liquid film formation. $Re \equiv \rho q/\eta_0$, $B_0 \equiv t^2\rho g/\sigma = 0\cdot1$, $N_p \equiv \eta_0^4 g/\rho\sigma^3 = 10^{-4}$, $l/t = 4$, where l is thickness of slot enlargement, t is thickness of slot upstream and q is flowrate per unit width (Kistler, 1983).

the solution shown in Fig. 8.12. It should be noted that the direction of the spines that control the tessellation changed smoothly from being normal to the slide to being normal to the substrate, but were positioned so that the nodal mesh was substantially refined in the bead zone where the flow turns rapidly. Moreover, one spine was forced to translate as the position of the wetting line shifted during Newton iteration, the other spines adjusting proportionally in accordion-like fashion between the static separation line at the slide lip where one spine is fixed, and the dynamic wetting line. Some searching was required to find a successful start-up approximation, but continuation sufficed for all the sequences of cases computed thereafter. It is important to bring to attention the fact that if fully developed flow is prescribed at the inflow plane by essential boundary conditions, and that plane is placed too close to the region of characteristic film thinning just above the coating bead, the standing wave or waves which are often detectable in reality may disappear entirely from the finite-element predictions. On the other hand, an asymptotic upstream solution is available (Ruschak, 1978) which can be used to good advantage as in eqn (8.41).

In curtain-coating a liquid film formed on a slide, as shown in Fig. 8.7(h),

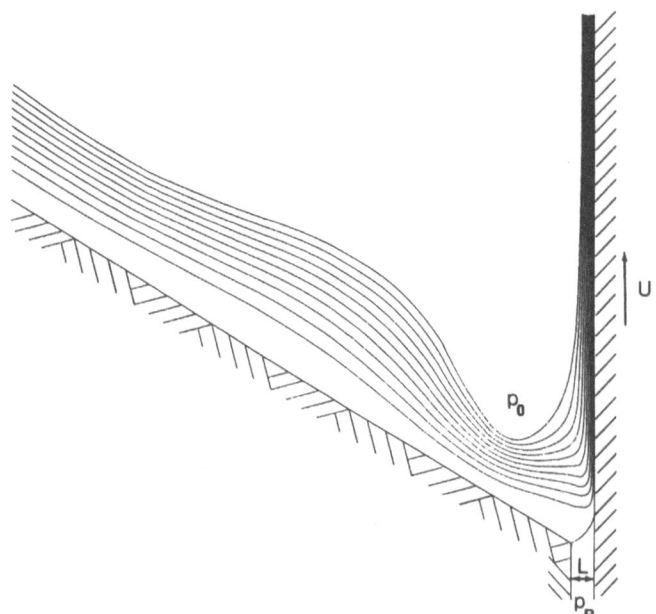

Fig. 8.12. Flow in slide-coating bead. $Re \equiv \rho UL/\eta_0 = 47$, $Q \equiv q/UL = 0.5$, $St \equiv \rho L^2 g/\eta_0 U = 0.04$, $Ca = 0.276$, $\Delta p = -2.61$, $\theta_{contact} = 165°$, where L is gap clearance, q is flowrate per unit width and Δp is dimensionless pressure difference $p_B - p_0$ across bead (Kobayashi et al., 1982).

or extruded through a narrow vertical slot falls as an unsupported sheet onto a moving substrate. The sheet approaches free gravity fall much further downstream than the forming zone (where full two-dimensional analysis is required) extends, but still far above, typically, the impinging region where flow in the sheet is affected by the moving substrate. Similarly when an extruded curtain is a highly viscous sheet being drawn by a wind-up mechanism (cf. Fig. 8.5) rather than being coated, the distance from die-slot to take-up is typically much longer than the forming zone. These flows are prototypes of the many real flows for which it is advantageous to subdivide the entire flow field into several flow regions; to make fully two-dimensional calculations, which are relatively costly, only for the forming zones; and to rely on various cost-saving approximations in the flow developing zones. For the cases of viscous liquid sheet formation the downstream boundary conditions on the two-dimensional finite-element equations are, as described in Section 8.3, the conditions of matching with

288   S. F. KISTLER AND L. E. SCRIVEN

one-dimensional finite-element equations for the evolving curtain or sheet profile further downstream. Alternatively, the two-dimensional finite-element solution can be viewed as providing the proper upstream boundary condition for the approximate film profile equations. The appropriate downstream boundary conditions on the film profile equations are free gravity fall at a point sufficiently far downstream for falling liquid curtains, and the wind-up speed or tension at the wind-up mechanism for drawn viscous sheets. It is important to test the influence of the matching plane. In the results for symmetrical viscous sheet formation, exemplified by Fig. 8.5, there is no significant effect of relocating the matching plane further downstream.

The sheet formation results show among other things that shearing motion in the die-slot rapidly relaxes into extensional flow with the plug-like velocity distribution that is the basis of the one-dimensional asymptotic approximation (see Section 1.3 and Chapters 6 and 7). In symmetrical cases the spines coincide with lines of a constant Cartesian coordinate, whereas in the sequence of solutions in Fig. 8.13 (Kistler *et al.*, 1981), which shows the fall of a curtain off the specially shaped lip of a slide, the spine

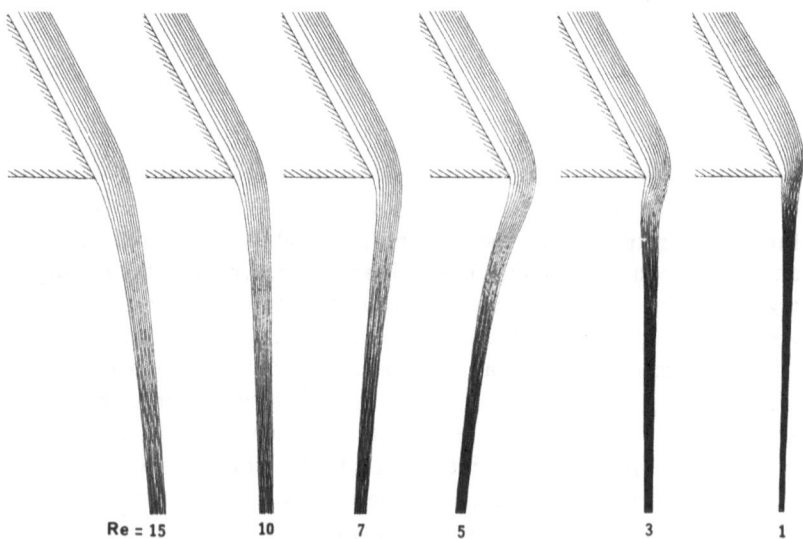

Re = 15   10   7   5   3   1

Fig. 8.13.   Deflection of liquid film falling over lip of inclined slide ('teapot effect'). $\sigma/(4\eta_0^4 g/\rho)^{1/3} = 1\cdot5$, $Re \equiv \rho q/\eta_0$, where q is flowrate per unit width (Kistler *et al.*, 1981).

that defines the outflow plane is forced during Newton iteration to be normal to the midsurface of the liquid sheet in order that the matching conditions, expressed in streamwise coordinates, can be imposed simply. The orientations of the spines between the one through the static separation line, which is fixed, and the outflow plane are interpolated or adjusted at each iteration in a prescribed manner.

At relatively high flowrates the trajectory of a curtain is 'ballistic', as for $Re = 15$ in Fig. 8.13. As the rate is lowered the trajectory turns 'anti-ballistic' and passes through a maximum deflection ($Re = 5$) while the forming zone outside the lip shortens. The mechanism of this so-called 'tea-pot' effect has been controversial but is settled by the finite-element analysis: at the root is lack of shear stress in the curtain downstream of the forming zone (Kistler, Dowd and Scriven, 1982). In Fig. 8.14 the theoretical predictions for a liquid curtain falling off the lip of a vertical plane are compared with precise measurements of top free surface profiles by means of a needle probe mounted on a micrometer which can be positioned vertically with a slide and dial indicator (Kistler *et al.*, 1981). The agreement with the computed predictions is excellent, except at intermediate Reynolds numbers where the deflection of the liquid sheet is greatest and most sensitive to the flow parameters, and hence where errors in measurement of flowrate, viscosity and surface tension are manifested most strongly (Kistler, Dowd and Scriven, 1982). However at $Re = 4$, for instance, even such details of the measured profile as the second inflection are correctly predicted. This example demonstrates that theoretical predictions based on the finite-element methods for viscous free surface flow outlined in the previous sections can account accurately for reality in coating flows. Furthermore, what can be predicted is far more than what can ordinarily be measured, for details of velocity, strain rate, vorticity, pressure, and viscous stress fields are virtually inaccessible on the small thickness scales of coating flows. Nevertheless, experimental results are generally indispensable, as for instance in the example just presented; for in the flow, quite apart from a purely hydrodynamic deflection of the sheet which is so well predicted by the finite-element calculations, there can be wetting and spreading over the underside of the lip. Indeed, in the laboratory it is found that once the liquid does wet the underside of the lip, the sequence in Fig. 8.14 cannot be reversed by simply increasing the flowrate. The theoretical analysis misses this hysteresis effect because the adhering of the meniscus to the corner is always enforced as an essential boundary condition, but does not account for the physics of contact line attachment or migration.

Flow in the impingement region where the freely falling sheet is coated

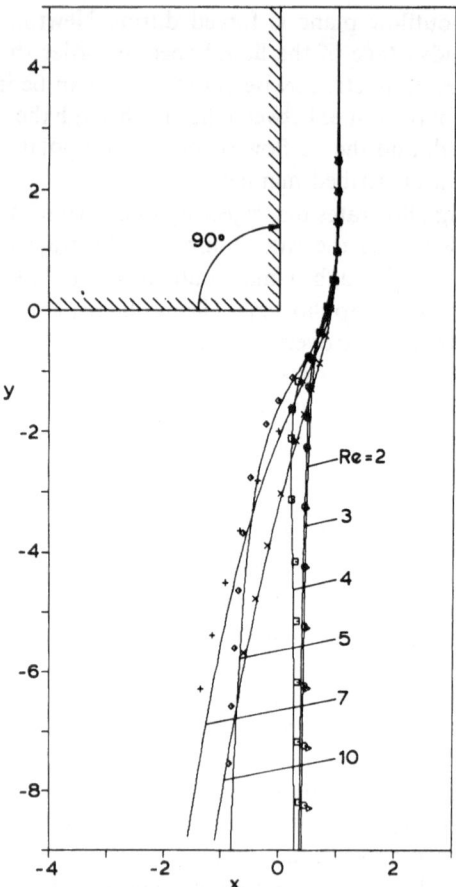

Fig. 8.14. Predicted top free surface profile of curtain compared with measured . points. $Re \equiv \rho q / \eta_0$, $\sigma / (4 \eta^4 g / \rho)^{1/3} = 1 \cdot 15$, where q is flowrate per unit width (Kistler *et al.*, 1981).

onto the moving substrate largely determines the operability of curtain-coating. The sequence of solutions in Fig. 8.15 illustrates performance at high enough capillary numbers for free surface shapes to be insensitive to surface tension (Kistler and Scriven, 1982). In this industrially relevant limiting regime the capillary pressure term and the end-point forces can be neglected in eqn (8.35); mathematically this means that the dynamic contact angle, rather than being an independent parameter to be specified, is a dependent variable and part of the solution just as is the wetting line

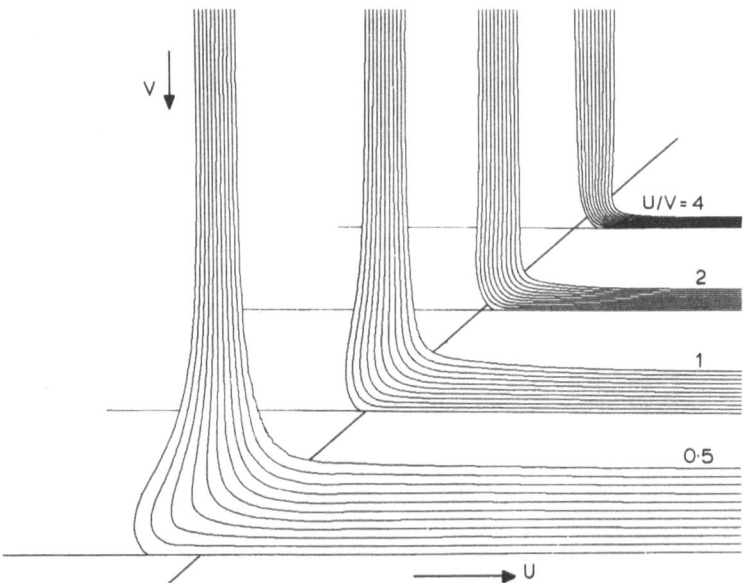

Fig. 8.15. Streamlines for liquid curtain impinging on moving substrate with increasing ratio of substrate speed U to impinging velocity V. $Re \equiv \rho q/\eta_0 = 2 \cdot 5$, where q is flowrate per unit width (Kistler and Scriven, 1982).

position. Finite-element analysis can thereby be used to explore the bounds of coating operability in this limiting regime. As the ratio of substrate speed U to impingement velocity V increases, the dynamic contact line shifts in the downstream direction and the dynamic contact angle approaches and finally reaches 180°—by hypothesis 180° is the upper limit for coating without catastrophic air entrainment. Other results show a similar progression as the Reynolds number decreases. As in the analysis of slide-coating, a slip velocity distribution limited to the first element adjoining the dynamic wetting line is used, the slip length remaining an empirical input parameter. The spines vary smoothly in orientation from being normal to the sheet to being normal to the substrate. The spine at the inflow plane is fixed in space; one spine shifts so that it always intersects the wetting line at 45°; and the intervening spines are spread or gathered accordingly. The spines downstream from the wetting line simply translate with the wetting line. Solutions for two-layered curtains, in which the ratio of viscosity in the top layer to that in the lower layer is increased, with the overall Reynolds number based on an averaged viscosity held constant, show that the global features of the flow do not change greatly. However, the details close to the dynamic wetting line indicate that it is advantageous for the layer of lower viscosity to make contact with the substrate.

8.6   FURTHER DEVELOPMENTS

These results illustrate the power of coating flow theory that makes use of the finite-element method as it is now developed for (Newtonian) viscous free surface flow. The general free surface parametrisation and finite-element representation presented in Section 8.3 are versatile enough to accommodate readily the complicated flow configurations in the forming zones encountered not only in slot-, roll-, slide-, and curtain-coating as described above, but also in kiss-, spin-, extrusion- and other coating operations. Full Newton iteration combined with asymptotic analysis of developing zones allows efficient iterative computations of the finite-element expansion coefficients. The computed results can be used to visualise velocity fields and associated strain-rate and vorticity fields—for example to reveal unwanted eddies and slow flow zones. They can be used to inspect strain, strain-rate, vorticity, and stress fields for rheological purposes. They can be employed to investigate the influence of the shape of confinement on the various fields, for instance to optimise a coating die design. They can furnish physical explanation of experimentally observed phenomena and insights into limits of operability of coating processes. All of these advantages are heightened by the difficulties of experimentally probing flows on the small scales that are common in coating processes.

The methods at hand can be, and increasingly are being, applied to other two-dimensional, steady, Newtonian, isothermal, incompressible flows with rigid and free boundaries. However there are many developments in the offing.

The finite-element methods are being developed in various directions. Already commonplace is the practice of coarsening and refining the tessellation—often referred to as refining the mesh of nodes—in order to confirm that refinement would lead ultimately to a finite-element solution that is independent of the sizes of the subdomains (Strang and Fix, 1973). What are needed are theory and systematic procedures for making such tests with only local or regional refinement. There are ongoing studies of the efficacy of various basis functions (e.g. Sani *et al.*, 1981). What are most promising are discontinuous basis functions for the pressure field, one of the reasons being that they open the way to so-called penalty formulations (see Hughes *et al.*, 1979 for a review), in which the pressure unknowns and continuity-derived equations are removed from the equation set, eqn (8.48), by introducing a synthetic equation of state. Penalty methods reduce the size of the global matrix problem and the front width needed; hence they can reduce computational costs. Kheshgi and Scriven (1982) developed a penalty method for viscous free surface flows and demonstrated that in the case of slot-coating (see Fig. 8.8) the computational time needed for frontal Gaussian elimination in the global matrix problem can be reduced by 30%.

Schemes of automatic spine placement and subsequent tessellation, or mesh generation, are needed to make the finite-element methods more accessible to non-specialist users. Forerunners of such schemes are methods currently being developed for adaptively relocating the reorienting spines and adjusting a tessellation as iteration proceeds, and also as parameters are changed. So far the most successful methods are for one-dimensional problems (Benner and Scriven, 1982; Gelinas *et al.*, 1981), but there is considerable incentive to extend these to two dimensions (Babuska and Rheinboldt, 1980; Davis and Flaherty, 1982). Another line of improvement is adaptive parameter step-size selection in continuation procedures (Kheshgi, 1982). Effects of the shape of confining boundaries can be found by varying the parameters in their representation. Quasi-Newton iteration methods, which are mentioned in Section 8.4, can reduce computational costs though often at some risk; their attractiveness will increase as sound adaptive variants of them are perfected.

Coating flow theory itself is being developed in several directions too, many of them with ever-deepening aid from finite-element mathematics. Central is the physics of air displacement and wetting at dynamic contact lines, as remarked in Section 8.2. The relevant scale is now known to be much smaller than that resolved by current finite-element solutions of the global flows and on that scale the local flows are surely three-dimensional and time-dependent. However, the air entrainment limit on coating speed evidently corresponds to a rapid inflation of that scale which is accompanied by approach to an apparent dynamic contact angle of 180°, as noted in Section 8.5.

An extremely important consideration, once the governing equations have been solved for a steady, two-dimensional coating flow, is its stability to disturbances. These can be themselves two-dimensional and likely to result in barring or surface waves that extend across the base flow. Or they can be three-dimensional and likely to result in ribbing or surface waves that extend downstream. In other cases instability manifests itself as less regular surface topography and in extreme cases as breakdown of the coated film, so that uncoated regions develop. The finite-element methods described in this chapter have been augmented by linear analysis of the stability of base flows to small disturbances that are two-dimensional (Bixler, 1982) or three-dimensional (Bixler, 1982; Ruschak, 1982a). The analysis leads to a large unsymmetric generalised eigenproblem, which Bixler (1982) solved by an efficient subspace iteration method (Stewart, 1976) that provides results needed to investigate the effect of the growing disturbance on the base flow, i.e. post-instability secondary flows, and other aspects of what is known as bifurcation behaviour. Ruschak (1982a) explored a quasi-steady approximation that reduces substantially the size of the eigenproblem so that it can be solved routinely, for instance with EISPACK routines, but with no clear saving in computational work and with some risk that the approximation is invalid at higher Reynolds numbers, in particular when the dominant mode is oscillatory. The successful predictions of ribbing instability in slot- or knife-coating (Bixler, 1982) and fingering of air displacing liquid from a slot (Ruschak, 1982a) show that linear stability analysis of viscous free surface flows by the finite-element method is not merely feasible but highly promising.

Unsteady flows are of great interest in connection with start-up, upsets such as substrate splices cause on passing through a coating bead, transitory control actions, and the like. Moreover an alternative way to study instabilities is to follow the transient response to selected disturbances by solving the complete time-dependent Navier–Stokes system. Progress has been made by Kheshgi and Scriven (1982), who merged an implicit time integrator used by Gresho et al. (1980) with the free surface parametrisation and Galerkin/finite-element method described in this chapter; they also incorporated a penalty method in order to reduce computational cost.

Flows in successful coating operations are in many cases two-dimensional except near the edges of the film being coated, where the situation is more or less three-dimensional—and entirely deserving of study. Finite-element analyses of three-dimensional viscous free surface flows have not yet appeared, chiefly because of the dismayingly large number of basis functions needed for reasonably accurate expansions of the pressure and velocity fields. Recent developments in short-cut methods and the growing availability of large scientific computers offer hope of extending coating flow theory to three-dimensional effects.

The effects of compliant substrate and confining surfaces, such as flexible blades and rubber rolls, are of considerable practical significance.

Finite-element methods have been perfected for elastohydrodynamic lubrication flows and are currently being developed for analysis of somewhat analogous coating flows. Another set of complications are thermal effects, and developments in non-isothermal analysis are likewise afoot; the same is true of mass transfer effects such as those induced by the surface-active agents and the volatile solvents that are so often used in coating technology.

Last but far from least are the rheological complications encountered in coating polymeric and other non-Newtonian liquids, for the flows can differ greatly from their Newtonian counterparts (Middleman and his group, e.g. Bauman *et al.*, 1982). The simplest non-Newtonian constitutive equation, namely the generalised Newtonian model that can account for the shear-rate dependence of viscosity (e.g. the Carreau model, cf. Bird *et al.*, 1977), can be incorporated without difficulty in the finite-element-based theory of viscous free surface flows (see Coyle *et al.*, 1982a). The non-linearity of the generalised viscous stress merely compounds that of the free surfaces and that of convective momentum transport (if it is present): the method of solving the algebraic equation set need not be changed at all. Viscoelastic liquids are another matter. Differential constitutive equations such as the convected Maxwell models (e.g. eqn (1.27)) and second-order fluid model (eqn (1.23)) are known to fail when elastic effects become important in steady, two-dimensional flow, that is, when the elastic contribution to the state of stress becomes comparable to the viscous contribution, the ratio being measured by the Deborah number (see Mendelson *et al.*, 1982; Tanner, 1982, or eqn (1.36)). There are indications of viscoelastic stress concentrations and so there are likely to be special material responses locally, quite possibly three-dimensional and unsteady, rather as in the case of the dynamic wetting line touched on in this chapter. The physics of viscoelastic free surface flows, the efficient representation of the rheological responses of the liquids, especially by integral or fading memory constitutive relations of the BKZ-type, and finite-element methods for accurately predicting important details of the flows, are all subjects of active research in the theory of coating flow.

REFERENCES

Babuska, I. and Rheinboldt, W. C. (1980). Reliable error estimation and mesh adaptation for the finite element method. In *Computational Methods in Nonlinear Mechanics* (Ed. J. F. Oden), North Holland Publishing Company, Amsterdam, pp. 67–108.

296   S. F. KISTLER AND L. E. SCRIVEN

Baumann, T., Sullivan, T. and Middleman, S. (1982). Ribbing instability of coating flows: Effect of polymer additives, *Chem. Eng. Commun.*, 14, 35.

Beguin, A. E. (1954). Method of Coating Strip Material. US Patent 2,681,294.

Benner, R. E. and Scriven, L. E. (1982). Adaptive subdomaining for solution of one-dimensional problems by the finite element method, SIAM 30th Anniversay Meeting, Stanford, 19–23 July, 1982.

Bird, R. B., Armstrong, R. C. and Hassager, O. (1977). *Dynamics of Polymeric Liquids*, Wiley, New York.

Bixler, N. E. (1982). Stability of a coating flow, Ph.D. Thesis, University of Minnesota.

Bixler, N. E. and Scriven, L. E. (1983a). Downstream development of three-dimensional viscocapillary film flow, *I & EC Fundam.* (to be submitted).

Bixler, N. E. and Scriven, L. E. (1983b). Robin conditions for open flow boundaries, *Phys. of Fluids* (to be submitted).

Brown, R. A. (1979). The shape and stability of three-dimensional interfaces, Ph.D. Thesis, University of Minnesota.

Brown, R. A., Scriven, L. E. and Silliman, W. J. (1980). Computer-aided analysis of nonlinear problems in transport phenomena. In *New Approaches to Nonlinear Problems in Dynamics* (Ed. P. Holmes), SIAM, Philadelphia, p. 298.

Chang, P. W., Patten, T. W. and Finlayson, B. A. (1979). Collocation and Galerkin finite element methods for viscoelastic fluid flow—II, *Comp. Fluids*, 7, 285.

Coyle, D. J., Macosko, C. W. and Scriven, L. E. (1982a). Computer simulation of nip flow in roll coating. In *Computer Applications in Coatings and Plastics*, ACS Symposium Series, pp. 251–64.

Coyle, D. J., Macosko, C. W. and Scriven, L. E. (1982b). Finite element analysis of reverse roll coating. AIChE Ann. Mtg, Los Angeles, 14–18 Nov., 1982, Paper 101h.

Davis, S. F. and Flaherty, J. E. (1982). An adaptive finite element method for initial-boundary value problems for partial differential equations, *SIAM J. Sci. Stat. Comput.*, 3, 6.

Dennis, J. E. and Moré, J. (1977). Quasi-Newton methods, motivation and theory, *SIAM Rev.*, 19, 46.

Dussan, E. B. (1976). The moving contact line: The slip boundary condition, *J. Fluid Mech.*, 77, 665.

Engelman, M. S., Strang, G. and Bathe, K.-J. (1981). The application of quasi-Newton methods in fluid mechanics, *Int. J. Num. Meth. Engng.*, 17, 707.

Frederiksen, C. S. and Watts, A. M. (1981). Finite-element method for time-dependent incompressible free surface flow, *J. Comp. Phys.*, 39, 282.

Gelinas, R. J., Doss, S. K. and Miller, K. (1981). The moving finite element method: Applications to general partial differential equations with multiple large gradients, *J. Comp. Phys.*, 40, 202.

Gifford, W. A. (1982). A finite element analysis of isothermal fiber formation, *Phys. Fluids*, 25, 219.

Gresho, P. M., Lee, R. L. and Sani, R. L. (1980). On the time-dependent solution of the incompressible Navier–Stokes equations in two and three dimensions. In *Recent Advances in Numerical Methods in Fluids*, Vol. 1 (Ed. C. Taylor and K. Morgan), Pineridge Press, Swansea, UK, pp. 27–81.

Higgins, B. G. (1980). Capillary Hydrodynamics and Coating Beads. Ph.D. Thesis, University of Minnesota.

Higgins, B. G. (1982). Downstream development of two-dimensional viscocapillary film flow, *I & EC Fundam.*, **21**, 168.

Higgins, B. G. and Scriven, L. E. (1979). Interfacial shape and evolution equations for liquid films and other viscocapillary flows, *I & EC Fundam.*, **18**, 208.

Higgins, B. G. and Scriven, L. E. (1980). Capillary pressure and viscous pressure drop set bounds on coating bead operability, *Chem. Eng. Sci.*, **35**, 673.

Higgins, B. G., Silliman, W. J., Brown, R. A. and Scriven, L. E. (1977). Theory of meniscus shape in film flow: A synthesis, *I & EC Fundam.*, **16**, 801.

Hocking, L. M. (1976). A moving fluid interface on a rough surface, *J. Fluid Mech.*, **76**, 801.

Hocking, L. M. (1977). A moving fluid interface: Part 2. The removal of the force singularity by a slip flow, *J. Fluid Mech.*, **79**, 209.

Hood, P. (1976). Frontal solution program for unsymmetric matrices, *Int. J. Num. Meth. Engng.*, **10**, 379.

Hood, P. (1977). Correction, *Int. J. Num. Meth. Engng.*, **11**, 1055.

Hughes, T. J. R., Liu, W. K. and Brooks, A. (1979). Finite element analysis of incompressible viscous flow by the penalty function formulation, *J. Comp. Phys.*, **30**, 1.

Huh, C. (1969). Capillary hydrodynamics, interfacial instability and the solid/liquid/fluid contact line, Ph.D. Thesis, University of Minnesota.

Huh, C. and Mason, S. G. (1977). The steady movement of a liquid meniscus in a capillary tube, *J. Fluid Mech.*, **81**, 401.

Huh, C. and Scriven, L. E. (1971). Hydrodynamic model of steady movement of a solid/liquid/fluid contact line, *J. Coll. Interf. Sci.*, **35**, 85.

Huyakorn, P. S., Taylor, C., Lee, R. L. and Gresho, P. M. (1978). A comparison of various mixed interpolation finite elements in the velocity–pressure formulation of the Navier–Stokes equations, *Comp. and Fluids*, **6**, 25.

Isaacson, E. and Keller, H. B. (1966). *Analysis of Numerical Methods*, Wiley, New York.

Jackson, N. R. and Finlayson, B. A. (1982). Calculation of hole pressure: I. Newtonian fluids, *J. Non-Newtonian Fluid Mech.*, **10**, 55.

Kheshgi, H. S. (1983). Motion of viscous liquid films, Ph.D. Thesis, University of Minnesota.

Kheshgi, H. A. and Scriven, L. E. (1982). Penalty-finite element analysis of time-dependent two-dimensional free surface film flows. Proc. Fourth International Symposium on Finite Element Methods in Flow Problems, Tokyo, 26–29 July, 1982.

Kistler, S. F. (1983). The fluid mechanics of curtain coating and related viscous free surface flows, Ph.D. Thesis, University of Minnesota.

Kistler, S. F. and Scriven, L. E. (1981). Liquid curtains and drawn sheets analysed. AIChE Ann. Mtg, New Orleans, 8–12 November, 1981, paper 20g.

Kistler, S. F. and Scriven, L. E. (1982). Finite element analysis of dynamic wetting for curtain coating at high capillary numbers. AIChE Nat. Mtg, Orlando, Feb. 28–March 3, 1982, paper 45d.

Kistler, S. F. and Scriven, L. E. (1983). Coating flow theory, *Int. J. Num. Meth. Fluids* (in press).

Kistler, S. F., Dowd, C. A. and Scriven, L. E. (1981). The teapot effect, *Bull. Amer. Phys. Soc.*, **26**, 1276.

Kistler, S. F., Dowd, C. A and Scriven, L. E. (1983). Finite element analysis and experimental study of the teapot effect (to be submitted).

Kobayashi, C., Saito, H. and Scriven, L. E. (1982). Study of slide coating by finite element method. AIChE Nat. Mtg, Orlando, Feb. 28–March 3, 1982, paper 45e.

Landau, L. D. and Levich, B. (1942). Dragging of a liquid film by a moving plate, *Acta Physicochimica URSS*, **17**, 41.

Lipschutz, M. M. (1969). *Differential Geometry. Schaum's Outline Series*, McGraw-Hill, New York.

Lowndes, J. (1980). The numerical simulation of the steady movement of a fluid meniscus in a capillary tube, *J. Fluid Mech.*, **101**, 631.

Mendelson, M. A., Yeh, P.-W., Brown, R. A. and Armstrong, R. C. (1982). Approximation error in finite element calculations of viscoelastic fluid flows, *J. Non-Newtonian Fluid Mech.*, **10**, 31.

Miyamoto, K. and Scriven, L. E. (1982). Breakdown of air film entrained by liquid coated on web, AIChE Ann. Mtg, Los Angeles, Nov. 14–19, 1982, paper 101g.

Nickell, R. E., Tanner, R. I. and Caswell, B. (1974). Solution of viscous incompressible jet and free surface flows using finite element methods, *J. Fluid Mech.*, **65**, 189.

Oliver, J. F., Huh, C. and Mason, S. G. (1977). Resistance to spreading of liquids by sharp edges, *J. Colloid Interf. Sci.*, **59**, 568.

Orr, F. M. (1976). Numerical simulation of viscous flow with a free surface, Ph.D. Thesis, University of Minnesota.

Orr, F. M. and Scriven, L. E. (1978). Rimming flow: Numerical simulation of steady viscous free-surface flow with surface tension, *J. Fluid Mech.*, **84**, 145.

Padday, J. F. (1961). Multiple Coating Apparatus. US Patent 3,005,440.

Rheinboldt, W. C. (1974). *Methods for Solving Systems of Nonlinear Equations*, SIAM, Philadelphia.

Richardson, S. (1970). A stick-slip problem related to motion of a free jet at low Reynolds number, *Proc. Cambridge Phil. Soc.*, **67**, 477.

Rivas, A. P. (1972). Meniscus computations: shapes of some technologically important liquid surfaces, M.S. Thesis, University of Minnesota.

Ruschak, K. J. (1974). The fluid mechanics of coating flows, Ph.D. Thesis, University of Minnesota.

Ruschak, K. J. (1976). Limiting flow in a pre-metered coating device, *Chem. Eng. Sci.*, **31**, 1057.

Ruschak, K. J. (1978). Flow of a falling film into a pool, *AIChE J.*, **24**, 705.

Ruschak, K. J. (1980). A method for incorporating free boundaries with surface tension in finite element fluid flow simulators, *Int. J. Num. Meth. Engng.*, **15**, 639.

Ruschak, K. J. (1982a). Linear stability analysis for free boundary flows by the finite-element method. AIChE Nat. Mtg, Orlando, Feb. 28–March 3, 1982, preprint 47f.

Ruschak, K. J. (1982b). Boundary conditions at a liquid/air interface in lubrication flows, *J. Fluid Mech.* (to appear).

Ruschak, K. J. and Scriven, L. E. (1976). Rimming flow of liquid in a rotating horizontal cylinder, *J. Fluid Mech.*, **76**, 113.

Saito, H. and Scriven, L. E. (1981). Study of coating flow by the finite element method, *J. Comp. Phys.*, **42**, 53.

Sani, R. L., Gresho, P. M., Lee, R. L. and Griffiths, D. F. (1981). |The|cause and cure of the spurious pressures generated by certain FEM solutions of the incompressible Navier–Stokes equations: Part I, *Int. J. Num. Meth. Fluids*, **1**, 17; Part II, ibid., **1**, 171.

Scriven, L. E. (1982). How does air entrain at wetting lines? AIChE Nat. Mtg, Orlando, Feb. 28–March 3, 1982, paper 46a.

Silliman, W. J. (1979). Viscous film flows with contact lines, Ph.D. Thesis, University of Minnesota.

Silliman, W. J. and Scriven, L. E. (1978). Slip of liquid inside channel exit, *Phys. Fluids*, **21**, 2115.

Silliman, W. J. and Scriven, L. E. (1980). Separating flow near a static contact line: slip at a wall and shape of a free surface, *J. Comp. Phys.*, **34**, 287.

Stewart, F. M. (1976). Simultaneous iteration for computing invariant subspaces of non-Hermitian matrices, *Numer. Math.*, **25**, 123.

Strang, G. and Fix, G. J. (1973). *An Analysis of the Finite Element Method*, Prentice-Hall, New Jersey.

Tanguy, P. and Choplin, L. (1981). Hydrodynamic study of the wire free coating problem: Numerical simulation using the finite element method. AIChE Ann. Mtg, New Orleans, November 8–12, 1981, paper 20f.

Tanner, R. I. (1982). The stability of some numerical schemes for model viscoelastic fluids, *J. Non-Newtonian Fluid Mech.*, **10**, 169.

Teletzke, G. F., Davis, H. T. and Scriven, L. E. (1982). Wetting hydrodynamics, *J. Fluid Mech.* (submitted).

Walters, R. A. (1980). The frontal method in hydrodynamics simulations, *Comp. Fluids*, **8**, 265.

Williamson, A. S. (1972). The tearing of an adhesive layer between flexible tapes pulled apart, *J. Fluid Mech.*, **52**, 639.

Wilson, S. (1969). The development of Poiseuille flow, *J. Fluid Mech.*, **38**, 793.

Zienkiewicz, O. C. (1977). *The Finite Element Method*, McGraw-Hill, London.

CHAPTER 9

Process Control

J. WORTBERG

Kunststoffverarbeitung,
Institut für Aachen, West Germany

9.1 INTRODUCTION

In the manufacture of finished or semi-finished parts in plastics the parts produced must be of a certain desired quality configuration. With plastics, the quality of a product is not dependent upon the raw material and its manufacture alone; it is also dependent to a large|extent on processing history. Reproducible process control that is specially tailored to the raw material and product is therefore particularly important. There will also be constant endeavour from an economic angle to satisfy quantitative and qualitative specifications with minimum outlay in terms of investment, raw material, energy and human labour. It is against this background that the use of computer systems for control in plastics processing has to be seen.

With the introduction of suitable measuring sensors, appropriate control loops and process control strategies, the human controller with his inevitably subjective assessment and his capacity for error, is being replaced by equipment that can carry out orders and monitor processes according to fixed-program principles. Following the generation of analogue control units, the development of suitable equipment for this led in the 1970s to digital technology, without which the complex process control concepts of present-day production plant would never have been economically feasible. Only with the coming of microcomputer technology has the control, regulation and monitoring of expensive production plant on the basis of economic criteria become a reality. This chapter therefore deals solely with

301

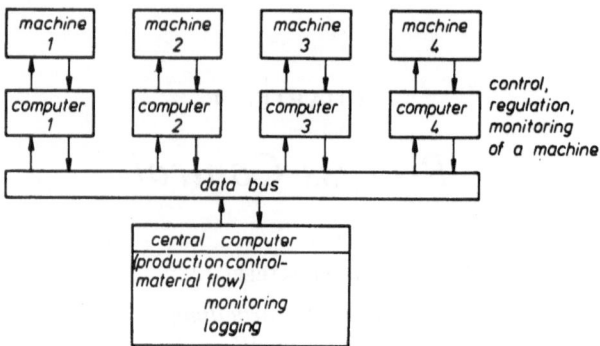

Fig. 9.1.   Computer configuration with decentralised machine computers and
centralised production master computer.

process control using computers, be these microcomputers or digital
process control computers.

The field of applications for a computer on a plastics processing machine
or processing plant can be divided into two parts. First, there are the
business management functions (production sequence, material flow, etc.),
which are not specifically process-related and essentially constitute control
engineering applications. Secondly, there are the specifically process-
related applications, which involve control and regulation functions in the
actual process (see Menges and Michaeli, 1976). We shall concentrate here
on the latter applications. Whereas up to just a few years ago there was talk
on cost grounds of having *central* process control computers to control a
number of production units simultaneously (see Koch, 1976; Menges and
Michaeli, 1976), development of the microcomputer has led to a configuration where *decentralised* computer systems, allocated to the individual machines, are envisaged for the specifically process-related
functions (see Upmeier, 1981; Wiegand, 1979). A central computer or
master computer, which is primarily assigned to production planning and
production monitoring, can then initiate a data transfer (target and actual
values) from the machine computer to the master computer or vice versa,
via appropriate interfaces (see Fig. 9.1).

9.2   CONTROL ENGINEERING FUNCTIONS FOR A PROCESS
CONTROL COMPUTER (MICROCOMPUTER) IN PLASTICS
PROCESSING

When assessing computer-controlled processing units one often wonders,
and rightly so, what it is that the computer can do that conventional

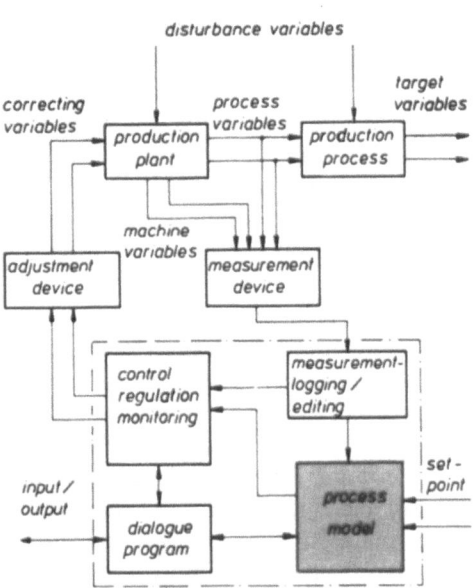

Fig. 9.2. Control by process computer.

systems are not able to do (see Hunkar, 1979). Figure 9.2 gives a diagram of specifically process-related applications for a computer (see Dormeier, 1977), which nowadays will generally mean a microcomputer.

9.2.1 Data acquisition and data editing

The measurement variables and process variables that are loggable in the process can readily be processed with the aid of a computer in such a way that conversion to traditional units or meaningful graphical representations of results is possible with little outlay (i.e. programming). The data storage facility in the computer is of particular interest here, since not only the current performance data but also past values can be incorporated in the representation of results (to highlight trends, etc.). In the extrusion process, for example, the following information is of interest: temperatures, pressures, rotational speeds and speeds by way of process variables; dimensions, throughputs, etc., by way of product-related target values. In hardware terms, the computer is connected to the process by special coupling elements. The information is transmitted either in analogue form (voltage, current) or digital form (frequency, switch status), with the analogue signals needing to be digitised prior to computer processing. The program is based on a time slot pattern which ensures that the process information is read in at set times. The so-called scan time, i.e. the interval between data inputs, is a function of the time behaviour of the process variables. Whereas slowly changing variables such as temperatures are only required at, say, 1-s or even 10-s intervals, more rapidly changing variables such as rotational speeds need to be scanned correspondingly more frequently at, say, 0·1-s intervals.

9.2.2 Control, regulation, monitoring

The characteristic feature of this application level is the closed signal loop. The computer not only records information but also feeds back control commands to the process or machine. From a technical program angle, the machine sequence control used on injection-moulding machines or hollow article blow-moulding units, for instance, consists of logic operations conducted with regard to temporal marginal conditions. In this respect, the computer has no advantage in process engineering terms; it merely enables the machine manufacturer to adjust more rapidly to customer-specific control modifications, through a simple program change. With the corresponding batch sizes, the computer also represents a cheaper alternative to the traditional control cabinet.

A higher application level is the use of a process control computer as a digital computer, also known as direct digital control or DDC (see Wiegand, 1979). Here, the computer takes over the function of analogue controllers in that it strives to attain optimum control results by means of a program containing one or more algorithms for calculating correcting variables from controlled variables.

A specific advantage of the computer here is that it is able to apply specially adapted control strategies for different plant-operating states, e.g. for 'start-up/shut-down' or 'continuous operation'. This it does by calling up tailor-made subprograms (see Anon., 1980; Kertscher, 1981; Schwab, 1980).

The constant comparison of target values with performance values also allows the computer to optimise its own controller function. By computing the control algorithm coefficients that characterise controller behaviour in accordance with theoretical optimisation rules, the computer will always ensure an optimum control result, even when there have been large-scale modifications in the point of operation (or set point). This self-adjustment of control system parameters, which is virtually impossible with conventional analogue technology, is known as 'adaptive control'.

9.2.3 Process optimisation

Probably the most interesting function of a computer in a processing machine is the possibility it offers for linking up a number of process variables or measurable variables by way of a prerequisite for optimised process control (see Parnaby *et al.*, 1975; 1978). Let us assume that a quality-determining target variable (or objective function) in a process (which may be either directly measurable or only indirectly calculable) may be influenced by several correcting variables. There are, however, limits on the range of adjustment as, otherwise, other target variables could not be controlled. The computer may therefore only intervene to obtain optimum process control on the basis of a preset *strategy*. If, for example, the dimensions or thickness of a pipe or profile are to be kept constant as a quality parameter, then take-up rate and screw speed are, in principle, available as correcting variables. Since, however, the screw speed can considerably influence melt temperature and melt homogeneity, this may only be used as a correcting variable providing that this marginal condition on the maximum possible throughput capacity is observed.

Other limitations could be set by a maximum permitted profile temperature upon exit from the cooling section, for instance. The strategy would now entail keeping dimensions constant by influencing take-up rate, whilst giving consideration to the process and product-conditioned marginal conditions and ensuring maximum permitted throughput (see Upmeier, 1981).

A further type of process control is that which uses statistical or mathematical–physical process models. Process models describe the inter-relationships between process variables (correlations) and permit the calculation of target variables, which either cannot be measured at all or can only be measured with a time delay, and which are then in turn available for controlled variables. Whilst both statistical and mathematical–physical process models are based on comprehensive process analyses (see Bergweiler, 1981; Buschhaus, 1982; Dormeier, 1977; Hellmeyer, 1977; Junk *et al.*, 1978; Michaeli, 1975; Parnaby and Kochhar, 1977; Predöhl, 1981; Ramm, 1981), the mathematical–physical model is to be given preference (in so far as it can be formulated simply). This is because the process inter-relationships are correctly described in qualitative terms in this model and extrapolation to points of operation not previously investigated is permissible, whereas the statistical methods are really only suitable for interpolations within a delimited field of operation. Exceptions are, however, justified in cases where the inter-relationship to be described is too complex or where adaption of the model is possible through the comparison of computed and measured target variables—possibly with a time lag (Ramm, 1981). In these cases the inter-relationships of interest can be described simply in mathematical terms, e.g. through linear or algebraic relationships:

$$\text{linear} \qquad y = k + ax_1 + bx_2 + cx_3 + \cdots \qquad (9.1)$$

$$\text{algebraic} \qquad y = k + ax_1^b + cx_2^d \ldots \qquad (9.2)$$

where y = target variable, x_i = influencing variables and $k, a, b, c, \ldots$ = operation-dependent coefficients to be adapted.

The transfer of process control and process optimisation from machine operators, who nowadays are often poorly trained and have little process engineering knowledge, to a process control computer naturally presupposes a corresponding process engineering background and a known strategy for optimum process control.

This fact, which is too often not fully appreciated, needs to be emphasised in this way because in a large number of cases the call for a process control computer represents a call for solutions to production difficulties, where not enough is known about causes and influencing variables. In this context reference is made to Chapters 2 to 8 of this book which describe simulation of polymer processes using mathematical–physical models. Although (at present, at least) simulation programs of this type are in many cases too complex and too large for process control at the machine (there can be problems with insufficiently known initial and boundary conditions in the process and problems with material data) and are also too computer intensive, they nonetheless show up the process inter-relationships correctly and this in turn leads to simplified control strategies for control at the machine. An example is process optimisation for injection moulding which is described below.

9.2.4 Input/output dialogue

It goes virtually without saying that a computer also opens up the way for easy input and output of set-up data, measurable data or computed data. The advantages of a central controlled data input, together with the facility for storing data that has been found to be right for a product on a previous occasion, not only makes for greater ease of operation but, above all, gives greater operational reliability and a reduction of what in some cases are considerable start-up times and hence start-up rejects.

One increasingly important advantage of the computer in a processing plant for the future is its capacity to *log data as required*. This applies not only to straightforward business management data, such as the quantities produced and production and stoppage times, but also covers the logging of process data, viewed in conjunction with process data control and monitoring systems. Quality-determining, process-relevant *performance values* need to be logged and documented without huge data banks being created. A sensible approach is only to log data when preset band widths are inadmissibly exceeded. (Who, after all, is interested in metre-long print-outs of quality production?) In this respect the computer supplies the processor not with more information but with less, though with information that has been edited and condensed. This will aid the processor not least in his dealings with customers, in the light of the growing development towards increased producer liability.

The data available on a machine or processing plant can naturally be transmitted to a higher-ranking process control master computer, or be called up by this, in the same way as input data can be transmitted from a central computer to the production unit if so desired. With a configuration of this type the logging and documentation functions can be carried out by the central computer.

9.3   COMPUTERS FOR TEMPERATURE CONTROL

The problem of temperature control is encountered in both extrusion and injection moulding (see Chapters 4 and 5). We shall therefore examine first digital control for mould and cylinder temperature and then the control of melt temperature.

9.3.1 Wall temperature control
In addition to the analogue systems employed for temperature control, digital temperature controllers are increasingly being introduced here. Figure 9.3 shows the block diagram of a scanned-data temperature control system using a microcomputer.

A characteristic feature of digital control is the discontinuous operating mode of the controller. The controlled variable is measured at constant time intervals and a correcting signal, such as the switch-in time for a strip heater, is computed from the control deviation using an algorithm in the computer.

In direct digital control it is possible to use algorithms that are adapted to the particular problem in question. Parameter-optimised control algorithms have primarily proved themselves here. As with analogue controllers,

DDC control calls for controller parameters to be optimally matched to the process in order for the desired control quality to be attained.

Since no setting rules exist for the controller parameters with higher-order algorithms, it is wise to carry out optimisation using a simulation program that describes the closed control loop (see Schwab, 1980). The DPID (Differential, Proportional, Integral, Differential closed loop control) algorithm will be taken as an illustration here. This has proved the best compromise in terms of optimum command behaviour and fault rectification behaviour.

$$
\begin{aligned}
y(k) = y(k-1) &+ K_Y(y(k-1) - y(k-2)) \\
&+ K_p(x_w(k) - x_w(k-1)) \\
&+ K_I x_w(k) \\
&+ K_{D1}(x_w(k) - 2x_w(k-1) + x_w(k-2)) \\
&+ K_{D2}(x_w(k) - 3x_w(k-1) + 3x_w(k-2) - x_w(k-3))
\end{aligned}
\tag{9.3}
$$

where

$y(i) =$ correcting variable for the ith scan period

$x_w(i) =$ controlled variable at scan time i

$$K_Y = \frac{T_v}{T_{AB} + T_v} \quad \text{digital filtering parameter}$$

$$K_P = \frac{K_R}{\left(1 + \dfrac{T_v}{T_{AB}}\right)} \quad \begin{array}{l}\text{proportional component parameter} \\ \text{(and controller amplification)}\end{array}$$

$$K_I = K_P \frac{T_{AB}}{T_n} \quad \text{integral component parameter}$$

$$K_{D1} = K_P \frac{T_{v1}}{T_{AB}} \quad \begin{array}{l}\text{parameter for 1st differentiation of} \\ \text{control difference}\end{array}$$

$$K_{D2} = K_P \frac{T_{v2}^2}{T_{AB}^2} \quad \begin{array}{l}\text{parameter for 2nd differentiation of} \\ \text{control difference}\end{array}$$

T_{AB} scan time

T_v, T_{v1}, T_{v2} delay times

T_n reset time

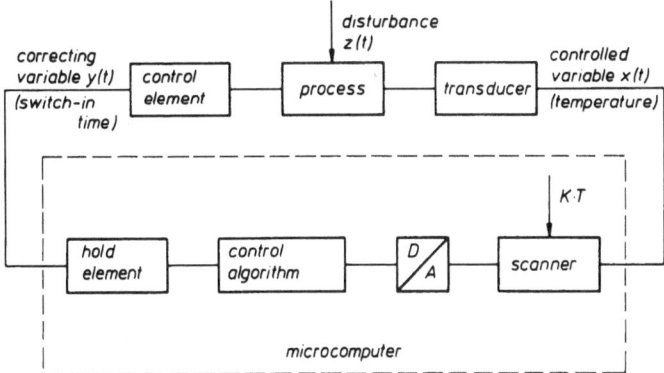

Fig. 9.3.   Block diagram of a scanned-data system for temperature control.
D/A = digital/analogue converter; K·T = scanning rate.

Figure 9.4 illustrates the control behaviour of this algorithm on a cylinder heating zone with a scanning interval of 30 s. The relatively high inertia of temperature control systems and the correspondingly long scanning times means that a single computer can process several temperature control systems simultaneously in multiplex mode.

9.3.2 Melt temperature control
Melt temperature, which constitutes an important parameter in the moulding process, is determined by the zone wall temperatures and by heating of the plastic in the machine through shear. If malfunctions occur during production this can lead to uncontrolled fluctuations in melt

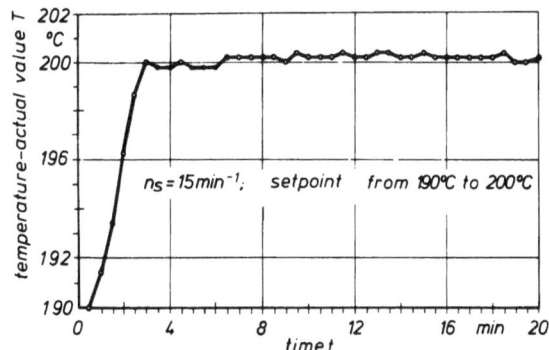

Fig. 9.4.   Control behaviour of the DPID algorithm on a cylinder heating zone.
n_S = screw speed.

temperature. There is thus a need for melt temperature to be controlled. By way of a restriction, it should be noted that this melt temperature control system cannot be applied in cases where heating or cooling only have slight influence on melt temperature on account of a prevailing high level of 'friction'.

The melt temperature control system, which works on the cascade control concept (see Schwab, 1980) is shown for analogue controllers in the block diagram in Fig. 9.5 with the melt temperature controller as the command controller and two zone controllers as auxiliary controllers. The command controller derives a correcting variable for its output from the difference between target melt temperature value and actual melt temperature value, which is then added to the zone temperature target values. These target value modifications lead on to a change in melt temperature through the zone temperature control.

A suitable extension of the zone wall temperature control system then opens up the possibility of using a microcomputer. Melt temperature is then scanned with the same scanning time as the zone wall temperature. This scan of melt temperature as a controlled variable follows at a characteristic distance from the cylinder wall on account of the permanent melt temperature profile at the screw tip.

A PID algorithm with self-optimising parameters was used for the melt temperature control. This will calculate favourable controller parameters upon request from the jump response given by the control system, following parameter adjustment rules and using the inflectional tangent method (see Schwab, 1980).

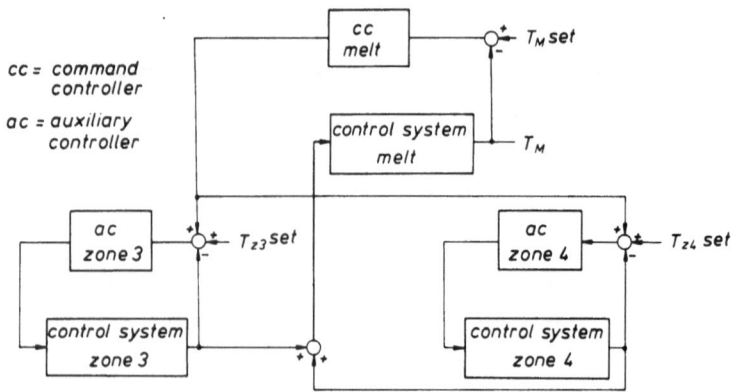

Fig. 9.5.   Block diagram of a cascade control for melt temperature.

The heating phase and plant start-up phase with a target melt temperature value of 200 °C is shown in Fig. 9.6. During the heating phase it was found advisable to switch off the melt temperature control and only to switch it on again after the screw started up. This avoided the excessive jump in melt temperature caused by 'frictional' energy when the screw was first switched on.

Instead of using heating capacity as the regulating element for melt temperature control it would be equally conceivable to apply the frictional energy of an adjustable choke.

Modification of melt temperature by means of a change in the friction heat per unit throughput takes effect within 2 to 3 min and is thus faster than using wall temperature to smooth out fluctuations. The consequences of fluctuations and modifications in screw speed and also the consequences of heating zone failure or target value modifications can be smoothed out quickly and reliably.

The disadvantages of this regulating element as compared with heating energy are the effects it has on throughput in extruders with counter-pressure dependent transport, potential repercussions on blend quality and greater outlay on technical equipment.

On injection-moulding machines the melt temperature can also be controlled by influencing back pressure. It should still be pointed out that measurement of melt temperature is necessary in all cases, either via thermocouples or similar devices, or from the travel time of an ultrasonic wave (see Recker *et al.*, 1976; Wiegand, 1979).

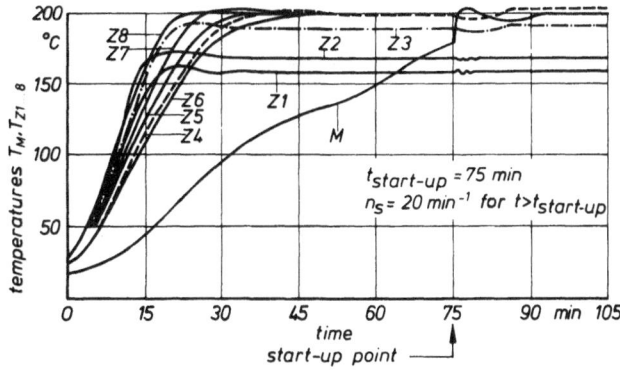

Fig. 9.6.   Heat-up and start-up with melt temperature control for $t > t_{\text{start-up}}$. The Z_i are temperature control zones: Z_1 to Z_3 are in the extruder, Z_4 is at the adapter and Z_5 to Z_8 are in the cross-head. M denotes the melt in front of the screw. n_S = screw speed.

9.4   COMPUTERS IN EXTRUSION

Computer control systems on extrusion plants not only serve to control the process parameters on the extruder itself but also take on functions beyond this, relating to one or more quality variables. In addition, the coordination and matching of individual process sections, such as during start-up, is an essential function. Descriptions will be given below of process control systems developed for different extrusion plants by the IKV (Plastics Processing Institute) at the University of Aachen, West Germany (see Wortberg, 1981). In this case rheological–thermodynamic process models provide the basis for improved process control. These establish *inter alia* an inter-relationship between the thermodynamic variables of pressure and temperature and the desired target variables of thickness and shrinkage, etc. They are also particularly suited (from a cost angle as well) to answering questions regarding plant layout and optimum process control.

9.4.1 Thickness control in film and sheet extrusion
In film and sheet extrusion thickness control constitutes a most important aspect of an overall control concept. The example of thickness control on extruded sheets will be taken to illustrate an adaptive control system that works with a mathematical–physical process model (see Bergweiler, 1981; Bergweiler *et al.*, 1980; Koch, 1976; Michaeli, 1975).

The process control concept is depicted in Fig. 9.7. The slow sheet speed in sheet extrusion means that there is a relatively lengthy delay time between molten plastic output from the die and thickness profile measurement on the cooled sheet. This delay time poses serious problems for

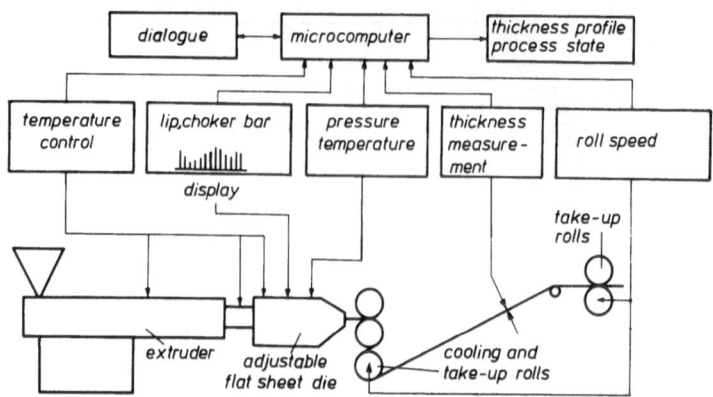

Fig. 9.7.   Process control in sheet extrusion.

automatic thickness control. Since no means has as yet been found of placing a thickness measurement sensor ¡nearer ¦to the die output, it was suggested that pressure and temperature readings be taken on the molten plastic over the width of the die and these correlated with the thickness profile (see Michaeli, 1975).

To this end, the die relief geometry of a flat sheet die defines a rheological measuring channel, with pressure and temperature readings and channel geometry supplying information on local melt throughput. If these pressure and temperature readings are taken at several points over the width of the die then this will indicate throughput distribution, and hence thickness distribution, over the width of the die.

For purposes of drawing up the process equation, the die gap was taken as the balance zone for the flow equation. For a slot channel of height H, width B and length L, the volumetric flow rate $\dot{V}$ is given by

$$\dot{V} = \frac{\Delta p H^3 B}{L 12 \bar{\eta}} \tag{9.4}$$

where Δp is the pressure drop in the channel and $\bar{\eta}$ is a representative viscosity for liquids with structural viscosity and is a function of shear rate $\dot{\gamma}$. Assuming a power law (which is always valid for certain regions of plastic melts) and with

$$\eta = K \dot{\gamma}^{v-1} \quad \text{and} \quad \dot{\gamma} = \frac{C \dot{V}}{B H^2} \tag{9.5}$$

this gives

$$\eta = K \left(\frac{C \dot{V}}{B H^2} \right)^{v-1} \tag{9.6}$$

The temperature dependence of the viscosity can be modelled thus (see eqn (1.18))

$$\eta(T) = \eta(T^*) \exp(-\zeta T) \tag{9.7}$$

and allowance can be made for stretching between the die and the rolls. This then gives the sheet thickness as

$$d = K^* H^n \exp(\zeta T) \left(\frac{\Delta p}{2L} \right)^m \frac{1}{V_{abz}} \tag{9.8}$$

This model equation provides the basis for the thickness control model (V_{abz} denotes the take-off velocity, m, n and K^* denote constants and T^*

denotes a reference temperature). The thickness formula is now programmed into the process control computer to give automatic thickness control. The pressure and temperature values are continuously read off by the computer and the thickness distribution calculated from the equation.

If there is any deviation between the actual thickness value and the target thickness value in the sheet the computer will proceed to correct the flow by adjusting the choker bars in the die. The success of this correction can then be checked in the following scan cycle.

Tests on the extrusion plant, however, revealed that control based solely on the process model equation is not sufficiently accurate. The principal reasons for this are as follows:

1. inaccuracies in pressure and temperature readings;
2. irregularities in flow channel geometry;
3. non-uniform die temperature control;
4. errors in describing the shirinkage effect;
5. errors in the process model.

A remedy for these difficulties is, however, provided by adaptive model correction. Here it is assumed that any errors that occur will be corrected by adaptive adjustment of the parameters in the model equation. Providing that only slight differences have to be smoothed out in this way, there will only be slight modification in the parameters and this will scarcely affect accuracy (see Bergweiler, 1981). A further advantage is that it is no longer absolute measurement accuracy that is decisive but only reproducible measurement accuracy.

The adaptation is intended not only to correct the steady-state model but also to correct temporary behaviour during changes in the point of operation and batch fluctuations. Since the model equation has been drawn up from a physical angle it may be assumed that the structure is correct. The adaptation thus consists in a parameter correction to be incorporated in the model equation with the correct timing and order of magnitude, Fig. 9.8.

In order to cut down the technical measurement outlay at the die, a means was also devised of calculating a number of the pressure values in the die, required by the model, through including a number of less accurate pressure values, Fig. 9.8. In calculating the pressure values, consideration is given both to local choker bar position and to information on the choker bar as a whole and its deflection (see Bergweiler, 1981).

In this way the inter-relationship between the individual die segments is implicitly established and taken into consideration. The rapidity of the

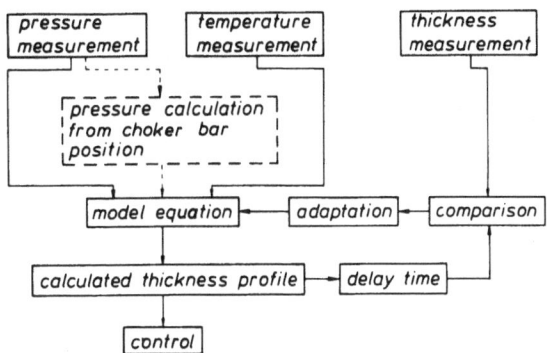

Fig. 9.8. Thickness control using a process model (adaptive).

control system is demonstrated by the reduction in standard deviation of the thickness tolerances during start-up (Fig. 9.9).

In tests with control circuits incorporating a delay time (Fig. 9.9) it is seen that adjustment to the final value attainable under the given marginal conditions of machine and production engineering follows after approximately four times the delay time.

In test (b) (Fig. 9.9), thickness control was carried out with the adaptive process model. Only three pressure transducers and two thermocouples were used here, along with measurement of the choker bar position. Through the measurement of pressure and temperature in the die, which allows the thickness to be calculated without any delay time, the thickness

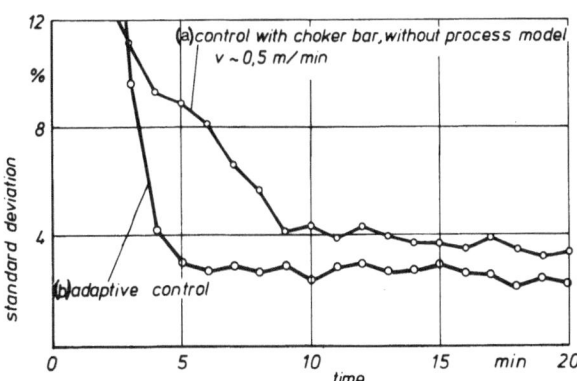

Fig. 9.9. Start-up phase: (a) Control via choker bar; (b) control with adaptive process model, three pressure transducers and two thermocouples, choker bar as the regulating unit.

tolerances are smoothed out very quickly. The tolerance band is attained after only one and a half times the delay time. With this method it is possible to achieve start-up times that are only a quarter to a fifth of the time taken for manual die setting.

It is equally possible to transpose this concept to all-round thickness control for extruded pipes, using automatic die centering. This is currently at the trial stage.

9.4.2 Dynamic throughput control in wire coating

The production processes for a large number of cable types (such as power cables and stranded conductors) involve frequent start-up phases or speed changes. This means that the plastics extruder runs through a lengthy unsteady phase due to large-scale changes in screw speed. Only when the new steady-state thermodynamic state is established is there a constant melt output for coating the cable. Conventional control systems for maintaining constant insulation thickness therefore do not work satisfactorily in the non-steady-state phases.

A forward strategy for thickness control, similar to that developed for sheet extrusion, was therefore worked out at the IKV (see Menges and Ramm, 1980; Ramm, 1981). Figure 9.10 shows the principle behind the system. The diameter of the wire or cable is measured both before and after insulating material is applied. These readings then serve to confirm or correct the model that runs the plant with the aid of a process control computer, in the same way as in sheet extrusion. The computer not only determines the control strategy but also assumes the actual process control and the control of all important control circuits (temperature, speed, etc.). If consideration is given to the influence of the chief process parameters

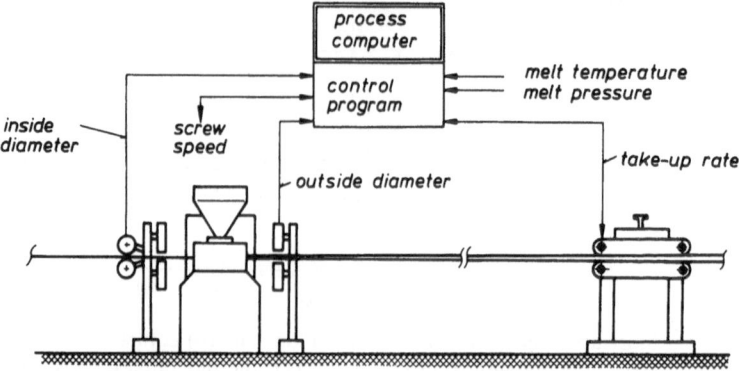

Fig. 9.10.   Principle of process control at a wire-coating plant.

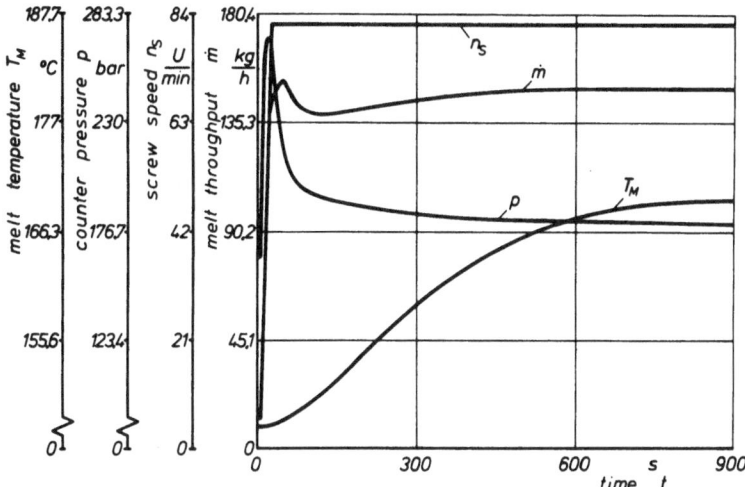

Fig. 9.11. Course of process output signals after change in screw speed.

(melt temperature, T_M, melt pressure, p, extruder speed, n_S) on the quality and quantity of the extruded melt, then a change in speed will give process signal output curves similar to those shown in Fig. 9.11. It is evident that this transient behaviour can no longer be described with steady-state models (see Parnaby and Kochhar, 1977). Hence, a similar approach to that adopted for sheet extrusion (see Section 9.4.1), for estimating melt throughput after speed changes, no longer produces satisfactory results. because the important process variable of screw speed, which has the greatest influence on throughput, is not contained in the model equation, Fig. 9.12.

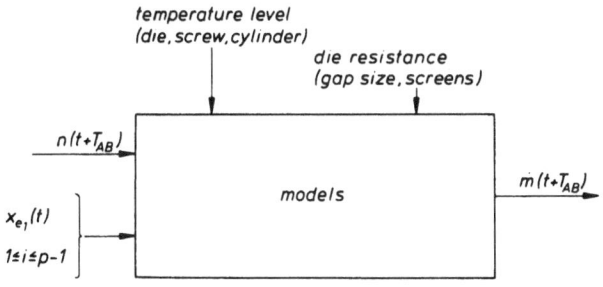

x_{e_i} = measurable process variables
$p-1$ = No. of additional process variables required for this model

Fig. 9.12. Model for estimating melt throughput during non-steady-state extruder operation.

After several empirical model formulations had been tried out, process control with the following formulation finally gave the best overall behaviour:

$$n_S = a_0 \dot{m}(a_1 p) + (a_2 T_M) \qquad (9.9)$$

where the a_i are constants, $\dot{m}$ denotes mass throughput, p denotes melt pressure and T_M denotes melt temperature. This equation is already given for the calculation of the necessary screw speed as the setting variable for melt throughput with respect to insulation thickness, when line speed adjustment (i.e. linear ramp) is given.

In order to calculate the expected screw speed at time $(k + 1)$ the coefficients a_i can be matched to the conditions actually prevailing, or where there is sufficiently good correlation between model and measurement, it may be taken from a model data record. The instantaneous values of pressure, p, and temperature, T_M, are introduced as measurements, together with the desired mass throughput, $\dot{m}$, so that the adjustment to the screw speed n_S can be made. By setting a tolerance range for permissible deviations between model and measurement it proved possible to achieve adaptive model correction. The line speed for the whole unit (which is proportional to $\dot{m}$) is also taken into consideration by the model. Where there are modifications in target values (start-up, shut-down) the extruder speed is brought into line accordingly.

The tests carried out showed that the adaptive method used permitted process control to within a close tolerance band on either side of the set target value. In other words, just a few metres after start-up the model had already 'learnt' the extruder's melt throughput behaviour so well that the wall thickness deviation (d) never strayed outside a tolerance band of 5% (Fig. 9.13)—not even when the plant was further accelerated once the end of the cable had reached the take-up reel (65 m away in this case). It can be seen how the screw speed (n_S) target value continues to increase after line speed (v_L) has been adjusted to production speed, so as to keep the wall thickness performance value within the tolerance band. This is even clearer in cases where large-scale adjustment in target values causes an even greater modification in extruder throughput, which is then compensated for by a further increase in screw speed.

Comparative investigations revealed that the quality of a model formulation is all the greater where:

1. the model formulation is built up from physical fundamental equations;

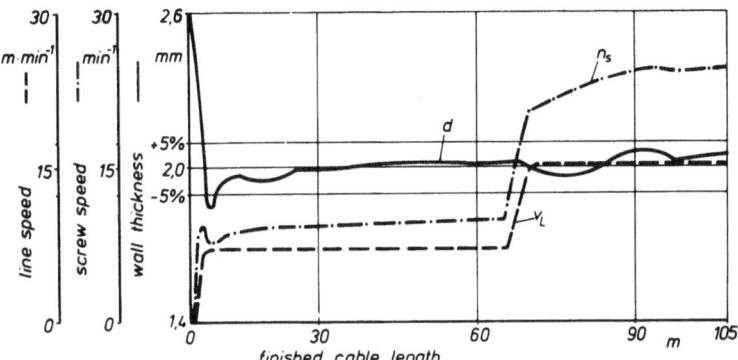

Fig. 9.13. Start-up: wall thickness control via model prediction.

2. the theory of gradually changing phases is abandoned and dynamic operating behaviour is included in the model formulation, or where several process variables that determine the next point of operation can be estimated in advance;

3. the interpretation of individual terms in the formulation (e.g. density, mean temperatures) is improved by using additional transducers and thermocouples.

A melt throughput calculation based on the flow calculation in the die is, in principle, also possible, though this has its drawbacks, stemming primarily from a lack of knowledge of the precise initial and boundary conditions required and a lack of material data. On-line die computation also takes up a great deal of computing time and hence compares unfavourably with a structurally simple model equation. Melt throughput control in the manner described may be transferred to any other extrusion process (see Predöhl, 1981).

9.4.3 Process control in extrusion blow moulding

The extrusion blow-moulding process may be divided into sub-processes in the direction of material transport, with each of these sub-processes having a certain degree of independence. These sub-processes are as follows: extrusion (extruder and die) with parison formation, the blowing and moulding process and the finishing work.

When it comes to process control for blow-moulding units, different process control concepts are available for controlling the process sequence

320   J. WORTBERG

in the individual stages. Work carried out to date and practical developments have produced the following systems:

1. parison wall thickness control;
2. parison length control;
3. parison speed control;
4. cycle time control;
5. weight control of blown piece.

Various cycle time control systems that work in conjunction with parison
length control are also in widespread use. Cycle time, t_Z, is made up of
blowing time, t_B, and machine time, t_M. The machine time is laid down by
the movement sequence open/close mould, lift/lower mould, etc. The
systems that are currently in general use require the blowing time, i.e. the
time in which the parison receives its definitive shape in the mould and
remains there cooling up to the point of ejection, to be given as the
empirical target value for cycle time control.

From a process engineering angle it is sensible to select the blowing time
that will allow perfect demoulding whilst giving the desired quality. This
time can be derived in much better form (than is calculable from measured
melt and mould wall temperatures) direct from the cooling pattern in the
moulded part and be characterised by a temperature, for instance. A
process model was thus developed at the IKV in the course of extensive
investigations, which allows cycle time control with readily manageable
equations, on the basis of a given demoulding temperature (see Dormeier,
1977; Junk et al., 1978; Menges et al., 1973). This involved the compilation
of a numerical computation procedure, based on energy and material
balances, to determine the melt temperature profile as a function of time
and place. This gives consideration to both parison cooling and the cooling
of the blown piece in the mould. The complexity of the necessary cooling
equation meant that on-line computation at the machine was not feasible.
The results of comprehensive simulation calculations on a large-scale
computer, with variation of initial and boundary conditions, are therefore
taken into consideration in a regression equation. The parameter sought
here is the cycle time, t_Z:

$$t_Z - \{A_1 \exp[A_2 + A_3 \exp(A_4 - A_5 t_Z)]\} + A_6 - t_M = 0 \qquad (9.10)$$

where t_Z denotes cycle time, t_M denotes machine time, and the A_i denote
coefficients that result from the different relevant variables, such as
temperatures, speed, times, material data and geometrical variables as in
Table 9.1.

TABLE 9.1
VALUES OF COEFFICIENTS FOR HDPE AND THE BLOW-MOULDING UNIT INVESTIGATED

Coefficients	Value
A_1	$6.92 h^{1.9627}$ (h in mm)
A_2	$0.00278 (T_U - 200\,°C) + 0.00998 (T_W - 10\,°C)$ (T_U, T_W in °C)
A_3	$0.00278 (T_{MD} - T_U)^{0.989}$ (T_{MD}, T_U in °C)
A_4	0.0546
A_5	$0.0084 \dfrac{x}{s_v . L_v}$ (x, s_v, L_v in mm)
A_6	$\dfrac{4h^2}{a_e} \ln\left(2 \dfrac{T_E - T_W}{T_K - T_W}\right)$ (h in mm; a_e in mm²/s; T_E, T_W, T_K in °C)

The following variables are required in order to determine coefficients A_1 to A_6: h = wall thickness of blown piece, x = position of computed demoulding temperature, s_v = parison wall thickness, L_v = parison length, a_e = effective thermal diffusivity, T_U = ambient temperature, T_W = mould wall temperature, T_K = crystallisation temperature, T_E = mean demoulding temperature, T_{MD} = melt temperature in die.

Coefficients A_1 to A_6 thus contain further regression coefficients which restrict the general validity range of the calculation, though it is only through these that the results can be used for process control at all. In a running blow-moulding unit, most of the values are constant so that the inter-relationship may also be described as:

$$t_{Blowing} = f(T, \text{constants}) \qquad (9.11)$$

With a given desired demoulding temperature T, the correct value for $t_{Blowing}$ can be derived iteratively from eqn (9.10). In the event of modification in one or more of the parameters regarded as constant (e.g. mould wall temperature, T_W, ambient temperature, T_U) the new blowing time to give the same demoulding temperature can be established during production. Process parameters can thus be selectively modified and any disturbance that occurs smoothed out. Direct allowance can be made for changes in material behaviour, temperature behaviour or cooling behaviour, so that the desired quality, in the form of a given demoulding temperature, is still observed.

Taking the cycle time, t_Z, calculated via the process model, and the

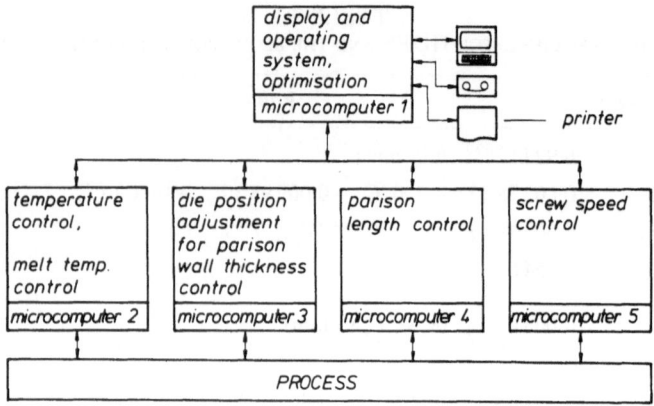

Fig. 9.14.   Microcomputer concept for extrusion blow moulding.

parison length, L_v, a mean target parison speed is determined and set using the screw speed.

The overall computer concept is built up on a modular basis for the individual task areas and comprises the following features (Fig. 9.14):

1. Central display and control system with optimisation facility.
2. DDC temperature control of ten zone wall temperatures including three zones for heating and cooling, cascade control for melt temperature.
3. Adaptive parison length control with subordinate screw speed control.
4. Screw speed control with AC and DC shunt motors.
5. Die position control to regulate parison wall thickness.

With a microcomputer system of this type it is possible to produce optimum hollow articles both during start-up of a blow-moulding unit and during any disturbance that may occur during steady-state production (see Menges and Schwab, 1980; Schwab, 1980).

9.5   COMPUTERS IN INJECTION MOULDING

Extensive investigations into process analysis of the injection-moulding process, both for thermoplastics and for elastomers and thermosets, have shown that it is wise for control to occur by setting those machine variables that have an influence on the actual thermodynamic moulding process (see

Buschhaus, 1982; Menges *et al.*, 1971; Stitz, 1973). The injection-moulding process can be divided up into two main phases to this end, |namely| the plasticisation phase (solid matter → molten mass) and the moulding formation phase (molten mass → finished part). During the first phase the plasticisation unit has the task of producing a homogeneous melt volume that remains constant from one shot to the next. The chief parameters here are cylinder temperature control, and screw speed and back pressure, taken in conjunction with screw geometry. This problem area will not be covered here, since the main process phase, which can be described with rheological–thermodynamic models, is the moulding formation phase with its quality-determining influence.

9.5.1 Process control for thermoplastics injection moulding

The moulding formation process in thermoplastics injection moulding will be looked at first, before details are given of the special features of reactive stock processing. If the process sequences are observed at the point where the moulding is actually formed, namely in the mould cavity, it is possible to divide the process up into three main sections, each of which requires different treatment in terms of process engineering:

1. filling phase,
2. compression phase,
3. holding pressure phase.

This breakdown is clearly recognisable in the melt pressure curve in the mould cavity, as shown in Fig. 9.15 (see Bongardt, 1982; Stitz, 1973). Each of these phases can be subsequently attributed with specific, striking properties in the moulded part, with the transitions being regarded as fluid in many cases. The main features of the filling phase can be identified with the surface quality of the moulding, the degree of orientation and crystallinity in the finished part and the thermal and mechanical stress on the melt. Complete filling of the mould cavities follows in the compression phase through the sharp increase in pressure, though this is also the stage when flash and moulding marks or even mould damage occur.

The holding pressure phase can essentially be said to influence the weight and shrinkage of the moulded part (see Stitz, 1973). The degree of crystallisation and orientation may also be influenced here in the same way as in the filling phase, though to a lesser extent and primarily in the inner region of the moulding.

In order to achieve optimum moulding quality it was necessary to

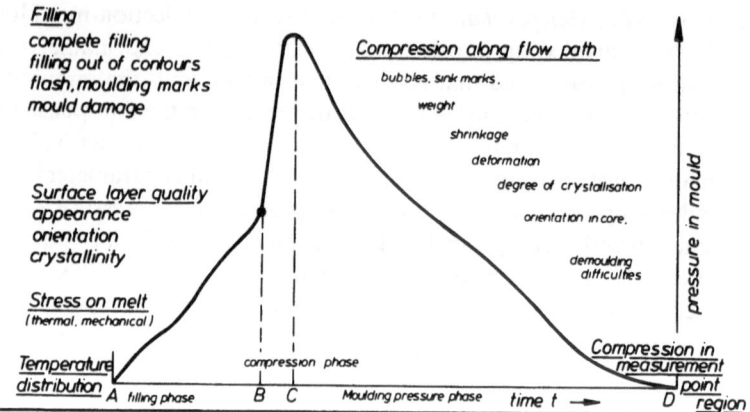

Fig. 9.15.   Identification of quality features with those sections of the pressure
curve that predominantly determine them.

develop process models which, with knowledge of pressure and temperature conditions, would permit the expected properties of the finished part to
be predicted during the production stage. Process control computers are
particularly suitable here (see Bongardt, 1982). An outline will be given
below of the basic principles behind the development of a process model
that can optimise the holding pressure phase of the injection-moulding
process when a process control computer is used.

The process model developed at the IKV makes use of the fact that the
density of plastic material can be influenced using pressure and temperature. The consequences, which differ from one material to another, are
expressed in the corresponding $p-v-T$ (pressure–specific volume–temperature) diagrams (cf. Fig. 9.16). These diagrams can be recorded with
suitable measuring equipment and in some cases it may be assumed that
different materials that have marked differences in flow behaviour (viscosity) will have very similar $p-v-T$ diagrams, as is the case with types of
polystyrene.

In order to process the $p-v-T$ diagrams further, such as in computers
assigned to machine control, it is appropriate to present them in the form of
an analytical description. The formulation used, which is empirical, has the
advantage of also describing the behaviour of semi-crystalline materials
correctly (see Bongardt, 1982). It reads as follows

$$v(p, T) = K_1 + (K_2 T) + K_3 \exp\left[(K_4 T) - (K_5 p)\right] + (K_6 p T^2) + \left(\frac{K_7}{K_8 + p}\right)$$

$$(9.12)$$

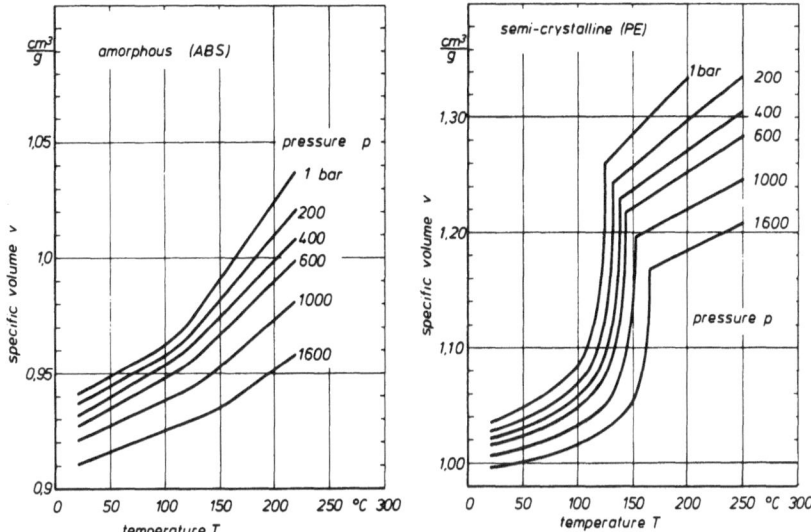

Fig. 9.16. p–v–T diagram for ABS and polyethylene.

where p denotes pressure, v denotes specific volume (reciprocal of density) and T denotes temperature; the K_i are constants. For amorphous materials, $K_3 = 0$. The K_i are determined by regression.

If the pressure pattern acting upon the moulding as it cools in the mould cavity is transposed to the p–v–T diagram, then it becomes possible to read off two properties of the subsequent finished part:

1.   the shrinkage, or weight and dimensions of the moulding,
2.   an estimate of the degree of orientation in the inner region of the moulding.

Thus, particularly for the precision parts that are being increasingly manufactured in plastics, an optimum pressure control during the holding pressure phase is absolutely essential (see Bongardt, 1982; Bongardt and Menges, 1980).

The first step in the procedure is to lay down the state curve of the molten plastic during the pressure holding phase in the p–v–T diagram (Fig. 9.17). Section I of the curve is laid down by the temperature of the melt in the screw ante-chamber, which represents the starting temperature for cooling in the mould, and by the maximum pressure inside the mould. The maximum pressure inside the mould is fixed by the maximum pressure permitted inside the mould or by the maximum pressure that the machine is

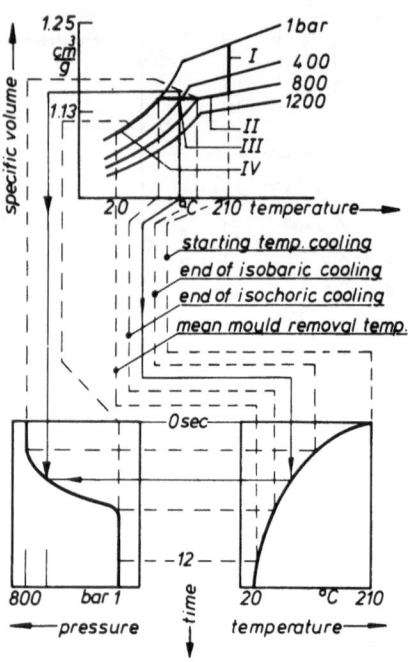

Fig. 9.17.   Establishing the optimum holding pressure curve.

able to apply during the holding pressure phase. Section I in the state diagram is run through within a very short time during the compression phase.

Section II of the curve is laid down by the maximum pressure inside the mould and by the target specific volume associated with the target shrinkage. The melt cools isobarically under constant maximum internal mould pressure until the target specific volume is reached. A constant pressure curve inside the mould at the start of the computed optimum holding pressure phase corresponds to this (Fig. 9.17, bottom left).

In section III of the state curve, suitable pressure control in the mould ensures that the target specific volume arrived at is kept constant up to the point where the 1 bar line is reached. The pressure values for the individual melt temperatures are obtained from the isobar in the p–v–T diagram that runs through the point where the isochorous state curve (in state section III) intersects the melt temperature under observation. The point in time at which this pressure has to be set in the mould is obtained from the computed cooling curve by correlating temperature and time (Fig. 9.17, bottom right). No more molten plastic flows into the mould cavity during state section III. The optimum pressure curve assigned to this section falls from the maximum pressure in the mould to 1 bar (Fig. 9.17, bottom left).

Section IV corresponds to isobaric cooling, at 1 bar, to the point where the mean demoulding temperature is reached, when the mould is opened and the moulded part ejected.

Laying down this state curve has the advantage that, with allowance for all the marginal conditions imposed by the mould or machine, all the molten plastic to which holding pressure has to be applied during cooling has already flowed into the mould at the earliest possible point in time. This is particularly important if the gate freezes rapidly. Where this is so, the full amount of melt required to give the specific final volume desired must already be inside the mould cavity at this stage.

The process model applied to control the holding pressure phase here not only requires the state diagram of the plastic to lay down the state curve but also requires the *measured pressure curve* and the *computed temperature curve* during cooling. In this case it is sufficient to calculate a mean temperature for the moulded part as a function of cooling time.

$$\bar{T}_{MW}(t) = T_W + \frac{8}{\pi^2}(T_M - T_W)\exp\left(-\frac{\pi^2 a_{eff}t}{S^2}\right) \qquad (9.13)$$

where $\bar{T}_{MW}$ = mean melt temperature in the mould; T_M = melt temperature in screw ante-chamber (measured); T_W = mould wall temperature (measured); S = wall thickness; a_{eff} = effective thermal diffusivity, and t = time during cooling.

In the event of modification in the thermal starting conditions or marginal conditions (T_M or T_W), the optimum holding pressure curve will always be calculated by a holding pressure optimisation program. An adaptive digital process control system then ensures that the computed pressure curve is realised on the machine.

The concept described constitutes part of a comprehensive computer control system on an injection-moulding machine, that has been developed at the IKV (see Bongardt, 1982; Bongardt and Menges, 1980; Menges and Bongardt, 1980; Schwab, 1981). The overall concept will be briefly outlined below.

The microcomputer system is made up of three independent computers. The real-time computer is coupled to the injection-moulding machine using analogue and digital input and output units and assumes the control of the injection-moulding process. During the running process the optimisation computer continuously calculates the optimum pressure curve inside the mould, which is then adjusted on the machine by the real-time computer. The operating computer organises the data flow between the data display unit and the real-time computer. All machine setting data is fed in via the data unit and the process state is logged here. If required, a graphic display of function curves for measured process variables can be obtained on the screen. The user programs for the real-time computer cover sequence control, hydraulic control, the adaptation program and temperature controls with a limit value monitor, Fig. 9.18. The hydraulic controls include a screw advance speed control with target value profile, control of maximum pressure inside the mould, holding pressure control with target value profile and back pressure control. Maximum pressure inside the mould is governed not by an adaptive control from one shot to the next but by a control within each cycle which keeps the maximum pressure constant, independently of modifications in process conditions, e.g. pronounced changes in injection speed.

The adaptation program automatically adjusts the controller parameters for the two profile controls to their optimum values. The optimi-

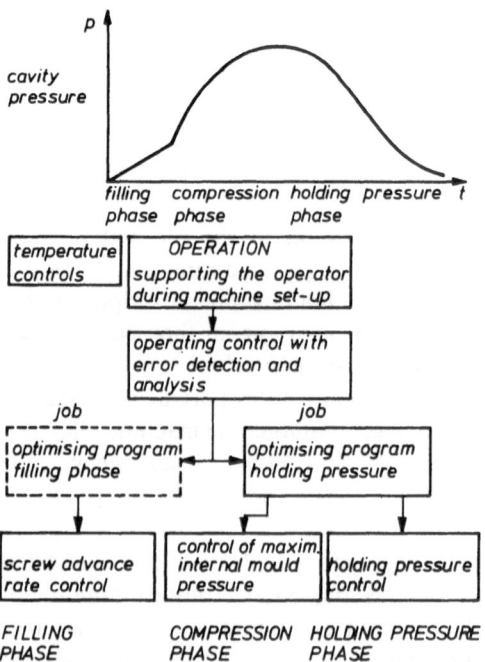

Fig. 9.18.   User program package for injection-moulding microcomputer control.

sation program computes the optimum pressure curve inside the mould in accordance with the p–v–T concept, as a function of time. Since this is a highly time-intensive computing operation and since rapid alignment of the pressure curve inside the mould to changes in process state is of particular importance during the start-up phase in injection moulding, an additional computer is employed for this task. The computed optimum pressure course inside the mould is transmitted to the holding pressure controller in the real-time computer as a target value profile.

This system thus marks the achievement of the essential target of computer control on injection-moulding machines, culminating from some 10 years'|development work—namely, a machine, which, provided it has a mould in working order and normal material, does not require any operator intervention. It can smooth out the usual disturbances automatically and keep to a pre-set point of operation, even under greatly modified conditions. In addition, it can speed up the running in of a new mould and, as a rule, conduct the start-up of an already known mould automatically.

9.5.2 Process control in the injection moulding of cross-linking moulding compounds

A process control concept for thermosets and elastomers has primarily to fulfil the following requirements:

1. logging,
2. elimination of fluctuations in moulding compound,
3. flash-free mouldings,
4. fully automatic demoulding,
5. constant product quality.

Fluctuations in the input data of moulding compounds (flow curing behaviour, viscosity) have a marked impact in the plasticisation phase and the injection phase (see Buschhaus, 1982; Egli, 1978; Menges and Buschhaus, 1981; Schönthaler and Niemann, 1981).

Technical measurement difficulties and inaccurate measurement readings make it impracticable to measure viscosity in the cylinder during the plasticisation phase. Data established during the injection phase supplies more reliable information here (see Frizelle and Norfleet, 1972). The viscosity in the cylinder influences the injection pressure required for a given injection velocity. If melt temperature is too low or cross-linking has begun the viscosity will be higher and more pressure will be required to fill the mould.

The viscosity of the moulding compound also has an influence on the

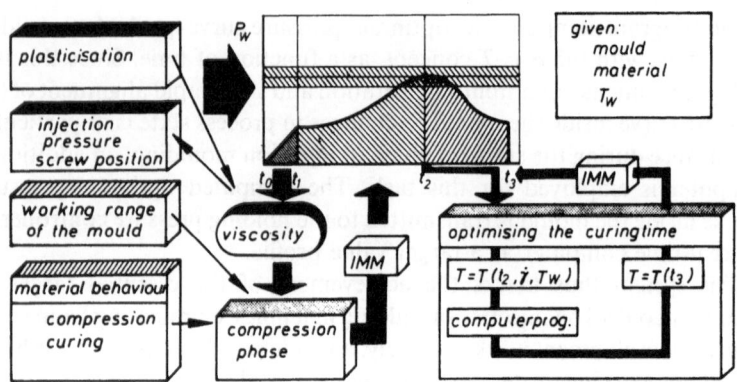

Fig. 9.19.   Process control for thermoset/elastomer injection moulding.

pressure increase in the mould during the filling phase. The value established when the viscosity state is monitored is used as a decision aid for the subsequent compression phase (Fig. 9.19). The switch-over from injection to compression follows as a function of pressure or force and follows in such a way that at the end of the filling phase there is no excess of pressure inside the mould at the points where outflow might occur.

The compression profile is controlled as a function of pressure inside the mould or as a function of force. The pressure profile over time is laid down on the basis of the working diagram for the mould and the viscosity behaviour of the moulding compound over time. The working diagram describes the operating behaviour of the mould, i.e. it shows which pressures are necessary in the mould to sufficiently compress a moulding compound at a particular temperature (see Buschhaus, 1982). An example of a working diagram for a mould found in practice is shown in Fig. 9.20. Providing that the shaded area (the 'operating window') is kept to, this will ensure sufficient compression on the one hand and either no flash or only minimal flash on the other hand.

During cross-linking the temperature and cross-linking profile over the cross-section of the moulding can be influenced by residence time, such that specific quality features can be obtained in this way. A feed-forward strategy is applied to optimise the curing phase (Fig. 9.19). Working from the temperature profile at the end of the compression phase, a computer program is used to determine the temperature equalisation process. A cross-linking profile over the cross-section of the moulding is calculated from the temperature field so that the time recorded to reach the required degree of cross-linking can be given as the demoulding time, Fig. 9.21.

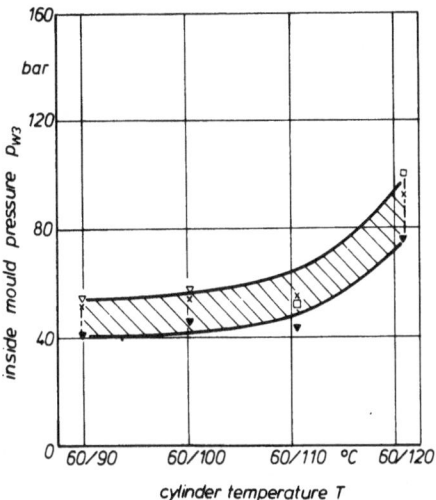

Fig. 9.20.   Working diagram for a mould (table station casing).

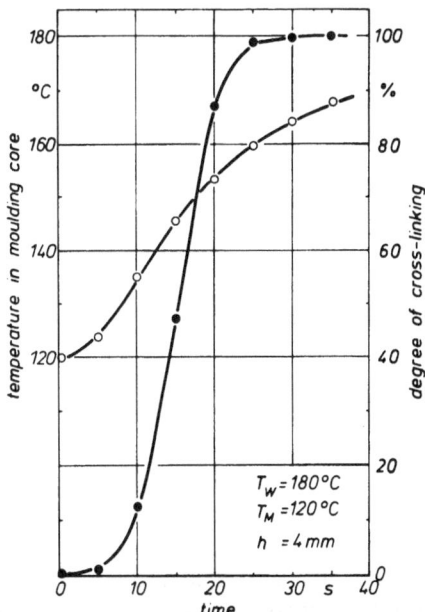

Fig. 9.21.   Temperature (—○—) and degree of cross-linking (—●—) as a function of time (computer program).

To this end it is sufficient to apply a closed solution for the time-dependent temperature field, since good correlation was obtained in the experimental comparison, although a constant melt starting temperature and constant material values must be assumed. It is likewise permissible to neglect the reaction heat is so far as the systems observed are only relatively mildly reactive (elastomers, phenolic resins). The resultant cross-linking profile can be determined by a difference method—generally just the inner layer. This is done by taking a reaction equation of, for example, the form

$$\Delta x = K(T)^n (1 - x)^n . \Delta t \qquad (9.14)$$

where

$$K(T) = K_0 \exp\left(-\frac{A}{RT}\right) \qquad (9.15)$$

x = degree (or fraction) of cross-linking, t = time, T = absolute temperature, n = reaction order, A = activation energy, R = universal gas constant, and K_0 = constant. If a temperature or time step size is given, it is then relatively easy to calculate the degree of cross-linking.

If an optimum, experimentally determined degree of cross-linking is given, this can then even be observed during disturbances in the moulding compound temperature or the mould temperature through adjustment of residence time.

The concept is built up in such a way that program modules are allocated to the individual process stages and these program modules are controlled by a microcomputer (see Menges and Bongardt, 1980).

9.6   PROBLEMS AND LIMITATIONS OF MICROCOMPUTERS

As has already been mentioned in Section 9.1, one of the chief limitations on the use of control systems is that the prevailing process inter-relationships must be formulable (see Fischer, 1978). Even the best microcomputer system is of no use if formulation is not feasible. This calls first of all for a detailed process analysis. Furthermore, even the best control concept can only be implemented in practice if the controlled variable is accessible to measuring equipment or can at least be derived from other variables. If it is planned to control, i.e. to keep constant, a variable that is not directly measurable in the process today, such as the degree of crystallisation or cross-linking or the shrinkage of an extrudate, then trouble will have to be taken in deriving the correlation between this quality feature and other, measurable process variables. The rheological–

mathematical description of processing history, or at least the relevant sections of it, takes on decisive importance in this concept. Work will have to be devoted here to bringing the process inter-relationships into a trimmed-down form that the microcomputer is able to manage. Apart from the problems mentioned here, there are also other problems associated with the development of microcomputer systems and their programming. The many different wishes of system users leads to the problem of optimum matching of both computer and programs to the processing unit in question. The traditional solution here is to use a process control computer and have the programs written in a high-level language. In this way at least the programs are relatively flexible and can be modified fairly easily.

REFERENCES

Anonymous. (1980). Outlook for extrusion controls, *Plastics Technol.*, 6, 57–9.
Bergweiler, E. (1981). Prozeßsteuerung bei der Flachfolien- und Tafelextrusion durch Mikrorechner, Thesis, RWTH, Aachen.
Bergweiler, E., Menges, G., Dierkes, A. and Michaeli, W. (1980). A new method of thickness control for flat film and sheet extrusion, *SPE Technical Papers*, 26 (38th ANTEC) New York, 50–2.
Bongardt, W. (1982). Verbesserung der Prozeßführung beim Spritzgießen durch selbsteinstellende Regelungen, Thesis, RWTH, Aachen.
Bongardt, W. and Menges, G. (1980). An injection moulding machine, which automatically finds its optimum working point, *SPE Technical Papers*, 26 (38th ANTEC) New York, 141–5.
Buschhaus, F. (1982). Beitrag zur Prozeßführung beim Spritzgießen von Duroplasten und Elastomeren, Thesis, RWTH, Aachen.
Dormeier, S. (1977). Ein Beitrag zur Automatisierung des Extrusionsblasformens, Thesis, RWTH, Aachen.
Egli, E. (1978). Automating the injection moulding process of Elastomers, *5th European Plastics and Rubber Conference*, Paris.
Fischer, P. (1978). Regelungs- und Steuerungstechnik bei Extrudern und Extrusionsanlagen für Kunststoffe, *Plastverarbeiter*, 29(5), 231–43.
Frizelle, W. G. and Norfleet, J. S. (1972). Viscosity measurements in the thermoset injection moulding, *PIA 9th Annual Research Conference*, Stevens Institute of Technology.
Hellmeyer, H. O. (1977). Ein Beitrag zur Automatisierung des Spritzgießprozesses, Thesis, RWTH Aachen.
Hunkar, D. (1979). The microprocessor: What it can really do, *Plastics Technology*, 4, 97–100.
Junk, P. B., Schwab, E., Dormeier, W. and Ramm, F. (1978). Von der Prozeßanalyse zur Rechnersteuerung—dargestellt am Beispiel des Extrusionsblasformens, *Kunststoffberater*, 4, 202–5; 5, 268–76; 6, 310–20.

Kertscher, E. (1981). Rechnergesteuerte Extrusionsanlagen für die Kabelherstellung, In *Rechnergesteuerte Extrusionsanlagen*, 'Ingenieurwissen', Series, VDI-Verlag, Düsseldorf.

Koch, W. (1976). Überwachung, Regelung und Steurung von Flachfolienproduktionsanlagen mit Rechnern, In *Rechnergesteuerte Extrusion?*, 'Ingenieurwissen' Series, VDI-Verlag, Düsseldorf.

Menges, G. and Bongardt, W. (1980). Der Einsatz von Mikrorechnern beim Spritzgießen, *Kunststoffe-Plastics* 27(2), 10–13.

Menges, G. and Buschhaus, F. (1981). Grundlagen für die Prozeßführung beim Spritzgießen von Duromeren und Elastomeren, *Plastverarbeiter*, ¦32(3), 322–5.

Menges, G. and Michaeli, W. (1976). Der Rechner als Hilfsmittel für Betrieb, Entwicklung und Forschung, In *Rechnergesteuerte Extrusion?* 'Ingenieurwissen' Series, VDI-Verlag, Düsseldorf.

Menges, G. and Ramm, F. (1980). Strategies for computer-controlled start-up of extruders, *SPE Technical Papers*, 26 (38th ANTEC), New York, 100–102.

Menges, G. and Schwab, E. (1980). Der Einsatz von Mikrorechnern beim Blasformen, *Kunststoffe-Plastics*, 27(1), 14–17.

Menges, G., Kulik, M. and Riehl, F. F. (1973). Entformungstemperatur und Kühlzeit bei der Herstellung extrusionsgeblasener Hohlkörper, *Plastverarbeiter*, 24(10), 621–3; 24(11), 685–90.

Menges, G., Stitz, S. and Vargel, J. (1971). Grundlagen der Prozeßsteuerung beim Spritzgießen, /*Kunststoffe*, 61(2), 74–80.

Michaeli, W. (1975). Zur Analyse des Flachfolien- und Tafelextrusionsprozesses, Thesis, RWTH, Aachen.

Parnaby, J. and Kochhar, A. K. (1977). Dynamical modelling and control of plastics extrusion processes, *Automatica*, 13, 177–83.

Parnaby, J., Battye, P. G., Hasson, G. A. and Hadwell, C. P. (1978). Computer-controlled injection moulding and extrusion, *Plastic and Rubber: Processing*, 3(3), 89–94.

Parnaby, J., Kochhar, A. K. and Wood, B. (1975). Development of computer control strategies for plastic extruders, *Polymer Engineering and Science*, 15(8), 594–605.

Predöhl, W. (1981). Optimierung des Betriebsverhaltens von Flachfolienanlagen mit Hilfe eines Rechners, In *Rechnergesteuerte Extrusionsanlagen*, 'Ingenieurwissen' Series, VDI-Verlag, Düsseldorf.

Ramm, F. (1981). Automatisierte Prozeßführung von Extrusionsanlagen mit adaptiven Prozeßmodellen, Thesis, RWTH, Aachen.

Recker, H. *et al.* (1976). *IKV-Dokumentation Messen, Steuern, Regeln (MSR) in der Kunststoffverarbeitung*, Carl Hanser Verlag, München.

Schönthaler, W. and Niemann, K. (1981). Optimales Spritzgießen von Duroplasten, *Kunststoffe* 71(6), 346–51.

Schwab, E. (1980). Mikrorechnereinsatz zur Prozeßführung beim Extrusionsblasformen, Thesis, RWTH, Aachen.

Schwab, E. (1981). Steuerung, Regelung und Prozeßführung beim Spritzgießen, *Plastverarbeiter*, 32(3), 315–21.

Stitz, S. (1973). Analyse der Formteilbildung beim Spritzgießen von Plastomeren als Grundlage der Prozeßsteuerung, Thesis, RWTH, Aachen.

Upmeier, H. (1981). Prozeßoptimieren, Steuern und Regeln der Blasfolienanlagen, In *Rechnergesteuerte Extrusionsanlagen*, 'Ingenieurwissen' Series, VDI-Verlag, Düsseldorf.

Wiegand, H.-G. (1979). *Prozeßautomatisierung beim Extrudieren und Spritzgießen von Kunststoffen*, Carl Hanser Verlag, München.

Wortberg, J. (1981). Der Rechner als Hilfsmittel für das Steuern und Regeln von Extrusionsanlagen, In *Rechnergesteuerte Extrusionsanlagen*, 'Ingenieurwissen' Series, VDI-Verlag, Düsseldorf.

Index

338 INDEX